Springer Theses

Recognizing Outstanding Ph.D. Research

Aims and Scope

The series "Springer Theses" brings together a selection of the very best Ph.D. theses from around the world and across the physical sciences. Each volume has been selected for its scientific excellence by an internationally renowned institute or department. For greater accessibility to non-specialists, the published versions include an extended introduction, as well as a foreword by the student's supervisor explaining the special relevance of the work for the field. The series as a whole provides a valuable resource both for newcomers to the research fields, and for other scientists seeking detailed background information on special questions. Finally, it serves as an accredited documentation of the valuable contributions made by today's younger generation of scientists.

How to Apply

Theses prepared at internationally leading universities and institutes may be nominated for publication by supervisors with the endorsement of the head of department. Each such department is entitled to nominate one thesis per year. Enquiries and requests for a nomination form should be sent, together with the thesis, to Ms. Nathalie Berg (see link on right).

Eligibility Criteria

To qualify for consideration, theses should fulfil the following criteria:

- They must be written in good English.
- The topic should fall within the confines of Chemistry, Physics, Earth Sciences, Engineering and related interdisciplinary fields such as Materials, Nanoscience, Chemical Engineering, Complex Systems and Biophysics.
- The work reported in the thesis must represent a significant scientific advance.
- A maximum of 30% of the text may be a verbatim reproduction from the author's previous publications.
- They must have been examined and passed during the 12 months prior to nomination.
- The theses should have a clearly defined structure including an introduction accessible to new PhD students and scientists not expert in the relevant field.
- After acceptance, theses should be supplemented by a "Supervisor's Foreword" outlining the wider scientific relevance of the work.

Indexed by zbMATH.

Kotaro Shimizu

Spin Moiré Engineering and Emergent Electromagnetism in Topological Spin Crystals

Doctoral Thesis accepted by
The University of Tokyo, Tokyo, Japan

Springer

Author
Dr. Kotaro Shimizu
Department of Applied Physics
The University of Tokyo
Tokyo, Japan

Supervisor
Prof. Yukitoshi Motome
Department of Applied Physics
The University of Tokyo
Tokyo, Japan

ISSN 2190-5053 ISSN 2190-5061 (electronic)
Springer Theses
ISBN 978-981-96-5991-3 ISBN 978-981-96-5992-0 (eBook)
https://doi.org/10.1007/978-981-96-5992-0

This Springer imprint is published by the registered company Springer Nature Singapore Pte Ltd.
The registered company address is: 152 Beach Road, #21-01/04 Gateway East, Singapore 189721, Singapore

Supervisor's Foreword

Topology, a branch of mathematics concerned with the properties of space that are preserved under continuous transformations, has become increasingly significant in modern condensed matter physics. It provides a powerful framework for understanding and classifying different phases of matter based on their global characteristics. Topological defects, such as vortices, domain walls, and skyrmions, represent disruptions in the ordered state of a system that cannot be eliminated by continuous transformations. These defects play crucial roles in various physical phenomena, from phase transitions to the dynamics of materials. In particular, topological spin textures in magnets, including chiral solitons, skyrmions, and hedgehogs, exhibit unique magnetic properties and quantum transport phenomena. Additionally, these structures can form periodic arrangements known as topological spin crystals, exhibiting macroscopic electromagnetic phenomena due to their noncollinear and noncoplanar spin textures. Consequently, understanding and controlling these topological properties is a significant challenge for developing novel functionalities in materials, with potential applications in spintronics, quantum computing, and advanced technologies.

This thesis by Kotaro Shimizu delves into a comprehensive theoretical investigation into the engineering of spin structures and their emergent physical properties by generating, annihilating, and controlling topological spin textures. A central concept introduced in this thesis is the "spin moiré" picture, which views topological spin crystals as superpositions of multiple spin density waves. Utilizing this perspective, Kotaro systematically explores the control of spin textures, their topological properties, and the resultant emergent electromagnetic responses, establishing a framework for engineering magnetic functionalities. Through extensive theoretical analysis grounded in variational methods, the Landau-Lifshitz-Gilbert equation, and linear spin-wave theory, this work demonstrates the control of these properties across a variety of topological spin textures, spanning one- to three-dimensional structures.

The discoveries in this thesis, rooted in the spin moiré picture, establish a versatile framework for investigating topological magnetism and pave the way for further exploration of emergent electromagnetic phenomena and functionalities of magnets. In addition, the insights gained from this study will stimulate further research into topological spintronics and beyond, offering a new route toward designing next-generation magnetic materials and devices with tailored properties.

Tokyo, Japan

Yukitoshi Motome

Preface

Topology plays an important role in modern condensed matter physics. Among many topological states of matter, magnetic states with topological spin structures such as magnetic skyrmions and hedgehogs have attracted considerable attention. These topological spin structures often form periodic arrangements, which are called topological spin crystals. They manifest nontrivial quantum transport and optical phenomena through emergent electromagnetic fields associated with the noncollinear and noncoplanar spin textures. One of the challenging problems in this field is flexible generation, annihilation, and control of the spin textures with different topology, leading to their novel functionality. Although this problem has been discussed in several specific cases, the previous studies lack a generic viewpoint. In this thesis, we propose the "spin moiré" picture to facilitate the systematic discussion on the control of the topological spin textures. Moiré is an interference fringe generated by a superposition of waves, which can be modulated in many ways by changing fundamental parameters in the superposition. In analogy with such control of conventional moirés, we theoretically explore the ways to control the spin moirés and their functionalities—spin moiré engineering.

We conduct a comprehensive study of spin moiré engineering for several different topological spin crystals spanning from one to three dimensions. In one dimension, we investigate the spin moiré engineering for a superposition of higher harmonic waves, the chiral soliton lattice (CSL). We study the emergent electric field at magnetic resonance and show that the higher frequency resonance modes are more clearly visible than the magnetic ones. We also find that the emergent electric field from the edges of the system becomes comparable to or larger than the bulk responses in realistic system sizes. In addition, we reveal that additional chiral solitons can be introduced into the system by selectively activating the edge mode with a circularly polarized AC magnetic field. Next, in two dimensions, we study the spin moiré engineering for the skyrmion lattices (SkLs). Through the phase shift between superposed waves and the uniform magnetization, we elucidate the topological phase diagram including the SkLs with the skyrmion number ranging from -2 to 2. We also clarify that the phase shift and the topological transition can be caused by an external magnetic field. Finally, in three dimensions, we study the spin moiré engineering for the hedgehog lattices (HLs). We clarify that spatial anisotropy can drive the modulation of the amplitudes of constituent waves of spin moirés, leading to the topological phase transition. Furthermore, by changing the twist angle of superposed waves, we find topological transitions with successive changes in the number of hedgehogs and antihedgehogs. We also systematically clarify the effect of the phase shift in the HLs and find the total number of hedgehogs and antihedgehogs unprecedentedly ranges up to 48. Furthermore, we show that the HL can cause a drift motion by an electric current and the threshold current density reflects the real-space configurations of the hedgehogs and antihedgehogs. Notably, we also discover the perfect hedgehog Hall effect that is a transverse motion of the hedgehogs without any longitudinal motion along the current direction.

To summarize, in this thesis, proposing the spin moiré picture, we systematically study the control of the spin textures, their topological properties, and the emergent electromagnetic fields for various topological spin textures. The spin moiré provides a novel platform for intriguing physics distinct from the structural moiré in the layered materials with several advantages: wide dimensionality, vector (spinor) degree of freedom, flexible control through the external fields, nontrivial topological properties, and the emergent electromagnetic phenomena. Our findings based on the spin moiré picture not only establish the versatile tool to study the control of topological magnetism but also pave the way for further exploration of topological spin textures and the resultant emergent electromagnetic phenomena.

Tokyo, Japan Kotaro Shimizu

Parts of this thesis have been published in the following journal articles

1. **"Phase transitions between helices, vortices, and hedgehogs driven by spatial anisotropy in chiral magnets"**
   K. Shimizu, S. Okumura, Y. Kato, and Y. Motome
   Phys. Rev. B **103**, 054427 (2021).

2. **"Spin moiré engineering of topological magnetism and emergent electromagnetic fields"**
   K. Shimizu, S. Okumura, Y. Kato, and Y. Motome
   Phys. Rev. B **103**, 184421 (2021).

3. **"Phase degree of freedom and topology in multiple-Q spin textures"**
   K. Shimizu, S. Okumura, Y. Kato, and Y. Motome
   Phys. Rev. B **105**, 224405 (2022).

4. **"Phase shift, ellipticity, angle, and topological number in skyrmion lattices"**
   K. Shimizu, S. Okumura, Y. Kato, and Y. Motome
   J. Phys.: Conf. Ser. **2164**, 012069 (2022).

5. **"Emergent electric field from magnetic resonances in a one-dimensional chiral magnet"**
   K. Shimizu, S. Okumura, Y. Kato, and Y. Motome
   Phys. Rev. B **108**, 134436 (2023).

6. **"Soliton penetration from edges in a monoaxial chiral magnet"**
   K. Shimizu, S. Okumura, Y. Kato, and Y. Motome
   Phys. Rev. B **111**, 024411 (2025).

7. **"Current-induced motion of nanoscale magnetic torons over the wide range of the Hall angle"**
   K. Shimizu, S. Okumura, Y. Kato, and Y. Motome
   Commun. Phys. **8**, 69 (2025).

Acknowledgements I would like to express my sincere gratitude to my supervisor Prof. Yukitoshi Motome for his instructive advice, continual encouragement, stimulating discussions, and critical readings through this manuscript. I would like to express my special thanks to Prof. Ryotaro Arita, Prof. Naoya Kanazawa, Prof. Masahito Mochizuki, and Prof. Rina Takagi for their valuable comments in the review process of this dissertation. I would also be grateful to Dr. Yasuyuki Kato and Dr. Shun Okumura for constructive discussions and helpful support in every aspect of my laboratory life.

I would like to thank all the former and present members of the Motome and Kato group, especially Mr. Motoki Amano, Dr. Troels Arnfred Bojesen, Dr. Kiyu Fukui, Dr. Koji Inui, Mr. Akito Iwano, Dr. Seong-Hoon Jang, Mr. Shoya Kasai, Mr. Kaito Kobayashi, Mr. Zhang Lingzhi, Dr. Petr Andreevich Mishchenko, Dr. Kazuki Nakazawa, Mr. Simon Nitsch, Mr. Takuma Ogihara, Dr. Ryo Ozawa, Dr. Rico Pohle, Mr. Shusei Wadashima, Dr. Ryota Yambe, Dr. Zhao Yi-Feng, and Mr. Seiya Yoshitomi for constant discussion and beneficent comments through not only laboratory meetings but also daily conversations.

My research has also been benefited from my stay in the group of Prof. Cristian Batista in the University of Tennessee and Prof. Gia-Wei Chern in the University of Virginia. I would like to express my special thanks to the group members, especially Dr. David Dahlbom and Dr. Sheng Zhang.

I acknowledge the excellent discussions on my research with Mr. Rintaro Eto, Dr. Yukako Fujishiro, Prof. Satoru Hayami, Dr. Koujiro Hoshi, Dr. Masataka Kawano, Prof. Junichiro Kishine, Dr. Jan Masell, Prof. Naoto Nagaosa, Prof. Eiji Saitoh, Dr. Yusuke Shimamoto, Dr. Hidemaro Suwa, Prof. Yoshihiko Togawa, and Dr. Katsuhiro Tanaka. I would also like to express my thanks to all the members of the research groups of Prof. Ryotaro Arita, Prof. Zongping Gong, Prof. Motoaki Hirayama, Prof. Takahiro Morimoto, Prof. Takahiro Sagawa, and Prof. Haruki Watanabe for fruitful discussions and constructive questions in the joint seminar. I would like to thank the secretary Ms. Yasuko Tanaka for her assistance in clerical work.

This work was supported by JSPS KAKENHI Grant Number No. JP21J20812 and the Program for Leading Graduate Schools (MERIT-WINGS). Parts of the numerical calculations were performed in the supercomputing systems in ISSP, the University of Tokyo.

Finally, I would like to express my sincere gratitude to my family and friends for their support and encouragement.

Competing Interests The author has no competing interests to declare that are relevant to the content of this manuscript.

Contents

Chapter 1
Introduction

1.1 Topology in Magnets

Topology plays an important role in not only mathematics but also physics. One of the key concepts in topology for magnetism is topological defects. The topological defects are classified by the homotopy groups. In magnets, localized spin moments can be described by a vector field whose parameter space is the M-sphere S^M, and whose topology can be classified by the homotopy group $\pi_n(S^M)$, where n depends on the spatial dimension and the types of the topological defects. The topological defects described by the homotopy groups $\pi_n(S^M)$ can be classified into two categories based on whether they include singularities where the spin cannot be defined. Regarding topological defects with singularities, they exhibit a relationship $n = d - m - 1$, where m is the dimension of the singular region and d is the spatial dimension. Topological defects without singularities adhere to the relationship $n = d$. Table 1.1 shows the homotopy groups of spheres and the corresponding topological defects that are useful in this thesis [1–4].

In general, the homotopy groups with $n < M$ are trivial, i.e., $\pi_n(S^M) = 0$. Hence, any spin textures can be continuously transformed into a trivial one such as the fully polarized state. Meanwhile, the homotopy groups with $n = M$ are given by $\pi_n(S^n) = \mathbb{Z}$, and hence these classes provide nontrivial examples of the topological defects, as exemplified in the table. For example, point defects in two-dimensional (2D) XY spin systems called vortices, which play a crucial role in the Berezinskii-Kosterlitz-Thouless (BKT) transition [5–7], are described by $\pi_1(S^1) = \mathbb{Z}$. This means that the topological defects are characterized by integers called the topological numbers. Since an operation of the group in $\mathbb{Z}$ is additive, topological defects with different topological numbers cannot be continuously transformed into each other. This is "topological protection", which makes the topological defects stable against perturbations. The homotopy groups with $n > M$ are complex and can be nontrivial. One representative and important example is a three-dimensional (3D) topological spin texture, hopfion, described by $\pi_3(S^2) = \mathbb{Z}$.

K. Shimizu, *Spin Moiré Engineering and Emergent Electromagnetism in Topological Spin Crystals*, Springer Theses, https://doi.org/10.1007/978-981-96-5992-0_1

Table 1.1 The homotopy groups of spheres relevant to this study and the corresponding topological spin textures

parameter space	π_1	π_2	π_3
S^1 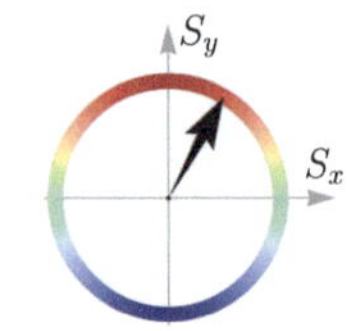	$\mathbb{Z}$ 	0	0
S^2 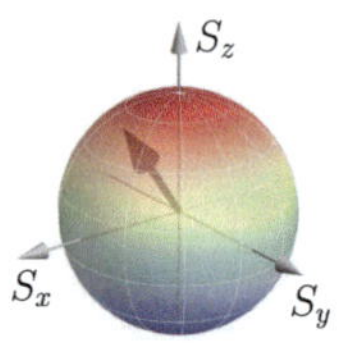	0	$\mathbb{Z}$ 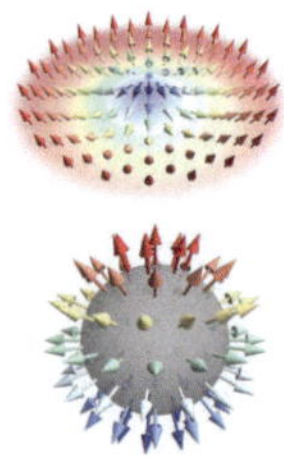	$\mathbb{Z}$ 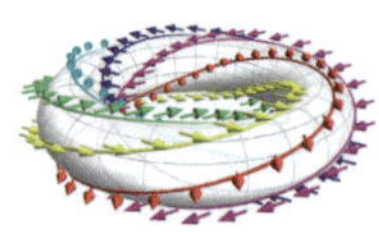

In this thesis, we study periodic arrays of the topological defects in magnetic materials. Before going into such spin textures, we briefly review several topological defects in one-, two-, and three-dimensional space in Sects. 1.1.1, 1.1.2, and 1.1.3, respectively [1, 3, 4]. Although the spin is originally the quantum degree of freedom, for simplicity, we here consider a three-component classical vector $\mathbf{S}(\mathbf{r}) = (S_x(\mathbf{r}), S_y(\mathbf{r}), S_z(\mathbf{r}))$ at position $\mathbf{r}$, define the normalized vector field described by

$$\mathbf{n}(\mathbf{r}) = \frac{\mathbf{S}(\mathbf{r})}{|\mathbf{S}(\mathbf{r})|} = (n_x(\mathbf{r}), n_y(\mathbf{r}), n_z(\mathbf{r})), \tag{1.1}$$

and introduce the variables describing the zenith angle $\theta(\mathbf{r})$ and the azimuthal angle $\phi(\mathbf{r})$ as

$$\mathbf{n}(\mathbf{r}) = (\sin\theta(\mathbf{r})\cos\phi(\mathbf{r}), \sin\theta(\mathbf{r})\sin\phi(\mathbf{r}), \cos\theta(\mathbf{r})). \tag{1.2}$$

1.1.1 *One Dimension*

Chiral Soliton

Let us first consider a topological defect in one-dimensional (1D) space, where topologically nontrivial spin textures are described by $\pi_1(S^M)$. A unique nontrivial example in Table 1.1 is a chiral soliton described by $\pi_1(S^1) = \mathbb{Z}$. Since the parameter space is S^1, we consider a 1D spin system where spins are confined in the xy plane, i.e., $S_z(\mathbf{r}) = 0$. Such a situation can be realized in the limit of strong easy-plane anisotropy. The topological spin texture allowed in this case is shown in Fig. 1.1, wrapping a unit circle (S^1) once without any singular points. The arrows represent the real-space configuration of $\mathbf{n}(\mathbf{r})$ and the colors represent the y component as shown in the inset. This is found in domain walls in ferromagnets and a bulk of monoaxial chiral magnets. These spin textures are characterized by an integer called the topological number given by [4, 8, 9]

$$N_{\text{soliton}} = \frac{1}{2\pi} \int_{-\infty}^{\infty} dz \frac{d\phi(z)}{dz}, \tag{1.3}$$

where z is the real-space coordinate along the swirling spin texture. This quantity counts how many times the spins wrap a 2D unit circle.

1.1.2 *Two Dimensions*

Vortex

Next, we consider the topological defects in 2D space ($d = 2$). For a vector field confined in the xy plane, a point defect ($m = 0$) called the vortex, which is described by the homotopy group $\pi_1(S^1) = \mathbb{Z}$, is allowed. Three types of such a point defect are exemplified in Fig. 1.2a–c. In each figure, the spin length vanishes at the center, which corresponds to the point defect. The existence of the singular points manifests

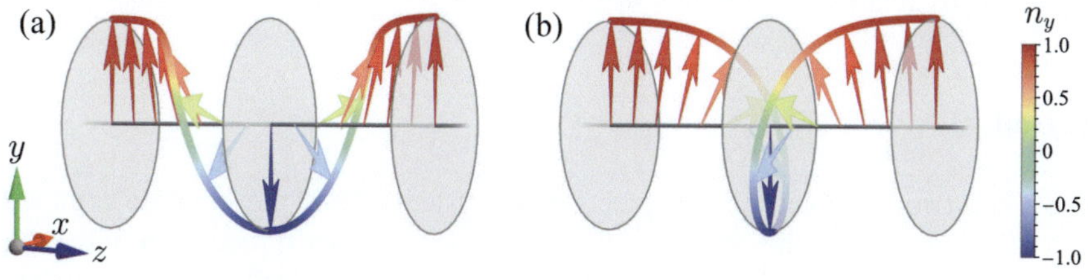

Fig. 1.1 Real-space spin configuration of a chiral soliton with **a** $N_{\text{soliton}} = 1$ and **b** $N_{\text{soliton}} = -1$. The arrows represent the normalized spins $\mathbf{n}(\mathbf{r})$ on the S_xS_y plane and their color displays the y component as shown in the inset

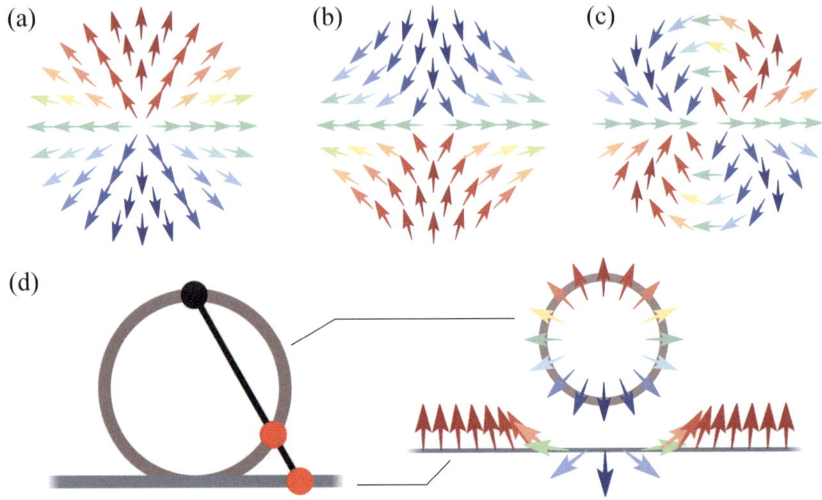

Fig. 1.2 Real-space spin configurations of **a** a vortex with $\zeta = 1$, **b** an antivortex with $\zeta = -1$, and **c** a vortex with $\zeta = 2$. **d** A stereographic projection from the vortex in 2D space to the chiral soliton in 1D space. The notations are common to those in Fig. 1.1

the difference between the vortex and chiral soliton, sharing the identical homotopy group. These defects are characterized by the topological number defined by [1, 3, 8]

$$\zeta = \frac{1}{2\pi} \int d\mathbf{l} \cdot \nabla\phi(\mathbf{r}), \tag{1.4}$$

where the integral is taken along a closed path surrounding the defect. This quantity also counts how many times the spins on the closed path wrap a 2D unit circle (S^1). It takes the values of $\zeta = 1, -1$, and 2 for Fig. 1.2a, b, and c, respectively. The defects with positive (negative) ζ are called (anti)vortices. The relation between the chiral soliton and the vortex, both of which are classified by $\pi_1(S^1) = \mathbb{Z}$, is schematically shown by a stereographic projection in Fig. 1.2d. The chiral soliton with N_{soliton} is topologically related to the vortex with $\zeta = N_{\text{soliton}}$.

Skyrmion

The above two examples are the defects in the systems where the spins are confined in a 2D plane. When the spins are allowed to point in any direction in 3D space ($M = 2$), the vortices and the chiral solitons become topologically trivial since they can be continuously modified into trivial structures such as a uniformly polarized state. Even in such systems, there are topological defects that are characterized by the homotopy groups $\pi_n(S^2)$ with $n \geq 2$ (see Table 1.1). First, we consider a 2D

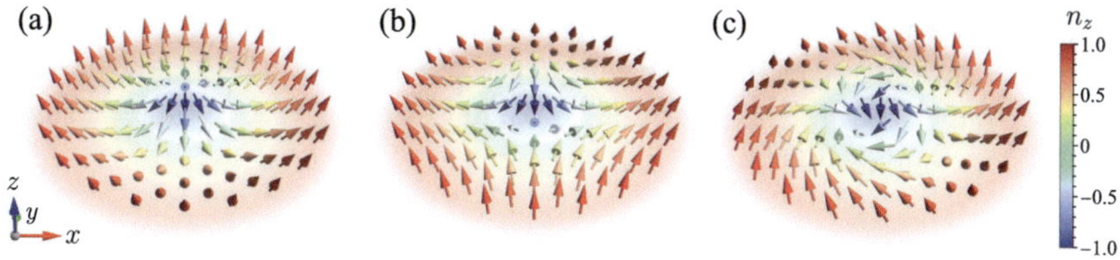

Fig. 1.3 Schematic pictures of **a** a skyrmion with $N_{\rm sk} = -1$ and $\zeta = 1$, **b** an antiskyrmion with $N_{\rm sk} = 1$ and $\zeta = -1$, and **c** a skyrmion with $N_{\rm sk} = -2$ and $\zeta = 2$. The helicity γ is set to 0. The arrows represent the normalized spins $\mathbf{n}(\mathbf{r})$ and their color displays the z component as shown in the inset

topological defect without singularities, called the skyrmion, which is described by $\pi_2(S^2) = \mathbb{Z}$. The skyrmions are topological solitons first proposed to explain hadrons in nuclear physics [10]. Later, their realization in spin textures was pointed out [11–14], and indeed, they have been recently discovered in magnetic materials as introduced in Sect. 1.4.2. The real-space spin configurations are exemplified in Fig. 1.3a–c. In these figures, the color of the arrows represents the z component of $\mathbf{n}(\mathbf{r})$. They are characterized by the topological number called the skyrmion number, which is defined as [2, 8, 15]

$$N_{\rm sk} = \frac{1}{4\pi} \int dxdy\, \mathbf{n}(\mathbf{r}) \cdot \left(\frac{\partial \mathbf{n}(\mathbf{r})}{\partial x} \times \frac{\partial \mathbf{n}(\mathbf{r})}{\partial y} \right), \tag{1.5}$$

where the integral is taken over the 2D xy plane. The integrand $b_z(\mathbf{r}) = \mathbf{n}(\mathbf{r}) \cdot \left(\frac{\partial \mathbf{n}(\mathbf{r})}{\partial x} \times \frac{\partial \mathbf{n}(\mathbf{r})}{\partial y} \right)$ represents the solid angle spanned by the neighboring spins, and hence it measures the spin noncoplanarity. This quantity is called the scalar spin chirality and is also termed the emergent magnetic field, which acts as a fictitious magnetic field on conduction electrons (see Sect. 1.3). The skyrmion number counts how many times the spins on the plane wrap a 3D unit sphere.

To see the relationship between $N_{\rm sk}$ and the spin texture, we directly calculate Eq. (1.5) by assuming a spin configuration given by Eq. (1.2) with $\theta(\mathbf{r}) = \theta(r)$ and $\phi(\mathbf{r}) = \phi(\psi)$, where r and ψ is the radial distance and polar angle defined by $\mathbf{r} = r(\cos\psi, \sin\psi)$, respectively. Then, Eq. (1.5) reads

$$N_{\rm sk} = \frac{1}{4\pi} \int_0^{\infty} dr \int_0^{2\pi} d\psi\, \frac{d\theta(r)}{dr} \frac{d\phi(\psi)}{d\psi} \sin\theta(r) = -\frac{1}{4\pi} \left[\cos\theta(r)\right]_0^{\infty} \left[\phi(\psi)\right]_0^{2\pi} . \tag{1.6}$$

Suppose that $\theta(0) = \pi$, $\lim_{r\to\infty} \theta(r) = 0$, and $\phi(\psi) = \zeta\psi + \gamma$, $N_{\rm sk}$ is then given by $N_{\rm sk} = -\zeta$, where ζ is the vorticity of spins projected onto the xy plane and γ is called the helicity. Hence, the skyrmion number is determined from the vorticity once the spins at the center of skyrmions and infinity are fixed. The skyrmions with positive (negative) vorticity are called (anti)skyrmions. The topological defects in Fig. 1.3a, b, and c are given by $\gamma = 0$ with $\zeta = 1, \zeta = -1$, and $\zeta = 2$, corresponding to the skyrmion with $N_{\mathrm{sk}} = -1$, the antiskyrmion with $N_{\mathrm{sk}} = 1$, and the skyrmion with $N_{\mathrm{sk}} = -2$, respectively.

1.1.3 *Three Dimensions*

Hedgehog

Next, we consider the topological defects in 3D space ($d = 3$). The first example is a hedgehog, which is characterized by the homotopy group $\pi_2(S^2) = \mathbb{Z}$, same with the skyrmions. Real-space spin configurations of the hedgehogs are exemplified in Fig. 1.4a–c. In all cases, the spin length vanishes at the center of the spheres, and the spins surrounding the point defects wrap a 3D unit sphere (S^2). The topological number for these defects is called the monopole charge, which is defined as [8, 16]

$$Q_{\mathrm{m}} = \frac{1}{4\pi} \int d\tilde{\mathbf{S}} \cdot \mathbf{b}(\mathbf{r}), \tag{1.7}$$

where the integral is taken on a closed surface surrounding the defect. Here, $\mathbf{b}(\mathbf{r}) = (b_x(\mathbf{r}), b_y(\mathbf{r}), b_z(\mathbf{r}))$ is the scalar spin chirality or the emergent magnetic field, whose component is given by

$$b_i(\mathbf{r}) = \frac{1}{2}\varepsilon^{ijk}\mathbf{n}(\mathbf{r}) \cdot (\partial_j \mathbf{n}(\mathbf{r}) \times \partial_k \mathbf{n}(\mathbf{r})), \tag{1.8}$$

where ε^{ijk} is the Levi-Civita symbol; in Eq. (1.8) and hereafter, the Einstein summation convention is used for repeated indices. The defects with positive (negative) Q_{m} are called (anti)hedgehogs. Equation (1.7) indicates that the monopole charge is given by a surface integral of the emergent magnetic field, and hence the hedgehog (antihedgehog) is also termed the emergent magnetic monopole (antimonopole). The monopole charge takes the values of $Q_{\mathrm{m}} = 1, -1$, and 2 for Fig. 1.4a, b, and c, respectively. Figure 1.4d shows a stereographic projection from a hedgehog with $Q_{\mathrm{m}} = 1$, which has a different spin texture from Fig. 1.4a, to the skyrmion with $N_{\mathrm{sk}} = -1$. Thus, there is a one-to-one correspondence between the topology of the hedgehog with Q_{m} and the skyrmion with $N_{\mathrm{sk}} = -Q_{\mathrm{m}}$, similar to the relationship between the vortex and the chiral soliton in Fig. 1.2d.

Hopfion

There is another 3D topological defect without any singularities called the hopfion. It is a topological soliton which was first studied in relativistic field theories [17–19]. The topology is described by the homotopy group $\pi_3(S^2) = \mathbb{Z}$, as the hopfions can be

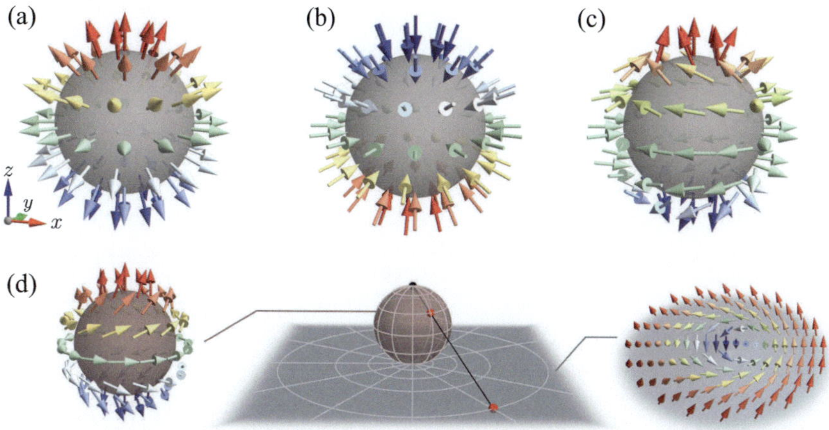

Fig. 1.4 Real-space spin configurations of **a** a hedgehog with $Q_{\rm m} = 1$, **b** an antihedgehog with $Q_{\rm m} = -1$, and **c** a hedgehog with $Q_{\rm m} = 2$. The gray spheres are guides for eyes. **d** A stereographic projection from the hedgehog with $Q_{\rm m} = 1$ in 3D space to the skyrmion with $N_{\rm sk} = -1$ in 2D space

defined by a projection of point defects in four dimensions to 3D space. A schematic picture of the real-space spin configuration of a hopfion is shown in Fig. 1.5: (a) a bird's-eye view, where the spins in the same directions are connected by the colored lines, (b) the xz cut at $y = 0$, and (c) the xy cut at $z = 0$. The topological number to characterize the hopfions is called the Hopf index, which is defined as [20–22]

$$N_{\rm H} = -\frac{1}{(4\pi)^2}\int d^3\mathbf{r}\, \mathbf{b}(\mathbf{r}) \cdot \mathbf{a}(\mathbf{r}), \tag{1.9}$$

where $\mathbf{a}(\mathbf{r})$ is the gauge potential related with $\mathbf{b}(\mathbf{r})$ in Eq. (1.8) as

$$\mathbf{b}(\mathbf{r}) = \nabla \times \mathbf{a}(\mathbf{r}). \tag{1.10}$$

1.2 Topological Spin Crystal

The topological defects introduced in the previous section can exist as a single object. Nonetheless, they often appear in magnets by forming periodic arrays called the topological spin crystals. Such periodic spin textures can be approximately represented by the superpositions of simple spin structures, such as spin helices and sinusoidal waves. When each composed wave is characterized by the wave vector $\mathbf{Q}_\eta$, the general form of the superposition is represented by

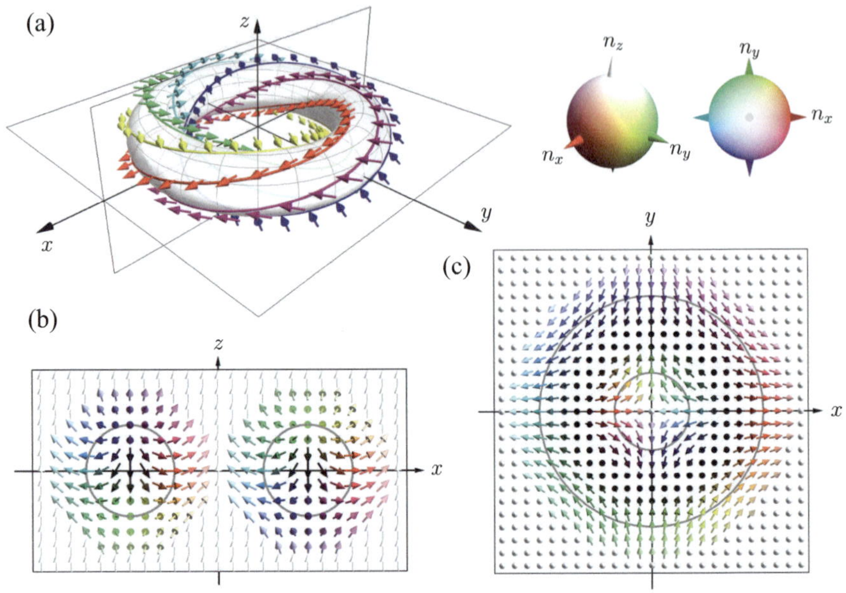

Fig. 1.5 **a** Schematic picture of the real-space spin configuration of a hopfion with $N_{\mathrm{H}} = 1$. The color of the arrows represents the xy components of the spins. The colored lines connect the spins with the same directions on the torus. **b** and **c** display the xz cut at $y = 0$ and the xy cut at $z = 0$, respectively

$$\mathbf{S}(\mathbf{r}) \propto \sum_{\eta=1}^{N_Q} \left(\psi_\eta^{\mathrm{c}} \mathbf{e}_\eta^1 \cos \mathcal{Q}_\eta + \psi_\eta^{\mathrm{s}} \mathbf{e}_\eta^2 \sin \mathcal{Q}_\eta\right) + \mathbf{m}, \tag{1.11}$$

where N_Q is the number of superposed waves, ψ_η^{c} and ψ_η^{s} are the amplitudes of cosinusoidal and sinusoidal waves with the wave vector $\mathbf{Q}_\eta$, respectively, $\mathbf{e}_\eta^1$ and $\mathbf{e}_\eta^2$ are the orthogonal unit vectors, $\mathcal{Q}_\eta = \mathbf{Q}_\eta \cdot \mathbf{r} + \varphi_\eta$ (φ_η is the phase), and $\mathbf{m}$ represents the uniform magnetization. We call such superposed spin textures the multiple-Q spin structures, and specifically, that with N_Q waves is called the $N_Q Q$ spin structure. The representative example is a superposition of the proper-screw spin helices with $\psi_\eta^{\mathrm{c}} = \psi_\eta^{\mathrm{s}}$ whose helical planes are perpendicular to the propagating directions, i.e., $\mathbf{Q}_\eta \times (\mathbf{e}_\eta^1 \times \mathbf{e}_\eta^2) = 0$. In the following, we introduce several topological spin crystals represented by a superposition of multiple spin helices in 1D, 2D, and 3D space. Specifically, we introduce the 1D chiral soliton lattice (CSL), 2D skyrmion lattice (SkL), and 3D hedgehog lattice (HL) in Sects. 1.2.1, 1.2.2, and 1.2.3, respectively.

1.2.1 Chiral Soliton Lattice

Let us first discuss the topological spin crystal in 1D space. Here, we consider a superposition of multiple helical spin structures, as exemplified in Fig. 1.6a; these spin structures exhibit elliptical modulations, as discussed later. The superposition of these elliptical helices brings about a spin structure with a periodic arrangement of chiral solitons, which is called the CSL. The spin configuration of the CSL is analytically given by $\theta(x) = 2\text{am}(\bar{x}, \kappa)$ and $\phi(x) = -\frac{\pi}{2}$, where $\bar{x} = \frac{K(\kappa)Q_{\text{CSL}}}{\pi}x$, Q_{CSL} is the ordering wave number, and $\text{am}(x, \kappa)$ and $K(\kappa)$ is the Jacobi amplitude function and the elliptic integral of the first kind with the elliptic modulus κ satisfying $0 \leq \kappa \leq 1$, respectively [23]. The parameters κ and Q_{CSL} depend on the system (see Sect. 1.4.1). Thus, $\mathbf{S}(x)$ reads

$$\mathbf{S}(x) = \left(0, -2\text{sn}(\bar{x}, \kappa)\text{cn}(\bar{x}, \kappa), 1 - \frac{2}{\kappa^2} + \frac{2}{\kappa^2}\text{dn}^2(\bar{x}, \kappa)\right), \tag{1.12}$$

where sn, cn, and dn are the Jacobi elliptic functions. Figure 1.6b–d represent the real-space spin configuration of the CSL given by Eq. (1.12) for several values of κ with a fixed Q_{CSL}. The spin texture with $\kappa = 0$ is shown in Fig. 1.6b, exhibiting the helical spin structure. When $\kappa = 0.9$, the spin texture acquires the solitonic feature and forms the CSL, as shown in Fig. 1.6c. By further increasing κ, the spin texture gets more nonlinear modulations, as exemplified by the CSL with $\kappa = 0.9999$ in Fig. 1.6d. Hence, the solitonic feature of the spin texture is prominent when κ approaches 1 and vanishes for $\kappa = 0$.

By Fourier transforming Eq. (1.12), the CSL can be represented as a superposition of multiple waves as

$$\mathbf{S}(x) = \sum_{\eta=1}^{\infty} \psi_\eta \left(\hat{\mathbf{z}}\cos(\eta Q_{\text{CSL}}x) - \hat{\mathbf{y}}\epsilon_\eta \sin(\eta Q_{\text{CSL}}x)\right) + m\hat{\mathbf{z}}, \tag{1.13}$$

where $\hat{\mathbf{y}}$ and $\hat{\mathbf{z}}$ are the unit vectors in the y and z directions, respectively,

$$\psi_\eta = \left(\frac{\pi}{\kappa K(\kappa)}\right)^2 \frac{2\eta}{\sinh(\eta\pi K(\kappa')/K(\kappa))}, \tag{1.14}$$

$$\epsilon_\eta = \tanh\left(\frac{\eta\pi K(\kappa')}{K(\kappa)}\right), \tag{1.15}$$

$$m = 1 - \frac{2}{\kappa^2}\left(1 - \frac{E(\kappa)}{K(\kappa)}\right) (\leq 0), \tag{1.16}$$

$$\kappa' = \sqrt{1 - \kappa^2}, \tag{1.17}$$

and $E(\kappa)$ is the elliptic integral of the second kind. Thus, the CSL is represented as a superposition of multiple elliptical helices with a wave number of ηQ_{CSL} as well as

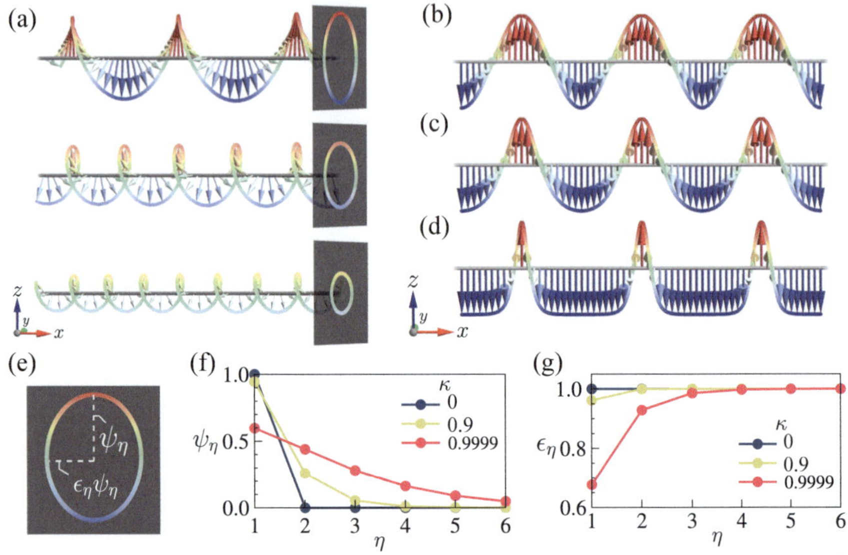

Fig. 1.6 **a** Schematic illustration of three elliptical waves propagating in the x direction with different wave numbers. The colored lines on the black planes represent the trajectories of spins on the $S_y S_z$ plane. Real-space spin configurations of the CSL in Eq. (1.12): **b** $\kappa = 0$, **c** $\kappa = 0.9$, and **d** $\kappa = 0.9999$. **e** Schematic illustration of the amplitude ψ_η and the ellipticity ϵ_η. η dependence of **f** ψ_η and **g** ϵ_η for several values of κ

the uniform magnetization m. Each constituent wave exhibits an elliptical trajectory in spin space, characterized by the major axis of ψ_η and the ellipticity denoted by ϵ_η, as schematically shown in Fig. 1.6e. We show ψ_η and ϵ_η for several values of κ in Fig. 1.6f and g, respectively. By increasing κ, ψ_η for $\eta \geq 2$ increases, whereas ψ_1 decreases. Hence, the stronger solitonic feature arises from the larger contributions from the higher harmonic waves with $\eta \geq 2$, leading to rapid spatial modulations of the spin configuration. In addition, ϵ_η decreases by increasing κ, as the majority of spins point in $\pm\hat{\mathbf{z}}$ directions due to the strong solitonic feature.

1.2.2 *Skyrmion Lattice*

Next, we show two examples of the multiple-Q spin structures in two dimensions. Hereafter, we consider the superposition of helices without elliptical modulations for simplicity. Figure 1.7a and b show the superpositions of two helices: (a) a schematic illustration of the two helices propagating in the orthogonal directions of $\mathbf{Q}_1$ and $\mathbf{Q}_2$, and (b) their superposition (the spin length is normalized after the superposition). When $\mathbf{m} = 0$, this double-Q structure has singular points periodically where the spin length vanishes, and the spin textures surrounding them look like vortices and

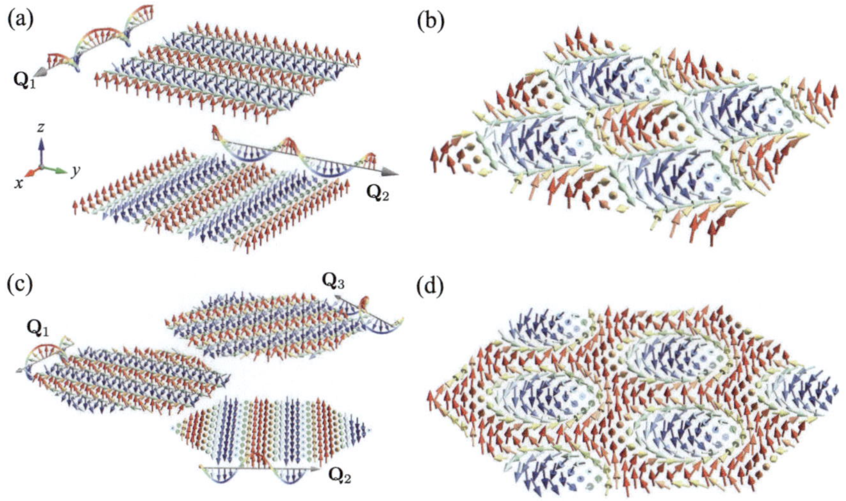

Fig. 1.7 **a** Schematic illustration of two spin helices propagating in the orthogonal directions denoted by the wave vectors $\mathbf{Q}_1$ and $\mathbf{Q}_2$. **b** Real-space spin configuration obtained by a superposition of the two helices in (**a**). The spin structure is called the double-Q vortex crystal (2Q-VC). **c** Schematic illustration of three spin helices propagating in the 120° directions with $\mathbf{Q}_1$, $\mathbf{Q}_2$, and $\mathbf{Q}_3$, and **d** their superposition called the triple-Q skyrmion lattice (3Q-SkL). The color of the arrows represents the z component of the spins, and the spin length is normalized in the superpositions in (**b**) and (**d**)

antivortices. Hence, this is called the 2Q vortex crystal (2Q-VC). In contrast to the vortices and antivortices confined in the 2D plane, $\mathbf{n}(\mathbf{r})$ spans the spherical space of S^2 in the present case, and therefore, the 2Q-VC is topologically trivial (see Table 1.1). Alternatively, this structure can be regarded as the periodic array of the meron and antimeron, each of which wraps the half of the unit sphere in spin space and shows the skyrmion number $N_{\mathrm{sk}} = 1/2$ and $-1/2$, respectively. Thus, this structure is also dubbed as the meron-antimeron lattice (MAML) [24]. On the other hand, when $\mathbf{m} \neq 0$, this 2Q structure turns into a topologically nontrivial one as the singular points vanish and the spins in the magnetic unit cell (MUC) wrap a unit sphere once. This is regarded as a square lattice of skyrmions, and hence called the 2Q skyrmion lattice (2Q-SkL). This structure has been found in several materials as the stable and metastable states (see Sect. 1.4.2).

Figure 1.7c and d show the other example with a superposition of three helices propagating in the 120° directions. In this case, the triple-Q structure in Fig. 1.7d is regarded as a hexagonal array of the skyrmions, and hence it is called the triple-Q skyrmion lattice (3Q-SkL). This structure has been discovered in many materials as will be discussed in Sect. 1.4.2.

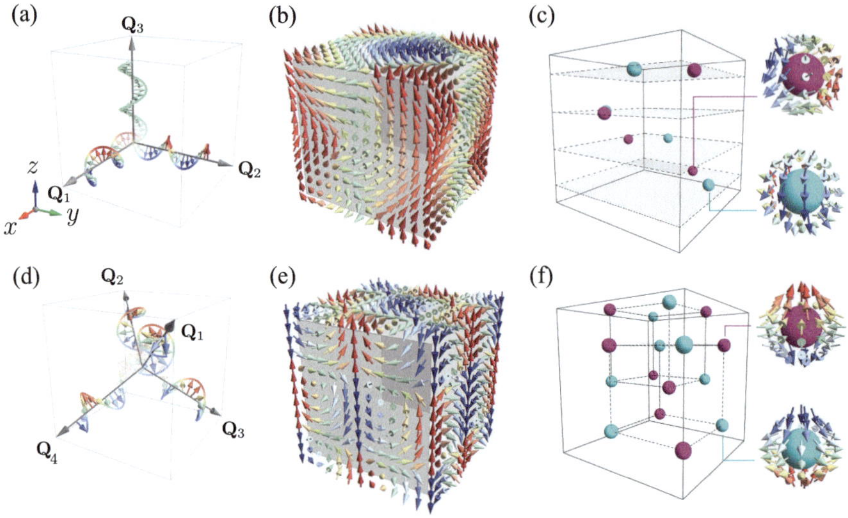

Fig. 1.8 **a** Schematic illustration of three spin helices along the orthogonal cubic directions $\mathbf{Q}_1$, $\mathbf{Q}_2$, and $\mathbf{Q}_3$, and **b** real-space spin configuration of their superposition in the MUC. The spin texture is called the $3Q$ hedgehog lattice ($3Q$-HL). **c** Real-space distribution of the hedgehogs (magenta spheres) and the antihedgehogs (cyan spheres) in the $3Q$-HL. The gray planes denote the xy planes on each of which a pair of hedgehog and antihedgehog lies. **d**, **e**, and **f** are the corresponding figures in the case of four spin helices along the tetrahedral directions. This is called the $4Q$-HL. In **f**, the dashed cubes connecting the hedgehogs and antihedgehogs are guides for eyes. The insets in (**c**) and (**f**) show the spin configurations surrounding the hedgehog and antihedgehog

1.2.3 Hedgehog Lattice

Finally, we show two examples in three dimensions. The first one is a superposition of three helices propagating in the orthogonal cubic directions, as shown in Fig. 1.8a. The resultant spin texture is shown in Fig. 1.8b. This $3Q$ spin structure has a periodic array of hedgehogs and antihedgehogs; there are four hedgehogs and four antihedgehogs in the MUC, whose configuration forms helices along the x, y, and z directions, as shown in Fig. 1.8c. Hence, this is called the $3Q$ hedgehog lattice ($3Q$-HL). The second example is the case with four helices propagating in the tetrahedral directions, called the $4Q$-HL, as shown in Fig. 1.8d. A resultant spin texture is exemplified in Fig. 1.8e.[1] The $4Q$-HL has a different array of hedgehogs and antihedgehogs, in which eight hedgehogs and eight antihedgehogs form two NaCl-type structures interpenetrating into each other, as shown in Fig. 1.8f. These HLs have been recently found in actual materials as discussed in Sect. 1.4.3.

[1] The positions and the number of hedgehogs and antihedgehogs in the spin texture depend on the phases φ_η in Eq. (1.11), as will be discussed in Sect. 7.3. Presented here is the result with $\sum_{\eta=1}^{4} \varphi_\eta = \frac{\pi}{3}$.

1.3 Emergent Electromagnetism

The topological spin structures introduced thus far exhibit noncollinear and noncoplanar spin textures. Such spin textures are known to generate an emergent electromagnetic field (EEMF) through the Berry phase mechanism [15, 16, 25–28], which acts on conduction electrons coupled with the spins. Furthermore, the EEMF gives rise to intriguing quantum transport phenomena. To see the emergence of this fictitious electromagnetic field, let us consider a simple model for spin-charge coupled metals in continuous space, whose Hamiltonian and Lagrangian are respectively given by

$$\mathcal{H}(\mathbf{r},t) = \mathbf{c}^\dagger(\mathbf{r},t)\left(-\frac{\hbar^2}{2m}\nabla^2 - \varepsilon_\mathrm{F} - J_\mathrm{sd}\mathbf{S}(\mathbf{r},t)\cdot\boldsymbol{\sigma}\right)\mathbf{c}(\mathbf{r},t), \tag{1.18}$$

$$\mathcal{L} = \int d^3\mathbf{r}\, i\hbar\mathbf{c}^\dagger(\mathbf{r},t)\frac{\partial\mathbf{c}(\mathbf{r},t)}{\partial t} - \mathcal{H}(\mathbf{r},t), \tag{1.19}$$

where $\mathbf{c}(\mathbf{r},t) = \big(c_\uparrow(\mathbf{r},t), c_\downarrow(\mathbf{r},t)\big)$, $c_\sigma^\dagger(\mathbf{r},t)$ $[c_\sigma(\mathbf{r},t)]$ is a creation (annihilation) operator of an itinerant electron at position $\mathbf{r}$ and time t with spin $\sigma =\uparrow$ or $\downarrow$, $\hbar$ is the Dirac constant, m is the electron mass, and ε_F is the Fermi energy. The first term in Eq. (1.18) describes the kinetic energy of electrons and the third term represents the coupling between the itinerant electron spin denoted by the vector of the Pauli matrices $\boldsymbol{\sigma} = (\sigma_x, \sigma_y, \sigma_z)$ and the localized spin $\mathbf{S}(\mathbf{r},t)$ with $\mathbf{S}(\mathbf{r},t) = S\mathbf{n}(\mathbf{r},t)$. By introducing a new electron operator $\mathbf{C}(\mathbf{r},t)$ given by

$$\mathbf{C}(\mathbf{r},t) = \mathsf{U}(\mathbf{r},t)\mathbf{c}(\mathbf{r},t), \tag{1.20}$$

with

$$\mathsf{U}(\mathbf{r},t) = \mathbf{m}(\mathbf{r},t)\cdot\boldsymbol{\sigma}, \tag{1.21}$$

where

$$\mathbf{m}(\mathbf{r},t) = \left(\sin\frac{\theta}{2}\cos\phi, \sin\frac{\theta}{2}\sin\phi, \cos\frac{\theta}{2}\right), \tag{1.22}$$

the spin quantization axis for itinerant electrons becomes parallel to the localized spin moment at each $\mathbf{r}$. When the spin-charge coupling is much larger than the bandwidth, one can neglect electrons whose spin is antiparallel to $\mathbf{n}(\mathbf{r},t)$, which is called the adiabatic approximation. In this case, by substituting $\mathbf{C}(\mathbf{r},t)$ with $(C(\mathbf{r},t), 0)$, the Lagrangian in Eq. (1.19) reads

$$\begin{aligned}\mathcal{L} = &\int d^3\mathbf{r}\, i\hbar C^\dagger(\mathbf{r},t)\frac{\partial C(\mathbf{r},t)}{\partial t} + |e|A_0^\mathrm{em}(\mathbf{r},t)C^\dagger(\mathbf{r},t)C(\mathbf{r},t)\\ &+\frac{\hbar^2}{2m}C^\dagger(\mathbf{r},t)\left(\nabla + i\frac{|e|}{\hbar}\mathbf{A}^\mathrm{em}(\mathbf{r},t)\right)^2 C(\mathbf{r},t) + (\varepsilon_\mathrm{F} + J_\mathrm{sd}S)\,C^\dagger(\mathbf{r},t)C(\mathbf{r},t),\end{aligned} \tag{1.23}$$

where

$$A_0^{\rm em}(\mathbf{r}, t) = -\frac{\hbar}{2|e|}(1 - \cos\theta(\mathbf{r}, t))\frac{\partial\phi(\mathbf{r}, t)}{\partial t}, \tag{1.24}$$

$$\mathbf{A}^{\rm em}(\mathbf{r}, t) = \frac{\hbar}{2|e|}(1 - \cos\theta(\mathbf{r}, t))\nabla\phi(\mathbf{r}, t). \tag{1.25}$$

Here, e is the elementary charge. Therefore, under the adiabatic approximation, electrons coupled to localized spin moments can be regarded as free electrons with the scalar potential $A_0^{\rm em}(\mathbf{r}, t)$ and the vector potential $\mathbf{A}^{\rm em}(\mathbf{r}, t)$.

The quantum mechanical phases acquired by itinerant electrons result in the scalar potential and the vector potential, and accordingly, these potentials generate the EEMF, $\mathbf{B}^{\rm em}(\mathbf{r}, t) = \nabla \times \mathbf{A}^{\rm em}(\mathbf{r}, t)$ and $\mathbf{E}^{\rm em}(\mathbf{r}, t) = -\nabla A_0^{\rm em}(\mathbf{r}, t) - \partial_t \mathbf{A}^{\rm em}(\mathbf{r}, t)$. The explicit forms are given by

$$\mathbf{B}^{\rm em}(\mathbf{r}, t) = \frac{\hbar}{2|e|}\mathbf{b}(\mathbf{r}, t), \tag{1.26}$$

$$\mathbf{E}^{\rm em}(\mathbf{r}, t) = \frac{\hbar}{2|e|}\mathbf{e}(\mathbf{r}, t), \tag{1.27}$$

where $\mathbf{b}(\mathbf{r}, t)$ is defined in Eq. (1.8) except for the explicit time dependence and

$$e_i(\mathbf{r}, t) = \mathbf{n}(\mathbf{r}, t) \cdot (\partial_i \mathbf{n}(\mathbf{r}, t) \times \partial_t \mathbf{n}(\mathbf{r}, t)). \tag{1.28}$$

Note that Eq. (1.26) has the same functional form as Eq. (1.8), and thus spin noncoplanarity is essential to have nonzero $\mathbf{B}^{\rm em}$. Meanwhile, Eq. (1.27) does not require the spin noncoplanarity, but the spin noncollinearity is enough to have nonzero $\mathbf{E}^{\rm em}$. In the following, we discuss the relation between the noncoplanar spin textures and the EEMF. Specifically, by taking the hedgehogs and the skyrmion as illustrative examples, we discuss the emergent magnetic field (EMF) and the emergent electric field (EEF) in Sects. 1.3.1 and 1.3.2, respectively. Then, we introduce the resultant quantum transport phenomena and optical phenomena in Sect. 1.3.3.

1.3.1 Emergent Magnetic Field

In order to understand the relation between the noncoplanar spin textures and the EMF, let us consider a 3D spin texture given by $\mathbf{S}(\mathbf{r}) = \frac{1}{r}(x, y, z)$, where $r = |\mathbf{r}|$. This represents the spin texture around the hedgehog in Fig. 1.4a. By straightforward calculations, one can obtain the corresponding EMF by Eq. (1.8) as

$$\mathbf{b}(\mathbf{r}) = \frac{\mathbf{r}}{r^3}. \tag{1.29}$$

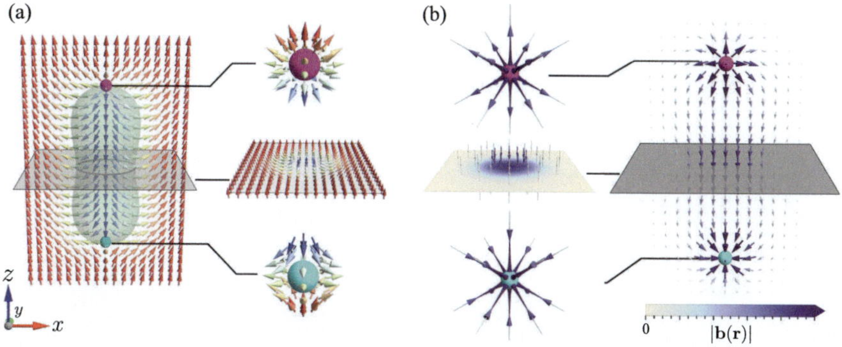

Fig. 1.9 **a** Real-space spin configuration of the hedgehog, antihedgehog, and skyrmion in 3D space. In the left panel, the spins on the xz plane are shown for better visibility. The green closed surface represents the isosurface with $S_z(\mathbf{r}) = 0$. The positions of the hedgehog and antihedgehog are denoted by the magenta and cyan spheres, respectively. **b** Real-space distribution of the stream of $\mathbf{b}(\mathbf{r})$ around the topological defects: hedgehog (left top), skyrmion (left middle), and antihedgehog (left bottom). Real-space distribution of $\mathbf{b}(\mathbf{r})$ on the xz plane (right). The arrows and the color represent the direction and norm of $\mathbf{b}(\mathbf{r})$, respectively

This indicates that the hedgehog can be regarded as a source of the EMF, i.e., a magnetic monopole. In the same way, the antihedgehog can be regarded as a sink of the EMF, i.e., a magnetic antimonopole. The schematic illustration for the correspondence between the spin textures and the EMF for the hedgehog and antihedgehog is shown in Fig. 1.9. The hedgehog with $Q_\mathrm{m} = 1$ and antihedgehog with $Q_\mathrm{m} = -1$ are connected by the Dirac string [29] composed of the spins pointing downward, whose intersection gives the core of the skyrmion with $N_\mathrm{sk} = -1$, as shown in Fig. 1.9a (see Sect. 6.1). Real-space distribution of the EMF arising from the spin configuration is shown in Fig. 1.9b. As we see above, the hedgehog and antihedgehog correspond to the magnetic monopole and antimonopole, respectively. Furthermore, one can show that a skyrmion carries the magnetic flux

$$\frac{\hbar}{2|e|}\int dxdyb_z(\mathbf{r}) = \frac{h}{|e|}N_\mathrm{sk}, \tag{1.30}$$

where h is the Planck constant. Hence, the skyrmion can be regarded as a carrier of the flux quantum.

1.3.2 Emergent Electric Field

Next, we consider the EEF. To see the relation between the topological spin texture and the EEF, let us consider the time evolution of the spin texture given by $\mathbf{S}(\mathbf{r}, t) = (\mathbf{r} - \mathbf{v}t)/\,|\mathbf{r} - \mathbf{v}t|$, which represents the spin configuration around the hedgehog moving with the velocity $\mathbf{v}$. One can again easily obtain the corresponding EEF by Eq. (1.28) as

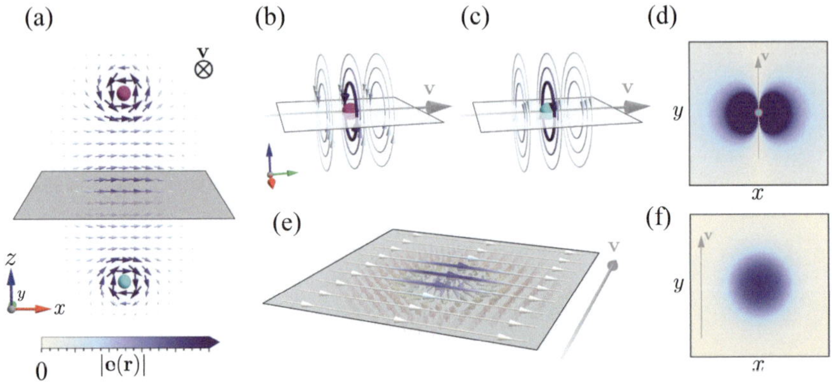

Fig. 1.10 **a** Real-space distribution of $\mathbf{e}(\mathbf{r})$ on the xz plane (see the text for the details). Stream of the EEF around **b** the hedgehog and **c** the antihedgehog. **d** Real-space distribution of $|\mathbf{e}(\mathbf{r})|$ on the xy planes intersecting the hedgehog and antihedgehog. The positions of the hedgehog and antihedgehog are common and denoted by the dots. **e** Stream of the EEF and the real-space spin configuration of the skyrmion on the xy plane. **f** Real-space distribution of $|\mathbf{e}(\mathbf{r})|$ on the skyrmion plane

$$\mathbf{e}(\mathbf{r}, t) = \frac{\mathbf{r} \times \mathbf{v}}{r^3}. \tag{1.31}$$

Around the hedgehog, the EEF is perpendicular to $\mathbf{v}$[2] and its amplitude vanishes along the axis parallel to $\mathbf{v}$. Real-space distribution of the EEF for the motion of the spin texture given in Fig. 1.9a with $\mathbf{v} = (0, v, 0)$ is shown in Fig. 1.10a. The stream of the EEF around the hedgehog and antihedgehog are given in Fig. 1.10b and c, respectively. Indeed, the EEF is circulating around the hedgehog and its y component is zero, as represented by the stream of the EEF exhibiting the closed loops on the xz plane. Figure 1.10d shows the distribution of $|\mathbf{e}(\mathbf{r})|$ on the xy intersection of the hedgehog, showcasing the vanishing EEF along the y axis; the distribution is common to the result for the antihedgehog. The EEF can also be nonzero around the skyrmion. The stream of the EEF and real-space distribution of $|\mathbf{e}(\mathbf{r})|$ around the skyrmion on the xy plane are shown in Fig. 1.10e and f, respectively. On this plane, the EEF only has the x component and its amplitude is stronger around the skyrmion core. This behavior can be understood from conventional electromagnetic induction by regarding the skyrmion as a magnetic flux (see Sect. 1.3.1).

Although the spin noncoplanarity is required to have nonzero EMF, the EEF only requires the spin noncollinearity in compensation for the time evolution of the spin texture. Thus, the EEF can be nonzero even in 1D spin textures such as the helical or conical spin texture. Let us consider a spin configuration given by

[2] We note that the EEF arising from the rigid translational motion of spin textures is generally perpendicular to the direction of the velocity. The real-space and real-time evolution of such a spin texture with velocity $\mathbf{v}$ is described by $\mathbf{n}(\mathbf{r}, t) = \mathbf{n}(\mathbf{r} - \mathbf{v}t)$. Under this condition, the EEF is given by the relation $e_i(\mathbf{r}, t) = -\mathbf{n} \cdot (\partial_i \mathbf{n} \times (\mathbf{v} \cdot \nabla)\mathbf{n})$, which directly leads to $\mathbf{e} \cdot \mathbf{v} = 0$.

$$\mathbf{S}(x,t) = \left(m(t), \sqrt{1-m^2(t)}\cos(Qx), \sqrt{1-m^2(t)}\sin(Qx)\right), \quad (1.32)$$

which represents the helical spin structure with the wave number Q showing the uniform spin canting with a time-dependent magnetization $m(t)$. The EEF is calculated as $\mathbf{e}(x,t) = (Q\partial_t m(t), 0, 0)$, and indeed, the EEF can be nonzero in 1D space.

1.3.3 Emergent Electromagnetic Phenomena

The EEMF can induce novel quantum transport phenomena. The representative example is the Hall responses due to the EMF, which is called the topological Hall effect, as schematically shown in Fig. 1.11a. The topological Hall effect has been experimentally observed in many materials. An example is shown in Fig. 1.11b for the experimental data of MnSi exhibiting the 3Q-SkL [30] (see Sect. 1.4.2). $\Delta\rho_{xy}$ is the topological Hall component estimated by the decomposition of the Hall resistivity into three parts: the normal Hall part $\rho_{xy}^{\mathrm{N}} = R_0 B$, the anomalous Hall part $\rho_{xy}^{\mathrm{A}} = \mu_0 R_s M$, and the additional topological Hall part $\Delta\rho_{xy} = \rho_{xy} - \rho_{xy}^{\mathrm{N}} - \rho_{xy}^{\mathrm{A}}$, where R_0 and R_s are the fitting parameters, B is the magnetic field, μ_0 is the permeability in vacuum, and M is the magnetization. The data clearly indicate that the additional topological Hall response appears in the temperature and field range corresponding to the 3Q-SkL phase. It is also known that the skyrmion shows a Hall motion called the skyrmion Hall effect, as shown in Fig. 1.11a, which is the counteraction of the electron's Hall response. Similar to the topological Hall effect, the EMF can induce the Nernst effect and the magneto-optical Kerr effect, and they are indeed experimentally observed. The Nernst effect has been observed in Gd_2PdSi_3 [33] and MnSi [34]. The magneto-optical Kerr effect has been observed in the noncollinear antiferromagnets γ-Fe_xMn_{1-x} and K_xRhO_2 [35] as well as Gd_2PdSi_3 [36] and MnGe [37].

The EEF can bring about direct voltage for electrons. Experimental detection of the EEF has been done for field-induced motion of a magnetic domain wall [31], ferromagnetic resonance in a patterned ferromagnetic film [38], and gyrating motion of a magnetic vortex [39]. Field dependence of the measured voltage associated with the domain wall motion in a permalloy nanowire is shown in Fig. 1.11c. Above the Walker threshold at the peak of the wall velocity represented by the open circles, the magnetic field parallel to the wire direction induces the precessional motion along with the translation of the domain wall, and accordingly, the wall-induced voltage was clearly measured above the threshold, as represented by the closed circles. In addition to the direct voltage measurement, the EEF has attracted renewed interest in the dynamics induced by an electric current. It was pointed out that the EEF by a current-induced motion of swirling spin textures gives rise to an inductance called an emergent inductance [40], and such behavior was observed in helical magnets [32] even at room temperature [41]. An example is shown in Fig. 1.11d for the experimental data of $Gd_3Ru_4Al_{12}$. The inductance appears in the imaginary part of the complex

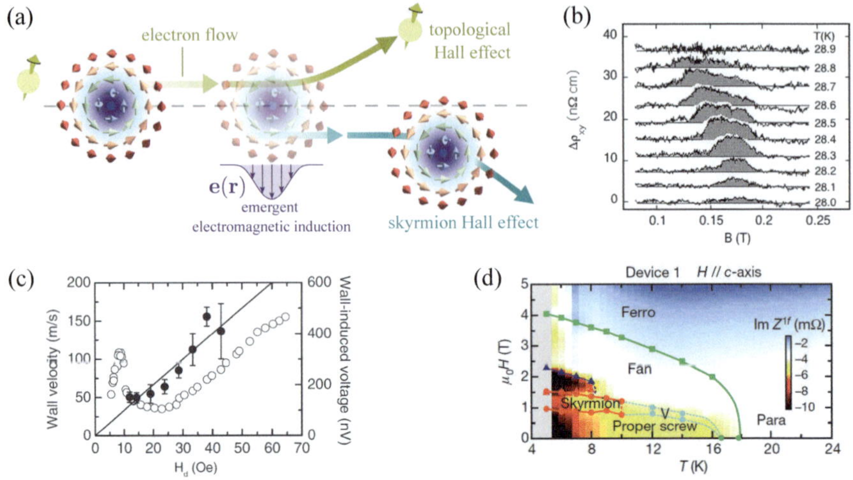

Fig. 1.11 **a** Schematic illustration of the emergent electromagnetic phenomena. **b** An experimental observation of the topological Hall effect in MnSi while varying the external magnetic field (B). $\Delta\rho_{xy}$ is the additional contribution to the normal and anomalous Hall responses; see the text for details. Reprinted figure with permission from Ref. [30]. Copyright (2009) by the American Physical Society. **c** Observation of the EEF in a permalloy while changing the magnetic field (H_d). Wall velocity and the voltage associated with the EEF arising from the motion of the domain wall are denoted by the open and closed circles, respectively. Reprinted figure with permission from Ref. [31]. Copyright (2009) by the American Physical Society. **d** The imaginary part of the impedance originating from the EEF for various magnetic phases in $Gd_3Ru_4Al_{12}$ as a function of the temperature (T) and the magnetic field ($\mu_0 H$). Reprinted figure from Ref. [32]. Reproduced with permission from Springer Nature

impedance of the fundamental components, $\mathrm{Im}Z^{1f}$, exhibiting the large value in several magnetic phases. Intriguingly, the amplitude of $\mathrm{Im}Z^{1f}$ takes the largest value in the proper screw and transverse conical phases located at the lower and higher field region of the SkL phase, respectively. The EEF can give rise to significant emergent electric phenomena with these spin textures much simpler than the SkL.

1.4 Experiments on Topological Spin Textures

In the previous sections, we introduced the topological spin crystals, which are periodic arrays of topological defects represented by a superposition of multiple spin density waves. We here briefly introduce the experimental observations of the topological spin textures, focusing on the CSLs, the SkLs, and the HLs.

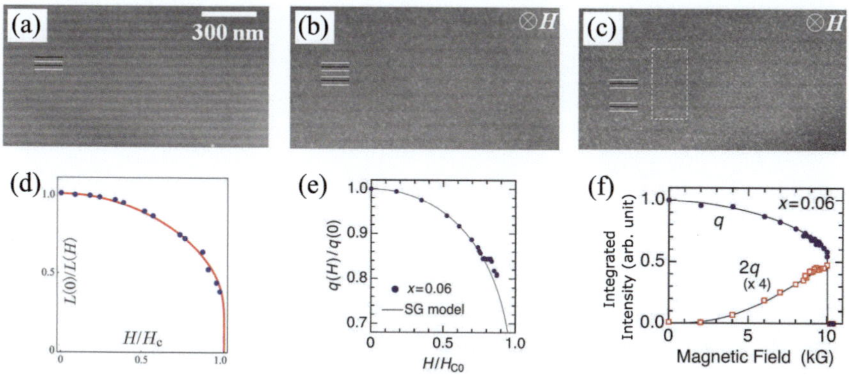

Fig. 1.12 Experimental results of LTEM for $CrNb_3S_6$ at 110 K under the magnetic field perpendicular to the plane: **a** 0 T, **b** 0.208 T, and **c** 0.224 T. The stripe patterns of the intensity are propagating in the direction of the crystallographic c axis. (d) Magnetic field dependence of $L(0)/L(H)$ in $CrNb_3S_6$. Reprinted figures with permission from Ref. [42]. Copyright (2012) by the American Physical Society. Magnetic field dependence of (e) the ordering wave number q and (f) the integrated intensities around the q and $2q$ peaks for the resonant X-ray diffraction experiment in $Yb(Ni_{1-x}Cu_x)_3Al_9$ with $x = 0.06$. Reprinted figures from Ref. [43] licensed under CC BY 4.0. Copyright (2017) T. Matsumura et al.

1.4.1 Chiral Soliton Lattice

A direct observation of the CSL was firstly made for a monoaxial chiral magnet $CrNb_3S_6$ [42]. It has also been reported in Ba_2CuGeO_7 [44–47], CuB_2O_4 [48, 49], and $Yb(Ni_{1-x}Cu_x)_3Al_9$ [43], and $Co_8Zn_{10}Mn_2$ [50]. $CrNb_3S_6$ has a layered hexagonal structure of 2H-type NbS_2 intercalated by Cr atoms. The CSL was directly observed by means of Lorentz transmission electron microscopy (LTEM) and small-angle electron diffraction [42], as shown in Fig. 1.12. The LTEM image at zero field is shown in Fig. 1.12a, exhibiting the equal widths of the bright and dark stripes with a period of $\simeq$48 nm. This means that a helical spin texture is realized with equal populations of spins with positive and negative S_z components. The result under the magnetic field perpendicular to the crystallographic c axis in $CrNb_3S_6$ at $H = 0.208$ T is shown in Fig. 1.12b. The width of the dark stripes decreases, whereas that of the bright stripes increases by increasing the magnetic field. This behavior indicates the formation of the CSL, where the positive and negative S_z populations are inequivalent [see Fig. 1.6b–d]. Furthermore, the magnetic modulation period increases. These trends maintain up to the critical field $H_c \simeq 0.230$ T, as shown in Fig. 1.12c.

The increase of the magnetic modulation period with the magnetic field is a common property of the CSL. This was theoretically predicted for a chiral magnet, whose Hamiltonian is given by the so-called chiral sine-Gordone model [51–53]. For such a model, the lowest-energy spin configuration under the magnetic field perpendicular to the chiral axis H is given by Eq. (1.12) with the parameters given by

$$\left(\frac{\kappa}{E(\kappa)}\right)^2 = \frac{H}{H_c}, \tag{1.33}$$

$$Q_{\mathrm{CSL}} = \frac{\pi^2}{4K(\kappa)E(\kappa)}Q_0, \tag{1.34}$$

with

$$H_c = \left(\frac{\pi Q_0}{4}\right)^2 J, \tag{1.35}$$

where H_c is the critical magnetic field, Q_0 is the ordering wave number at zero field, i.e., the wave number of the helical spin structure, and J is the coefficient of the exchange interaction. For the model details, see Sect. 3.2. By increasing H, κ representing the solitonic feature of the CSL increases, and accordingly, Q_{CSL} monotonically decreases, namely, the magnetic modulation period increases. Figure 1.12d shows the field dependence of the CSL period $L(H) = 2\pi/Q_{\mathrm{CSL}}$ in $CrNb_3S_6$ and the theoretical result, which are denoted by the blue dots and the red line, respectively, showing a good agreement with each other. This increasing behavior of $L(H)$ with H is one of the hallmarks of the CSL. A similar behavior was reported for the resonant X-ray diffraction experiment in $Yb(Ni_{1-x}Cu_x)_3Al_9$. Figure 1.12e shows the H dependence of the ordering wave number $q(H)$ denoted by the dots and the theoretical result denoted by the line, both of which are in good agreement for $H/H_{c0} \lesssim 0.8$, where H_{c0} is the critical field. The lock-in of q and the deviation from the theory for $H/H_{c0} \gtrsim 0.8$ arises from the interplay between conduction electrons and localized spin, which cannot be described by the chiral sine-Gordon model [54]. The integrated diffraction intensity around $q(H)$ and $2q(H)$ are shown in Fig. 1.12f. Due to the strong solitonic feature of the CSL, the contribution from the higher harmonics increases with H, as discussed in Sect. 1.2.1.

1.4.2 Skyrmion Lattice

An isolated skyrmion and its periodic array, i.e., SkL, were theoretically predicted to be stabilized in a magnet without inversion symmetry (see Sect. 1.5.1) [11, 12, 14]. Later, it was first observed in a B20 compound MnSi, which possesses a chiral crystal structure [58]. The phase diagram for the temperature and magnetic field is shown in Fig. 1.13a. At zero field, the compound exhibits a helical order below $\simeq$30 K, which turns into a conical order with spin canting in an applied magnetic field. The SkL was discovered in the A phase just below the critical temperature in the field-induced conical region whose nature has long evaded being identified. Figure 1.13b shows the small angle neutron scattering (SANS) data in the A phase, which indicates clear hexagonal Bragg spots evincing a superposition of three spin helices consistent with the $3Q$-SkL introduced in Sect. 1.2.2. A similar phase diagram, but with a

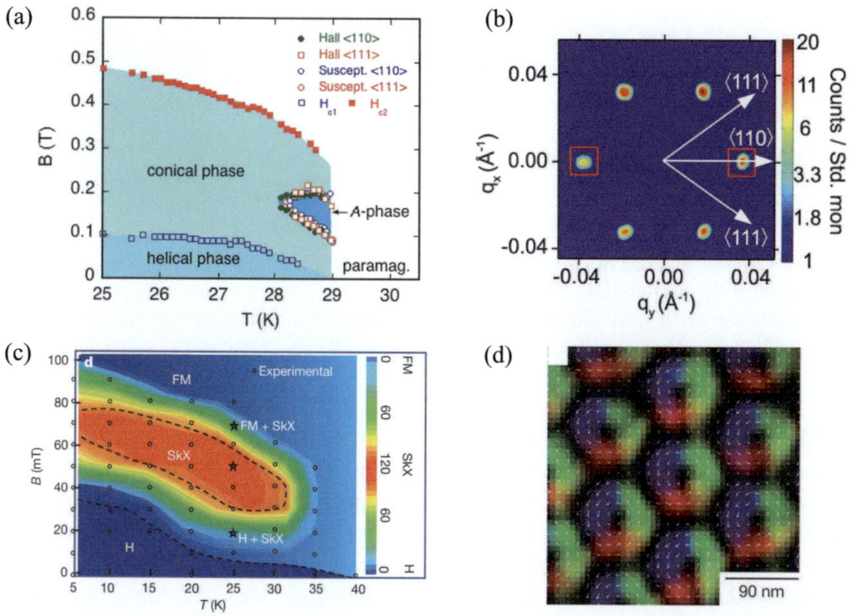

Fig. 1.13 **a** Phase diagram of MnSi as a function of the temperature (T) and magnetic field (B). A $3Q$-SkL was found in the A phase. The figure is taken from Ref. [30]. Copyright (2009) by the American Physical Society. **b** SANS data in the A phase, which indicates the formation of the $3Q$-SkL. The figure is taken from Ref. [55]. Copyright (2011) by the American Physical Society. **c** Phase diagram of a thin film of $Fe_{1-x}Co_xSi$. The red region indicates the $3Q$-SkL phase. **d** Real-space image of the periodic array of swirling spin textures in the $3Q$-SkL observed by the LTEM. Reprinted figures with permission from Ref. [56]. Reproduced with permission from Springer Nature

substantially extended region of the SkL, was obtained for a thin film of $Fe_{1-x}Co_xSi_x$, as shown in Fig. 1.13c [56]. In this case, a real-space image of the spin configuration was obtained by using LTEM as shown in Fig. 1.13d, which shows a hexagonal array of the swirling spin textures like skyrmions. After these discoveries, the SkL has been found in a wide range of magnets with similar phase diagrams to Fig. 1.13a and c, such as other B20 compounds $Mn_{1-x}Fe_xGe$ [59, 60], the multiferroic insulator Cu_2OSeO_3 [61, 62], the β-Mn type Co-Zn-Mn alloys [63], and the $4f$-electron compounds EuPtSi [64–66] and $EuNiGe_3$ [67]. The magnetic modulation period depends on materials and it typically ranges from 10 to 100 nm. These SkLs are usually observed as the $3Q$-SkL composed of three helices shown in Fig. 1.7d, which is the close-packing of skyrmions. Meanwhile, the square SkL like Fig. 1.7b can be realized as a metastable state [24, 68–70]. Notably, the square SkL was recently found in the $4f$-electron compound $EuNiGe_3$ even in an equilibrium state [67].

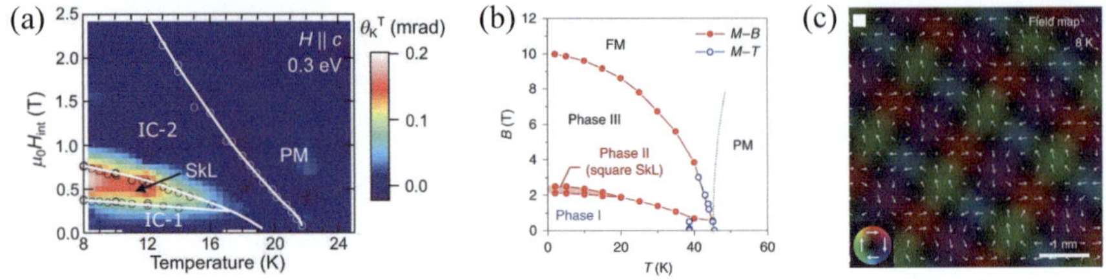

Fig. 1.14 **a** Phase diagram of Gd_2PdSi_3. The color represents the topological Kerr rotation angle θ_K^T. Reprinted figure from Ref. [36] licensed under CC BY 4.0. Copyright (2023) by Y. Kato et al. **b** Phase diagram of $GdRu_2Si_2$ and (c) real-space image of the square lattice SkL in Phase II. Reprinted figures with permission from Ref. [57]. Copyright (2023) by Nature Publishing Group

While all the above examples have noncentrosymmetric lattice structures with inversion symmetry breaking, it has also been predicted that the SkL can be stabilized even in a centrosymmetric magnet with the mechanisms discussed in Sects. 1.5.2 and 1.5.3. Indeed, new types of SkLs have been recently discovered in centrosymmetric magnets. The first candidate of this family is the triangular lattice compound Gd_2PdSi_3 [71]. Subsequently, the SkL was also found in the Gd-based breathing kagome compound $Gd_3Ru_4Al_{12}$ [72], the body-centered tetragonal lattice compound $GdRu_2Si_2$ [57], and the centrosymmetric binary compound $EuAl_4$ [73]. The phase diagram as well as the topological optical response of Gd_2PdSi_3 is shown in Fig. 1.14a. This material exhibits the $3Q$-SkL phase even at very low temperature with a nonzero magnetic field and accordingly, large topological Hall resistivity is observed in this phase [71]. The IC-1 and IC-2 phases respectively located at the lower and higher field region of the SkL phase show almost no topological Hall response. The proposed spin texture of the IC-1 phase at zero field is a topologically trivial $3Q$ state composed of three helices with different phases from the $3Q$-SkL, corresponding to the MAML (see Sect. 1.2.2). In these compounds, the magnetic modulation period of the SkL is much smaller than in the compounds introduced above. For instance, the SkL shows the period of ∼2.5 nm, ∼2.8 nm, ∼1.9 nm, and ∼3.5 nm in Gd_2PdSi_3, $Gd_3Ru_4Al_{12}$, $GdRu_2Si_2$, and $EuAl_4$, respectively. Such a small period brings about a larger emergent magnetic field and a larger topological Hall effect. Indeed, the topological Hall resistivity observed in Gd_2PdSi_3 is ∼3 μΩcm, which is three orders of magnitude larger than that in MnSi, ∼5 nΩcm [see Fig. 1.11b]. Also importantly, the square SkL, which is exceptional in noncentrosymmetric materials, has been observed in $GdRu_2Si_2$ and $EuAl_4$ as a stable phase. The square arrangement of tiny skyrmions has been directly observed in $GdRu_2Si_2$ exhibiting the phase diagram shown in Fig. 1.14b. Figure 1.14c shows the real-space image of the SkL observed by LTEM for the phase II. We will discuss the distinct stabilization mechanisms for these small-period SkLs in centrosymmetric materials in Sect. 1.5.

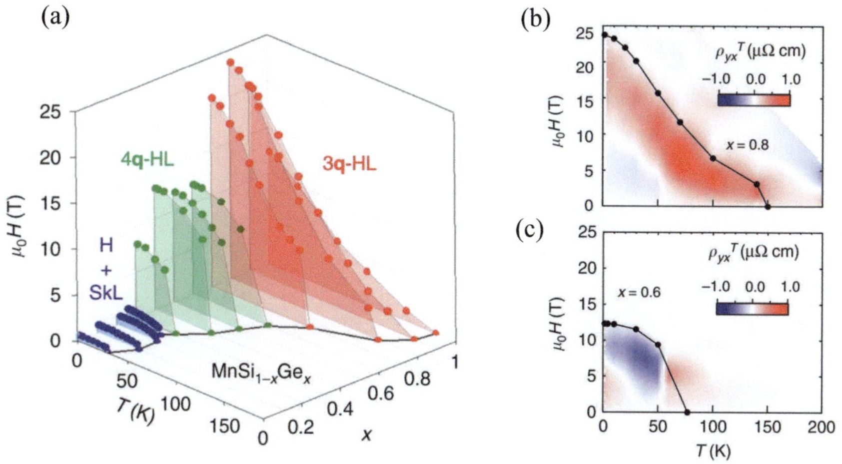

Fig. 1.15 **a** Phase diagram of $MnSi_{1-x}Ge_x$. The phase boundaries for the helical and the SkL states (H+SkL), the 4Q-HL, and the 3Q-HL are shown. **b** and **c** show the color plots of the topological Hall resistivity at $x = 0.8$ and 0.6, respectively. Reprinted figure from Ref. [74] licensed under CC BY 4.0. Copyright (2023) by Y. Fujishiro et al.

1.4.3 Hedgehog Lattice

A 3D HL was proposed to understand the peculiar magnetic and transport properties of another B20 compound MnGe [76, 77] (first referred to as the simple-cubic SkL in Ref. [77]). In stark contrast to the SkL in MnSi and other materials, the HL is stabilized even in the absence of the magnetic field, and the stable region is extended to a much wider region of the temperature and field. Combining the LTEM image of [001]-orientated MnGe under the magnetic field perpendicular to the sample plane [78] and the SANS data [74, 76, 77], the spin configuration is reasonably understood as a superposition of three helices propagating in the orthogonal cubic directions, as in the 3Q-HL introduced in Fig. 1.8a.

Another HL was discovered by doping Si to MnGe [74]. Substituting Ge with Si reduces the lattice constant, leading to an increase in bandwidth and reduced electron correlation effects. Through this band engineering, $MnSi_{1-x}Ge_x$ shows the rich magnetic phase diagram, as shown in Fig. 1.15a. In the region near the end members, MnSi ($x = 0$) and MnGe ($x = 1$), the system exhibits the SkL and the 3Q-HL, respectively [the former is denoted as the helical and SkL states (H+SkL) in Fig. 1.15a]. In the intermediate region for $0.3 \lesssim x \lesssim 0.7$, however, a different magnetic state from these two appears. In this region, the SANS experiment revealed the spots with the angles $\simeq \pm 70°$ and $\simeq \pm 110°$, consistent with the superposition of four tetrahedral waves shown in Fig. 1.8d. Combining this and anomalous transport phenomena suggesting a noncoplanar spin texture (see below), the new state is interpreted as a 4Q-HL similar to Fig. 1.8d.

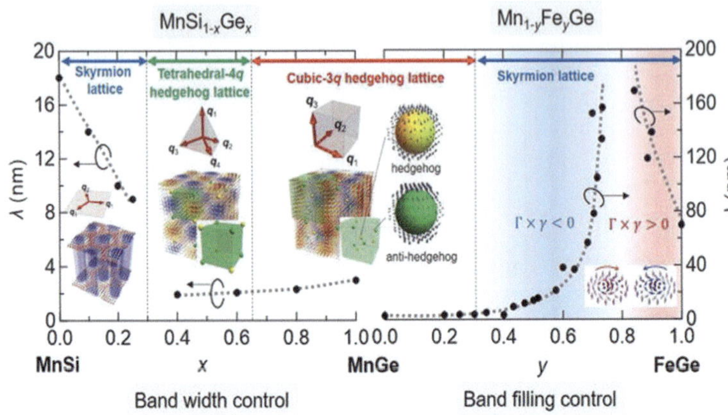

Fig. 1.16 Magnetic modulation period λ as a function of (left) x in $MnSi_{1-x}Ge_x$ and (right) y in $Mn_{1-y}Fe_yGe$. Reprinted figure with permission from Ref. [75]. Copyright (2020) by AIP Publishing

Figure 1.15b and c show the topological Hall resistivity on the plane of the temperature and the magnetic field for $MnSi_{1-x}Ge_x$ with $x = 0.8$ and $x = 0.6$, corresponding to the $3Q$- and $4Q$-HLs, respectively. In both cases, the results signal large topological Hall responses in the HL states below the critical fields denoted by the black lines, with complicated dependence on the temperature and field including the sign changes. In addition, an anomalously large Seebeck effect [79] and a topological Nernst effect [80] have been observed in MnGe. The emergent electromagnetic fields associated with the evolution of the noncoplanar spin textures in the HLs are speculated to play a crucial role in these anomalous behaviors, but the microscopic origin has not been fully clarified yet.

Thus, $MnSi_{1-x}Ge_x$ exhibits three different types of topological spin textures, the $3Q$-SkL, $3Q$-HL, and $4Q$-HL by controlling the band structure through x. Furthermore, the $3Q$-HL as well as the $3Q$-SkL has also been reported in $Mn_{1-y}Fe_yGe$, where the substitution of Mn with Fe leads to another band engineering through the electron filling and the spin-orbit coupling [75]. Interestingly, the magnetic period λ of the SkL and the HLs in these compounds shows characteristic x and y dependences, as plotted in Fig. 1.16. In $MnSi_{1-x}Ge_x$, the SkL exhibits $\lambda \simeq 18$ nm at $x = 0$ and λ decreases continuously as x increases. Meanwhile, λ is considerably small in the HL region for $0.3 \lesssim x \leq 1$; it is typically 2–3 nm and gradually increases with x through the regions of the two types of the HLs. In $Mn_{1-y}Fe_yGe$, λ takes the small value in the $3Q$-HL for $0 \leq y \lesssim 0.3$, which is similar to MnGe. Meanwhile, in the SkL phase, λ exhibits a divergent increase with y within the range $0.3 \lesssim y \lesssim 0.8$. Subsequently, it decreases accompanied by the helicity reversal due to the sign flip of the DM interaction. This behavior suggests that the stabilization mechanism is different between the SkL and the two HLs, while these spin textures can be controlled through band engineering. We discuss this issue in the next section. Although the above examples exhibiting the $3Q$- and $4Q$-HLs have noncentrosymmetric structures, recently, the $4Q$-HL similar to Fig. 1.8d as well as the topological $2Q$ state was reported in a centrosymmetric cubic perovskite $SrFeO_3$ at finite temperature even in the absence of the magnetic field [81]. Similar to the above examples, the $4Q$-HL exhibits a short magnetic modulation period of $\sim$1.8 nm and large topological Hall resistivity [82, 83].

1.5 Stabilization Mechanism

For the topological spin textures introduced in the previous sections, several different mechanisms have been argued for their stabilization [15, 84]. In this section, we introduce three of them: the Dzyaloshinskii-Moriya (DM) interaction, frustrated exchange interactions, and itinerant frustration.

1.5.1 Dzyaloshinskii-Moriya Interaction

The DM interaction is an antisymmetric exchange interaction originating from the spin-orbit coupling in noncentrosymmetric systems without spatial inversion symmetry [85, 86]. It is given in the form of $-\mathbf{D}_{ij} \cdot (\mathbf{S}_i \times \mathbf{S}_j)$, where $\mathbf{D}_{ij}$ is the vector determined by the symmetry of the lattice structure, and $\mathbf{S}_i$ denotes the spin at site i. This form indicates that the DM interaction favors a twist of spins in the perpendicular directions to the DM vector $\mathbf{D}_{ij}$.

A minimal Hamiltonian including the DM interaction for noncentrosymmetric magnets is given by

$$\mathcal{H} = \sum_{i,j} \left[-J \mathbf{S}_i \cdot \mathbf{S}_j - \mathbf{D}_{ij} \cdot (\mathbf{S}_i \times \mathbf{S}_j) \right], \tag{1.36}$$

where J is the coupling constant for the Heisenberg interaction. Suppose J is ferromagnetic ($J > 0$), the ground state of this Hamiltonian is given by a helical structure with the modulation period

$$\lambda = \frac{2\pi}{\arctan(D/J)} \simeq 2\pi \frac{J}{D}, \tag{1.37}$$

where $D = |\mathbf{D}_{ij}|$. The helical plane is perpendicular to the DM vector $\mathbf{D}_{ij}$. The approximation in the rightmost side of Eq. (1.37) holds when D is much smaller than J, which is realized in many materials introduced in Sect. 1.4.2. Hence, the magnetic period of swirling spin textures stabilized by the DM interaction tends to be large, typically several tens to hundreds nm. Indeed, the origin of the SkLs with relatively long period found in the wide range of materials introduced in Sect. 1.4.2 is ascribed to the competition between the ferromagnetic exchange interaction and the DM interaction in an external magnetic field at finite temperature.

1.5.2 Frustrated Exchange Interactions

Frustrated exchange interactions in insulating magnets may also stabilize topological spin textures. Typical examples are found in geometrically frustrated antiferromagnets whose lattice structures are composed of triangles. For instance, a SkL was found in an antiferromagnetic Heisenberg model on a 2D triangular lattice with the competition between the nearest-neighbor ferromagnetic and third-neighbor antiferromagnetic interactions under an external magnetic field [87]. Moreover, double-Q, quadruple-Q, and sextuple-Q spin structures were found on a 3D pyrochlore lattice with nearest-neighbor antiferromagnetic and second-neighbor ferromagnetic interactions [88]. In addition to these geometrically frustrated systems, frustration due to bond-dependent exchange interactions, such as compass and Kitaev type interactions, was shown to stabilize SkLs [89] and magnetic vortices [90–92] even on bipartite lattices without geometrical frustration. The size of the topological spin textures stabilized by this mechanism can be very small compared to that by the DM interaction. In addition, this mechanism works even for centrosymmetric systems. In such cases, the skyrmions and antiskyrmions are energetically degenerate.

Furthermore, it has also been proposed that the frustration between the two-spin and four-spin or higher-order exchange interactions brings about the noncollinear and noncoplanar spin textures [93–95]. The general form of such multiple-spin interactions is given by

$$J_{ijk\cdots}^{\alpha_i\alpha_j\alpha_k\cdots} S_i^{\alpha_i} S_j^{\alpha_j} S_k^{\alpha_k} \cdots, \tag{1.38}$$

where $S_i^{\alpha_i}$ denotes the α_i component of spin at site i. Considering the most fundamental multiple-spin interactions invariant under any rotations in spin space, the three-spin interaction takes the form of $\mathbf{S}_i \cdot (\mathbf{S}_j \times \mathbf{S}_k)$, while the four-spin interaction can be expressed as, e.g., $(\mathbf{S}_i \cdot \mathbf{S}_j)(\mathbf{S}_k \cdot \mathbf{S}_l)$ and $(\mathbf{S}_i \times \mathbf{S}_j) \cdot (\mathbf{S}_k \times \mathbf{S}_l)$. Thus, multiple-spin interactions naturally incorporate quantities representing the chirality, thereby favoring noncollinear and noncoplanar spin textures. An explicit form of the multiple-spin interactions can be obtained by a series expansion in the perturbation theory. For instance, in the Mott insulating state where electrons are spatially localized by strong correlation, the perturbation expansion in terms of the electron hopping generates systematically the multiple-spin interactions through higher-order processes beyond two-spin exchanges [96–99]. It was shown that the frustration between the two- and four-spin exchange interactions stabilize the $3Q$ noncoplanar spin texture [93]. Furthermore, cooperation of this exchange frustration and the DM interaction was proposed to explain the SkL with a very short magnetic modulation period in Fe/Ir(111) [94]. Recently, effects of higher-order exchange interactions have been also discussed as a mechanism to stabilize the HL observed in $MnSi_{1-x}Ge_x$ [95].

1.5.3 Itinerant Frustration

Itinerant frustration is a competition among electron-mediated interactions in metallic magnets, leading to the instability toward the different single-Q states with different wave vectors related to each other by symmetry [100–103]. Let us consider a simple model for spin-charge coupled metals, the Kondo lattice model (or s-d model), whose Hamiltonian is given by

$$\mathcal{H} = -t \sum_{ij,\sigma} \left(c^\dagger_{i\sigma} c_{j\sigma} + \text{h.c.} \right) - J_\text{K} \sum_i \sum_{i\sigma\sigma'} c^\dagger_{i\sigma} \boldsymbol{\sigma}_{\sigma\sigma'} c_{i\sigma'} \cdot \mathbf{S}_i, \tag{1.39}$$

where $c^\dagger_{i\sigma}$ ($c_{i\sigma}$) is a creation (annihilation) operator of an itinerant electron at site i with spin $\sigma = \uparrow$ or $\downarrow$. The first term describes the kinetic energy of electrons, and the second term represents the onsite coupling between the itinerant electron spin denoted by the vector of the Pauli matrices $\boldsymbol{\sigma}_{\sigma\sigma'}$ and the localized spin $\mathbf{S}_i$. The numerical calculation based on the Hamiltonian in Eq. (1.39) clarified that the noncoplanar spin structures including the 3Q-SkL are stabilized in the spin-charge coupled system even without an external magnetic field [100, 101, 104–107]. For instance, it was shown that the 3Q state exhibiting nonzero scalar spin chirality is stabilized on the triangular lattice near 1/4 and 3/4 fillings [104, 106]. Furthermore, the 3Q-SkLs, exhibiting $|N_\text{sk}| = 2$ at zero field and $|N_\text{sk}| = 1$ under magnetic field, were also reported to be stabilized on the triangular lattice [101] (see Sect. 3.3).

To understand the stabilization mechanism of these multiple-Q states, let us consider the weak spin-charge coupling limit where $J_\text{K} \ll t$, and derive an effective spin model by the perturbation expansion in terms of J_K/t [108] (see Sect. 3.3 for more details). The lowest-order contribution is a two-spin interaction known as the Ruderman-Kittel-Kasuya-Yosida (RKKY) interaction [109–111]. The RKKY interaction is a long-range exchange interaction oscillating with a period dictated by the Fermi surfaces of itinerant electrons. Since the interaction is isotropic in spin space, it prefers a 1Q state in general. Usually, the propagation direction of the 1Q state shows several candidates each of which is energetically degenerate due to crystalline symmetry, and one of them is spontaneously selected in this state. The next lowest order in the expansion gives four-spin interactions, which include many different types depending on the perturbation processes. A systematic study of the four-spin interactions for the model in Eq. (1.39) was done in Ref. [112], which revealed that a biquadratic interaction with positive coupling constant plays a crucial role in stabilizing a noncoplanar spin texture [106]. Later, a more elaborate derivation in Ref. [102] showed a fundamental form of the positive biquadratic interaction as

$$\tilde{K}\sum_{\eta}\left(\mathbf{S}_{\mathbf{Q}_{\eta}}\cdot\mathbf{S}_{-\mathbf{Q}_{\eta}}\right)^{2}, \tag{1.40}$$

where the summation is taken over the characteristic wave vectors $\mathbf{Q}_\eta$ dictated by the Fermi surfaces of itinerant electrons, as in the RKKY interaction. It was shown that the competition between the RKKY interaction favoring the $1Q$ state and this biquadratic interaction with $\tilde{K} > 0$ preferring noncollinear and noncoplanar spin configurations, reasonably explains the origin of topological spin crystals, such as SkLs in the model in Eq. (1.39) [100, 102, 112].

Similarly, further higher-order interactions can be derived from the perturbation expansion and its general form is given by

$$\tilde{J}^{\alpha_1\alpha_2\alpha_3\cdots}_{\mathbf{q}_1\mathbf{q}_2\mathbf{q}_3\cdots}S^{\alpha_1}_{\mathbf{q}_1}S^{\alpha_2}_{\mathbf{q}_2}S^{\alpha_3}_{\mathbf{q}_3}\cdots\delta_{\mathbf{q}_1+\mathbf{q}_2+\mathbf{q}_3+\cdots,l\mathbf{G}}, \tag{1.41}$$

where $S^{\alpha}_{\mathbf{q}}$ is the Fourier transform of S^{α}_{i}; δ is the Kronecker delta for the momentum conservation, l is an integer, and $\mathbf{G}$ is the reciprocal lattice vector. In Eq. (1.41), multiple wave vectors can contribute to energy gains, leading to the stability of the multiple-Q spin textures. We note that Eq. (1.41) is the Fourier transform of Eq. (1.38); the former and latter are derived for metallic and insulating systems, respectively. Although they share the same functional form, the significant difference lies in the $\mathbf{q}$ dependence of $\tilde{J}$. In the metallic system, dominant contributions arise from several sets of $\mathbf{Q}_\eta$, leading to long-range interactions in real space. Meanwhile, in the insulating system, $\tilde{J}$ varies gradually in Fourier space, leading to short-ranged interactions.

It is noteworthy that the magnetic period of the topological spin textures stabilized by this mechanism can be very short, compared to that by the DM interaction in Sect. 1.5.1: It is typically $\sim 1/k_{\mathrm{F}}$, where k_{F} is the magnitude of the Fermi wave vector, as $\mathbf{Q}_\eta$ are set by the Fermi surfaces. Furthermore, this mechanism can work even for centrosymmetric systems as the frustrated exchange interactions in Sect. 1.5.2. Indeed, the multiple-spin interactions arising from the spin-charge coupling have attracted attention for a possible scenario for the short-period SkLs in centrosymmetric metals introduced in Sect. 1.4.2. Moreover, this mechanism was also proposed to stabilize HLs observed in metallic magnets introduced in Sect. 1.4.3. It was shown that the itinerant frustration explained the $4Q$-HL as well as the $2Q$ state observed in the centrosymmetric $SrFeO_3$ [113]. Cooperation of the itinerant frustration and the DM interaction was also proposed to explain the $3Q$- and $4Q$-HLs in noncentrosymmetric metals $MnSi_{1-x}Ge_x$ and $Mn_{1-y}Fe_yGe$ [114].

1.6 Purpose of This Thesis

As introduced in the preceding sections, a variety of topological spin textures with different symmetry, dimensionality, and topology, have been proposed, and some of them were indeed discovered in real materials. Also, several different mechanisms were discussed for their stabilization. Such topological spin textures have great potential in their electronic functionality as well as the magnetic one. Indeed, they manifest nontrivial quantum transport and optical phenomena through the EEMF. Furthermore, these spin structures are nano- and micro-scale, robust against perturbations due to their topologically protected stability, and mobile in an electric current. Hence, the engineering significance of utilizing them as information carriers in magnetic memory devices has also become increasingly apparent. One of the challenging problems in these magnets is flexible generation, annihilation, and control of the spin textures with different topology, leading to their novel functionality controllable for next-generation devices. Although such a problem has been discussed in several specific cases, the previous studies lack a generic viewpoint.

In this thesis, we theoretically investigate the possibility of such generation and control of the topological spin textures in a comprehensive manner. For this purpose, we introduce the concept of "spin moiré". Moiré is a superstructure resulting from an interference of waves, widely emerging in physics. Importantly, a variety of moiré patterns are realized in many ways by changing a number of parameters, e.g., periods, amplitudes, phases, propagating directions, and the number of the superposed waves. As discussed in Sect. 1.2, the topological spin textures are composed of superpositions of multiple spin density waves such as helices, and hence they can be regarded as moiré patterns in the spin degree of freedom. Based on this spin moiré picture, we discuss the way to generate and control the topological spin textures by exploiting the lessons in the conventional moirés—spin moiré engineering. Specifically, considering CSLs in one dimension, SkLs in two dimensions, and HLs in three dimensions, we study the effects of several different parameters to control these moiré patterns, such as the number, amplitudes, twist angles, and relative phases of the superposed waves. We elucidate not only the modulations of the spin textures but also the evolution of the topological properties and the emergent electromagnetic fields. Modulating the parameters of the moiré can be achieved through external fields such as electric and magnetic fields. We also investigate possible ways to achieve such a spin moiré engineering through not only static external fields for equilibrium states but also dynamically oscillating fields and electric currents for nonequilibrium states. Furthermore, we also aim to investigate the effect of the incoherent superpositions in a spin moiré, which is an important issue in controlling the spin moiré.

1.7 Organization of This Thesis

The organization of this thesis is as follows. In Chap. 2, we introduce the key concept in this thesis, the spin moiré engineering. In Chaps. 3 and 4, we present the model and the numerical methods used in this thesis, respectively. In Chap. 5, we study the spin moiré engineering in one dimension. We specifically focus on the effect of the number of superposed waves and examine how the solitonic feature of the spin texture introduced by the higher-harmonic waves affects the topological properties and the EEF. To discuss the control associated with the circularly- and linearly-polarized AC magnetic fields, we conduct systematic analyses using the linear spin-wave theory and the Landau-Lifshitz-Gilbert (LLG) equation. In Chap. 6, we study the spin moiré engineering in two dimensions. In this chapter, we consider the phase degree of freedom for the two types of $3Q$-SkLs and investigate their effect on the spin textures, symmetry, and topological properties by using a general framework based on the hyperspace representation developed in Chap. 4. In Chap. 7, we discuss the spin moiré engineering in three dimensions. We first study the control of the topological properties through the amplitude of the superposed waves for the HLs by using the variational calculation. Next, we discuss the effect of the twist angle as well as the net magnetization by both through calculations of the topological properties and variational calculations based on the Ginzburg-Landau theory. Next, we discuss the effect of the phase shift for the two types of HLs by using the hyperspace representation developed in Chap. 4. Finally, we investigate the current-induced dynamics of the HLs by numerical simulations based on the LLG equation and discuss the effect from the amplitude of the superposed waves controlled by the external magnetic field. Chapter 8 is devoted to the summary of this thesis.

References

1. Mermin, N. D. (1979). *Reviews of Modern Physics, 51*, 591.
2. Rajaraman, R. (1987). *Solitons and instantons: An Introduction to Solitons and Instantons in Quantum Field Theory*. North Holland.
3. Nakahara, M. (2003). *Geometry, topology and physics*. CRC Press.
4. Manton, N., & Sutcliffe, P. (2004). *Topological solitons*. Cambridge University Press.
5. Berezinskii, V. (1971). *Journal of Experimental and Theoretical Physics, 32*, 493.
6. Kosterlitz, J. M., & Thouless, D. J. (1973). *Journal of Physics C: Solid State Physics, 6*, 1181.
7. Kosterlitz, J. M. (1974). *Journal of Physics C: Solid State Physics, 7*, 1046.
8. Braun, H.-B. (2012). *Advances in Physics, 61*, 1.
9. Tatara, G. (2019). *Physica E: Low-dimensional Systems and Nanostructures, 106*, 208.
10. Skyrme, T. H. R. (1962). *Nuclear Physics, 31*, 556.
11. Bogdanov, A. N., & Yablonskii, D. (1989). *Zhurnal Eksperimental'noi i Teoreticheskoi Fiziki 95*, 178.
12. Bogdanov, A., & Hubert, A. (1994). *Journal of Magnetism and Magnetic Materials, 138*, 255.
13. Bogdanov, A. (1995). *JETP Letters, 62*, 247.
14. Rößler, U. K., Bogdanov, A. N., & Pfleiderer, C. (2006). *Nature, 442*, 797.
15. Nagaosa, N., & Tokura, Y. (2013). *Nature Nanotechnology, 8*, 899.
16. Volovik, G. E. (1987). *Journal of Physics C: Solid State Physics, 20*, L83.
17. Korepin, V. E., & Faddeev, L. D. (1975). *Theoretical and Mathematical Physics, 25*, 1039.
18. Faddeev, L., & Niemi, A. J. (1997). *Nature, 387*, 58.
19. Gladikowski, J., & Hellmund, M. (1997). *Physical Review D, 56*, 5194.
20. Whitehead, J. (1947). *Proceedings of the National academy of Sciences of the United States of America, 33*, 117.
21. Sutcliffe, P. (2018). *Journal of Physics A: Mathematical and Theoretical, 51*, Article 375401.
22. Rybakov, F. N., Kiselev, N. S., Borisov, A. B., Döring, L., Melcher, C., & Blügel, S. (2022). *APL Materials, 10*, Article 111113.
23. Togawa, Y., Kousaka, Y., Inoue, K., & Kishine, J.-I. (2016). *Journal of the Physical Society of Japan, 85*, Article 112001.

24. Yu, X., Koshibae, W., Tokunaga, Y., Shibata, K., Taguchi, Y., Nagaosa, N., & Tokura, Y. (2018). *Nature, 564*, 95.
25. Berry, M. V. (1984). *Proceedings of the Royal Society A: Mathematical, Physical and Engineering Sciences, 392*, 45.
26. Xiao, D., Chang, M.-C., & Niu, Q. (2010). *Reviews of Modern Physics, 82*, 1959.
27. Nagaosa, N., & Tokura, Y. (2012). *Physica Scripta, T146*, Article 014020.
28. Nagaosa, N., Yu, X. Z., & Tokura, Y. (2012). *Philosophical Transactions of the Royal Society A, 370*, 5806.
29. Shimizu, K., Okumura, S., Kato, Y., & Motome, Y. (2021). *Physical Review B, 103*, Article 184421.
30. Neubauer, A., Pfleiderer, C., Binz, B., Rosch, A., Ritz, R., Niklowitz, P. G., & Böni, P. (2009). *Physical Review Letters, 102*, Article 186602.
31. Yang, S. A., Beach, G. S. D., Knutson, C., Xiao, D., Niu, Q., Tsoi, M., & Erskine, J. L. (2009). *Physical Review Letters, 102*, Article 067201.
32. Yokouchi, T., Kagawa, F., Hirschberger, M., Otani, Y., Nagaosa, N., & Tokura, Y. (2020). *Nature, 586*, 232.
33. Hirschberger, M., Spitz, L., Nomoto, T., Kurumaji, T., Gao, S., Masell, J., Nakajima, T., Kikkawa, A., Yamasaki, Y., Sagayama, H., Nakao, H., Taguchi, Y., Arita, R., Arima, T.-H., & Tokura, Y. (2020). *Physical Review Letters, 125*, Article 076602.
34. Oike, H., Ebino, T., Koretsune, T., Kikkawa, A., Hirschberger, M., Taguchi, Y., Tokura, Y., & Kagawa, F. (2022). *Physical Review B, 106*, Article 214425.
35. Feng, W., Hanke, J.-P., Zhou, X., Guo, G.-Y., Blügel, S., Mokrousov, Y., & Yao, Y. (2020). *Nature Communications, 11*, 118.
36. Kato, Y. D., Okamura, Y., Hirschberger, M., Tokura, Y., & Takahashi, Y. (2023). *Nature Communications, 14*, 5416.
37. Hayashi, Y., Okamura, Y., Kanazawa, N., Yu, T., Koretsune, T., Arita, R., Tsukazaki, A., Ichikawa, M., Kawasaki, M., Tokura, Y., & Takahashi, Y. (2021). *Nature Communications, 12*, 5974.
38. Yamane, Y., Sasage, K., An, T., Harii, K., Ohe, J., Ieda, J., Barnes, S. E., Saitoh, E., & Maekawa, S. (2011). *Physical Review Letters, 107*, Article 236602.
39. Tanabe, K., Chiba, D., Ohe, J., Kasai, S., Kohno, H., Barnes, S. E., Maekawa, S., Kobayashi, K., & Ono, T. (2012). *Nature Communications, 3*, 845.
40. Nagaosa, N. (2019). *Japanese Journal of Applied Physics, 58*, Article 120909.
41. Kitaori, A., Kanazawa, N., Yokouchi, T., Kagawa, F., Nagaosa, N., & Tokura, Y. (2021). *Proceedings of the National academy of Sciences of the United States of America, 118*, Article e2105422118.
42. Togawa, Y., Koyama, T., Takayanagi, K., Mori, S., Kousaka, Y., Akimitsu, J., Nishihara, S., Inoue, K., Ovchinnikov, A. S., & Kishine, J. (2012). *Physical Review Letters, 108*, Article 107202.
43. Matsumura, T., Kita, Y., Kubo, K., Yoshikawa, Y., Michimura, S., Inami, T., Kousaka, Y., Inoue, K., & Ohara, S. (2017). *Journal of the Physical Society of Japan, 86*, Article 124702.
44. Zheludev, A., Shirane, G., Sasago, Y., Kiode, N., & Uchinokura, K. (1996). *Physical Review B, 54*, 15163.
45. Zheludev, A., Maslov, S., Shirane, G., Sasago, Y., Koide, N., & Uchinokura, K. (1997). *Physical Review Letters, 78*, 4857.
46. Zheludev, A., Maslov, S., Shirane, G., Sasago, Y., Koide, N., & Uchinokura, K. (1998). *Physical Review B, 57*, 2968.
47. Mühlbauer, S., Gvasaliya, S. N., Pomjakushina, E., & Zheludev, A. (2011). *Physical Review B, 84*, Article 180406.
48. Roessli, B., Schefer, J., Petrakovskii, G. A., Ouladdiaf, B., Boehm, M., Staub, U., Vorotinov, A., & Bezmaternikh, L. (2001). *Physical Review Letters, 86*, 1885.
49. Kousaka, Y., Yano, S.-I., Kishine, J.-I., Yoshida, Y., Inoue, K., Kikuchi, K., & Akimitsu, J. (2007). *Journal of the Physical Society of Japan, 76*, Article 123709.

50. Brearton, R., Moody, S. H., Turnbull, L. A., Hatton, P. D., Štefančič, A., Balakrishnan, G., van der Laan, G., & Hesjedal, T. (2023). *Advanced Physics Research, 2*, 2200116.
51. Dzyaloshinskii, I. E. (1964). *Journal of Experimental and Theoretical Physics, 19*, 960.
52. Izyumov, Y. A. (1984). *Soviet Physics Uspekhi, 27*, 845.
53. Kishine, J.-I., Inoue, K., & Yoshida, Y. (2005). *Progress of Theoretical Physics Supplement, 159*, 82.
54. Okumura, S., Kato, Y., & Motome, Y. (2018). *Journal of the Physical Society of Japan, 87*, Article 033708.
55. Adams, T., Mühlbauer, S., Pfleiderer, C., Jonietz, F., Bauer, A., Neubauer, A., Georgii, R., Böni, P., Keiderling, U., Everschor, K., Garst, M., & Rosch, A. (2011). *Physical Review Letters, 107*, Article 217206.
56. Yu, X. Z., Onose, Y., Kanazawa, N., Park, J. H., Han, J. H., Matsui, Y., Nagaosa, N., & Tokura, Y. (2010). *Nature, 465*, 901.
57. Khanh, N. D., Nakajima, T., Yu, X., Gao, S., Shibata, K., Hirschberger, M., Yamasaki, Y., Sagayama, H., Nakao, H., Peng, L., Nakajima, K., Takagi, R., Arima, T., Tokura, Y., & Seki, S. (2020). *Nature Nanotechnology, 15*, 444.
58. Mühlbauer, S., Binz, B., Jonietz, F., Pfleiderer, C., Rosch, A., Neubauer, A., Georgii, R., & Böni, P. (2009). *Science, 323*, 915.
59. Grigoriev, S. V., Potapova, N. M., Siegfried, S.-A., Dyadkin, V. A., Moskvin, E. V., Dmitriev, V., Menzel, D., Dewhurst, C. D., Chernyshov, D., Sadykov, R. A., Fomicheva, L. N., & Tsvyashchenko, A. V. (2013). *Physical Review Letters, 110*, Article 207201.
60. Shibata, K., Yu, X., Hara, T., Morikawa, D., Kanazawa, N., Kimoto, K., Ishiwata, S., Matsui, Y., & Tokura, Y. (2013). *Nature Nanotechnology, 8*, 723.
61. Seki, S., Yu, X. Z., Ishiwata, S., & Tokura, Y. (2012). *Science, 336*, 198.
62. Adams, T., Chacon, A., Wagner, M., Bauer, A., Brandl, G., Pedersen, B., Berger, H., Lemmens, P., & Pfleiderer, C. (2012). *Physical Review Letters, 108*, Article 237204.
63. Tokunaga, Y., Yu, X. Z., White, J. S., Rønnow, H. M., Morikawa, D., Taguchi, Y., & Tokura, Y. (2015). *Nature Communications, 6*, 7638.
64. Kakihana, M., Aoki, D., Nakamura, A., Honda, F., Nakashima, M., Amako, Y., Takeuchi, T., Harima, H., Hedo, M., Nakama, T., & Ōnuki, Y. (2019). *Journal of the Physical Society of Japan, 88*, Article 094705.
65. Tabata, C., Matsumura, T., Nakao, H., Michimura, S., Kakihana, M., Inami, T., Kaneko, K., Hedo, M., Nakama, T., & Ōnuki, Y. (2019). *Journal of the Physical Society of Japan, 88*, Article 093704.
66. Ōnuki, Y., Kakihana, M., Iha, W., Nakaima, K., Aoki, D., Nakamura, A., Honda, F., Nakashima, M., Amako, Y., Gouchi, J., Uwatoko, Y., Nakamura, S., Sakakibara, T., Takeuchi, T., Haga, Y., Ikeda, H., Harima, H., Hedo, M., Nakama, T., & Conf, J. P. S. (2020). *Proc., 30*, Article 011008.
67. Singh, D., et al. (2023). *Nature Communications, 14*, 8050.
68. Karube, K., et al. (2016). *Nature Materials, 15*, 1237.
69. T. Nakajima, H. Oike, A. Kikkawa, E. P. Gilbert, N. Booth, K. Kakurai, Y. Taguchi, Y. Tokura, F. Kagawa, and T. hisa Arima, Sci. Adv. **3**, e1602562 (2017).
70. Takagi, R., Yamasaki, Y., Yokouchi, T., Ukleev, V., Yokoyama, Y., Nakao, H., Arima, T., Tokura, Y., & Seki, S. (2020). *Nature Communications, 11*, 5685.
71. Kurumaji, T., Nakajima, T., Hirschberger, M., Kikkawa, A., Yamasaki, Y., Sagayama, H., Nakao, H., Taguchi, Y., Arima, T.-H., & Tokura, Y. (2019). *Science, 365*, 914.
72. Hirschberger, M., et al. (2019). *Nature Communications, 10*, 5831.
73. Takagi, R., et al. (2022). *Nature Communications, 13*, 1472.
74. Fujishiro, Y., Kanazawa, N., Nakajima, T., Yu, X. Z., Ohishi, K., Kawamura, Y., Kakurai, K., Arima, T., Mitamura, H., Miyake, A., Akiba, K., Tokunaga, M., Matsuo, A., Kindo, K., Koretsune, T., Arita, R., & Tokura, Y. (2019). *Nature Communications, 10*, 1059.
75. Fujishiro, Y., Kanazawa, N., & Tokura, Y. (2020). *Applied Physics Letters, 116*, Article 090501.

76. Kanazawa, N., Onose, Y., Arima, T., Okuyama, D., Ohoyama, K., Wakimoto, S., Kakurai, K., Ishiwata, S., & Tokura, Y. (2011). *Physical Review Letters, 106*, Article 156603.
77. Kanazawa, N., Kim, J.-H., Inosov, D. S., White, J. S., Egetenmeyer, N., Gavilano, J. L., Ishiwata, S., Onose, Y., Arima, T., Keimer, B., & Tokura, Y. (2012). *Physical Review B, 86*, Article 134425.
78. Tanigaki, T., Shibata, K., Kanazawa, N., Yu, X., Onose, Y., Park, H. S., Shindo, D., & Tokura, Y. (2015). *Nano Letters, 15*, 5438.
79. Fujishiro, Y., Kanazawa, N., Shimojima, T., Nakamura, A., Ishizaka, K., Koretsune, T., Arita, R., Miyake, A., Mitamura, H., Akiba, K., Tokunaga, M., Shiogai, J., Kimura, S., Awaji, S., Tsukazaki, A., Kikkawa, A., Taguchi, Y., & Tokura, Y. (2018). *Nature Communications, 9*, 408.
80. Shiomi, Y., Kanazawa, N., Shibata, K., Onose, Y., & Tokura, Y. (2013). *Physical Review B, 88*, Article 064409.
81. Ishiwata, S., Nakajima, T., Kim, J.-H., Inosov, D. S., Kanazawa, N., White, J. S., Gavilano, J. L., Georgii, R., Seemann, K. M., Brandl, G., Manuel, P., Khalyavin, D. D., Seki, S., Tokunaga, Y., Kinoshita, M., Long, Y. W., Kaneko, Y., Taguchi, Y., Arima, T., & Tokura, Y. (2020). *Physical Review B, 101*, Article 134406.
82. Ishiwata, S., Tokunaga, M., Kaneko, Y., Okuyama, D., Tokunaga, Y., Wakimoto, S., Kakurai, K., Arima, T., Taguchi, Y., & Tokura, Y. (2011). *Physical Review B, 84*, Article 054427.
83. Chakraverty, S., Matsuda, T., Wadati, H., Okamoto, J., Yamasaki, Y., Nakao, H., Murakami, Y., Ishiwata, S., Kawasaki, M., Taguchi, Y., Tokura, Y., & Hwang, H. Y. (2013). *Physical Review B, 88*, Article 220405.
84. Tokura, Y., & Kanazawa, N. (2020). *Chemical Reviews, 121*, 2857.
85. Dzyaloshinsky, I. (1958). *Journal of Physics and Chemistry of Solids, 4*, 241.
86. Moriya, T. (1960). *Physical Review, 120*, 91.
87. Okubo, T., Chung, S., & Kawamura, H. (2012). *Physical Review Letters, 108*, Article 017206.
88. Okubo, T., Nguyen, T. H., & Kawamura, H. (2011). *Physical Review B, 84*, Article 144432.
89. Amoroso, D., Barone, P., & Picozzi, S. (2020). *Nature Communications, 11*, 5784.
90. Becker, M., Hermanns, M., Bauer, B., Garst, M., & Trebst, S. (2015). *Physical Review B, 91*, Article 155135.
91. Rousochatzakis, I., Rössler, U. K., van den Brink, J., & Daghofer, M. (2016). *Physical Review B, 93*, Article 104417.
92. Yao, X., & Dong, S. (2016). *Science and Reports, 6*, 26750.
93. Kurz, P., Bihlmayer, G., Hirai, K., & Blügel, S. (2001). *Physical Review Letters, 86*, 1106.
94. Heinze, S., Von Bergmann, K., Menzel, M., Brede, J., Kubetzka, A., Wiesendanger, R., Bihlmayer, G., & Blügel, S. (2011). *Nature Physics, 7*, 713.
95. Grytsiuk, S., Hanke, J.-P., Hoffmann, M., Bouaziz, J., Gomonay, O., Bihlmayer, G., Lounis, S., Mokrousov, Y., & Blügel, S. (2020). *Nature Communications, 11*, 511.
96. Takahashi, M. (1977). *Journal of Physics C: Solid State Physics, 10*, 1289.
97. Calzado, C. J., & Malrieu, J.-P. (2004). *Physical Review B, 69*, Article 094435.
98. Tanaka, K., Yokoyama, Y., & Hotta, C. (2018). *Journal of the Physical Society of Japan, 87*, Article 023702.
99. Hoffmann, M., & Blügel, S. (2020). *Physical Review B, 101*, Article 024418.
100. Ozawa, R., Hayami, S., Barros, K., Chern, G. W., Motome, Y., & Batista, C. D. (2016). *Journal of the Physical Society of Japan, 85*, Article 103703.
101. Ozawa, R., Hayami, S., & Motome, Y. (2017). *Physical Review Letters, 118*, Article 147205.
102. Hayami, S., Ozawa, R., & Motome, Y. (2017). *Physical Review B, 95*, Article 224424.
103. Hayami, S., & Motome, Y. (2021). *Journal of Physics: Condensed Matter, 33*, Article 443001.
104. Martin, I., & Batista, C. D. (2008). *Physical Review Letters, 101*, Article 156402.
105. Chern, G.-W. (2010). *Physical Review Letters, 105*, Article 226403.
106. Akagi, Y., & Motome, Y. (2010). *Journal of the Physical Society of Japan, 79*, Article 083711.
107. Batista, C. D., Lin, S.-Z., Hayami, S., & Kamiya, Y. (2016). *Reports on Progress in Physics, 79*, Article 084504.
108. Hayami, S., & Motome, Y. (2018). *Physical Review Letters, 121*, Article 137202.

109. Ruderman, M. A., & Kittel, C. (1954). *Physical Review, 96*, 99.
110. Kasuya, T. (1956). *Progress in Theoretical Physics, 16*, 45.
111. Yosida, K. (1957). *Physical Review, 106*, 893.
112. Akagi, Y., Udagawa, M., & Motome, Y. (2012). *Physical Review Letters, 108*, Article 096401.
113. Okumura, S., Hayami, S., Kato, Y., & Motome, Y. (2022). *Journal of the Physical Society of Japan, 91*, Article 093702.
114. Okumura, S., Hayami, S., Kato, Y., & Motome, Y. (2020). *Physical Review B, 101*, Article 144416.

Chapter 2
Spin Moiré Engineering

In this chapter, we introduce the main concept of this thesis, "spin moiré" picture. Moiré is an interference fringe generated by a superposition of waves [1, 2]. It has been extensively studied in optics and applied to many optical devices, taking advantage of the flexible control of the superstructures by various parameters, such as the periods, amplitudes, phases, and propagating angles of the superposed waves. In this thesis, we extend this idea to superpositions of spin helices in magnets for exploring a systematic control of the complicated spin structures and their topological properties.

In Sect. 2.1, we briefly review moiré phenomena in physics and introduce the spin moiré picture by extending the concept to magnetism. In Sect. 2.2, we discuss how the spin textures are modulated in analogy with the control of moirés in other fields such as optics.

2.1 Moiré Pattern

In Fig. 2.1, we present examples of moiré patterns. Figure 2.1a and b show the superpositions of two line gratings and two concentric gratings, respectively. In both cases, one can see superstructures with larger length scales and different symmetry compared to those in the original gratings. These types of superstructures generated by superpositions are called moirés. Importantly, a variety of moiré patterns are realized in many ways by changing a parameter in the superposed waves. Indeed, the moiré pattern in Fig. 2.1c is obtained by the superposition of the same concentric gratings shown in Fig. 2.1b but arranged in a slightly different configuration, exhibiting distinct shapes, periodicities, and symmetries.

To further demonstrate the moiré patterns flexibly modulated in different ways through various parameters, we show several moiré patterns composed of two waves in Fig. 2.2. Figure 2.2a shows the two sinusoidal waves A and B, and the moiré fringe

K. Shimizu, *Spin Moiré Engineering and Emergent Electromagnetism in Topological Spin Crystals*, Springer Theses, https://doi.org/10.1007/978-981-96-5992-0_2

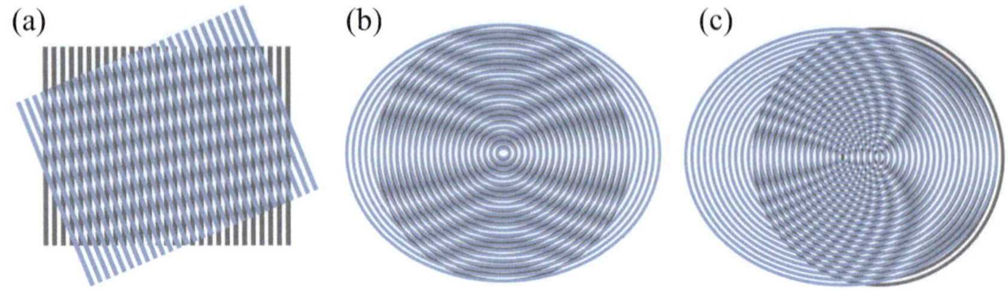

Fig. 2.1 Examples of the moiré patterns generated by superpositions of **a** two line gratings and **b**, **c** two concentric gratings. The moiré pattern in (**c**) is obtained by shifting the gray grating in (**b**)

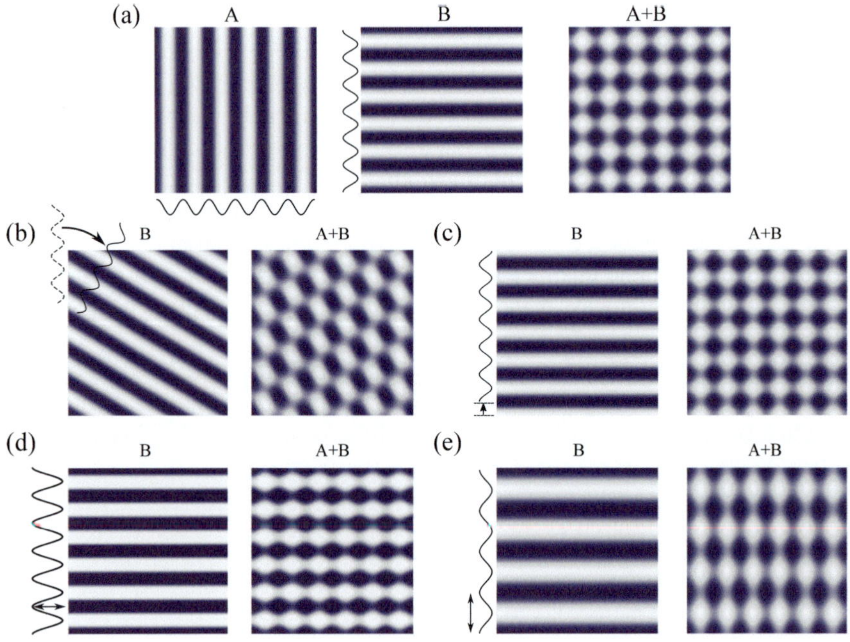

Fig. 2.2 **a** A moiré pattern A+B generated by superposing two sinusoidal waves A and B. Variations of moiré patterns obtained by changing the wave B in **b** the propagating direction, **c** the phase, **d** the amplitude, and **e** the pitch

A+B obtained by superposing the two waves. The variants of the moiré patterns by changing the wave B in the propagating direction, the phase, the amplitude, and the period are exemplified in Fig. 2.2b, c, d, and e, respectively. For each modulation of B, the moiré pattern A+B changes in a different way.[1] Besides these manipulations,

[1] The moiré pattern in (b) does not change from Fig. 2.2a as a phase shift in the linearly independent waves is equivalent to just a translation of the moiré. This is not the case for linearly dependent superpositions.

there are many other ways to change the moiré, e.g., by replacing the sinusoidal waves with other periodic waves. Thus, one can flexibly modulate the superstructure in the moiré patterns by many different parameters.

Recently, moirés have attracted renewed interest in condensed matter physics, especially in atomic layered materials. There, the 2D arrays of atoms cause interferences in their stacking structures, and the resultant superlattices bring up a new length scale, and consequently, an energy scale in the electronic structure. The highlight is a twisted bilayer graphene, where two graphenes are stacked, as illustrated in Fig. 2.3. In this system, the modulation of the propagating direction like in Fig. 2.2b, called the twist angle, is a pivotal parameter. The moiré superlattice structures with the twist angle $\theta = \frac{\pi}{36}$ and $\theta = \frac{\pi}{24}$ are shown in Fig. 2.3a and b, respectively. The twisted bilayer graphene exhibits the hexagonal moiré superlattice structure, whose unit cell is represented by the gray hexagon, and the moiré period can be drastically controlled through θ; the period is given by $l = \frac{a}{2\sin(\theta/2)}$, where a is the lattice constant of the graphene. Although the difference in the twist angle is tiny, the moiré period in Fig. 2.3a is ~1.5 times larger than that in Fig. 2.3b. Furthermore, the moiré superlattice exhibits an intriguing structure such as a twelvefold rotational symmetric quasicrystal structure, shown in Fig. 2.3c with $\theta = \frac{\pi}{6}$. Thus, the twist angle brings about a variety of superstructures and accordingly, affects the low-energy electronic state [3–6]. Through the diverse control through the twist angle, the twisted bilayer graphene exhibits a flat band with $\theta \sim 1.1°$, known as the magic angle [5, 7]. The reduction in bandwidth associated with flat band formation amplifies electronic correlation effects, leading to the emergence of various strong correlation phenomena in graphenes, while they are composed of a light element. In addition to the band engineering by the twist angle, the electron filling can also be controlled by the gate voltage. Through these control parameters in the twisted bilayer graphene, it has been reported that emergent quantum phenomena such as superconductivity [8] and correlated insulators [9] manifest around the magic angle.

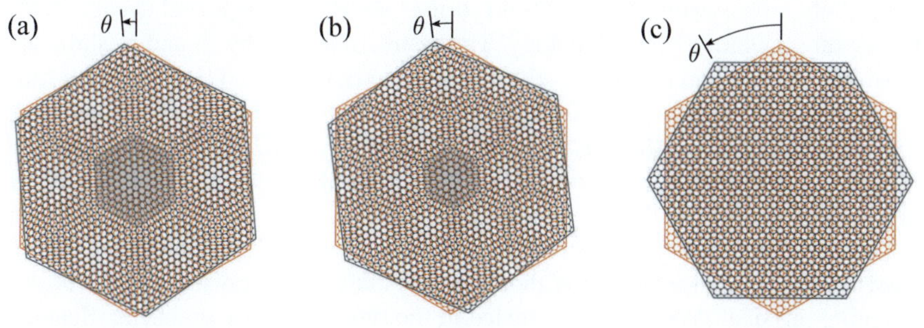

Fig. 2.3 Schematic illustrations of the moiré patterns of the twisted bilayer graphene with different twist angles: **a** $\theta = \frac{\pi}{36}$, **b** $\theta = \frac{\pi}{24}$, and **c** $\theta = \frac{\pi}{6}$. The gray hexagons in (**a**) and (**b**) represent the unit cells of the moiré superlattices. The moiré superlattice in (**c**) is a twelvefold rotational symmetric quasicrystal

2.2 Spin Moiré and Its Engineering

In this thesis, we extend the concept of moiré to magnetism. Topological spin crystals show complicated spin textures which are composed of multiple spin helices, such as the CSL, SkL, and HL introduced in Sect. 1.2. In these cases, the spin textures have superstructures with periodic arrays of topological defects whose periodicity and symmetry are different from the constituent waves. We regard such spin textures as the moiré patterns in magnetism and call them "spin moiré". By exploiting the lessons in conventional moirés which facilitate the flexible control of the superstructures, we discuss how to control the spin textures and resultant magnetic and electronic phenomena—spin moiré engineering.

To shape the idea more clearly, we present some examples of spin moirés and discuss their control, in analogy with the conventional moirés in Sect. 2.1. We here focus on the 2D cases as examples, while the extensions to one and three dimensions are straightforward. Let us consider an example of spin moirés composed of two spin helices A and B propagating in the orthogonal directions $\mathbf{Q}_1$ and $\mathbf{Q}_2$, respectively, as shown in Fig. 1.7a. For the resultant spin moiré A+B shown in Fig. 1.7b, we exemplify several manipulations of the spin structure in Fig. 2.4. Here, similar to in Fig. 2.2, we fixed the helix A and modulate the helix B by (a) the propagating direction, (b) the phase, (c) the amplitude, and (d) the period. In all the cases, the resultant spin moirés look different from the original one, as shown in Fig. 2.2. While the topological property is trivial (see Sect. 1.2.2) and intact through the moiré modulations in the above examples, we can control topological properties and symmetries as well as the spin textures through the spin moiré engineering. An illustrative example is shown in Fig. 2.5. Now, we consider phase shifts in the $3Q$-SkL shown in Fig. 1.7d composed of three helices. The spin texture comprises a hexagonal array of skyrmions with the skyrmion number $N_{\mathrm{sk}} = 1$ per MUC and has sixfold rotational symmetry. Let us consider a phase shift in the $\mathbf{Q}_1$ component from this state. Although a phase shift in the $2Q$-VC is reduced to a spatial translation of the spin texture, as shown in Fig. 2.4b, a phase shift in the $3Q$-SkL brings about an unexpected change in the spin texture, symmetry, and topology. The results obtained by $\frac{\pi}{2}$ and π shifts are shown in Fig. 2.5b and c, respectively. The symmetry is reduced to threefold for the $\frac{\pi}{2}$ shift, but recovered to sixfold for the π shift. Accordingly, the topological property is also changed: The $\frac{\pi}{2}$ shift gives the MAML with $N_{\mathrm{sk}} = 0$, which has a different arrangement of merons and antimerons from the MAML in Fig. 1.7b, while the π shift leads to the SkL with $N_{\mathrm{sk}} = -1$.

Thus, the spin moirés are defined as a straightforward extension of the conventional moirés, and hence they bring about not only unexpected superstructures but also intriguing physical properties as revealed in the twisted bilayer graphene. Besides, the spin moiré provides yet another novel platform for intriguing physics, with several advantages to the structural moiré in the layered materials, as listed below.

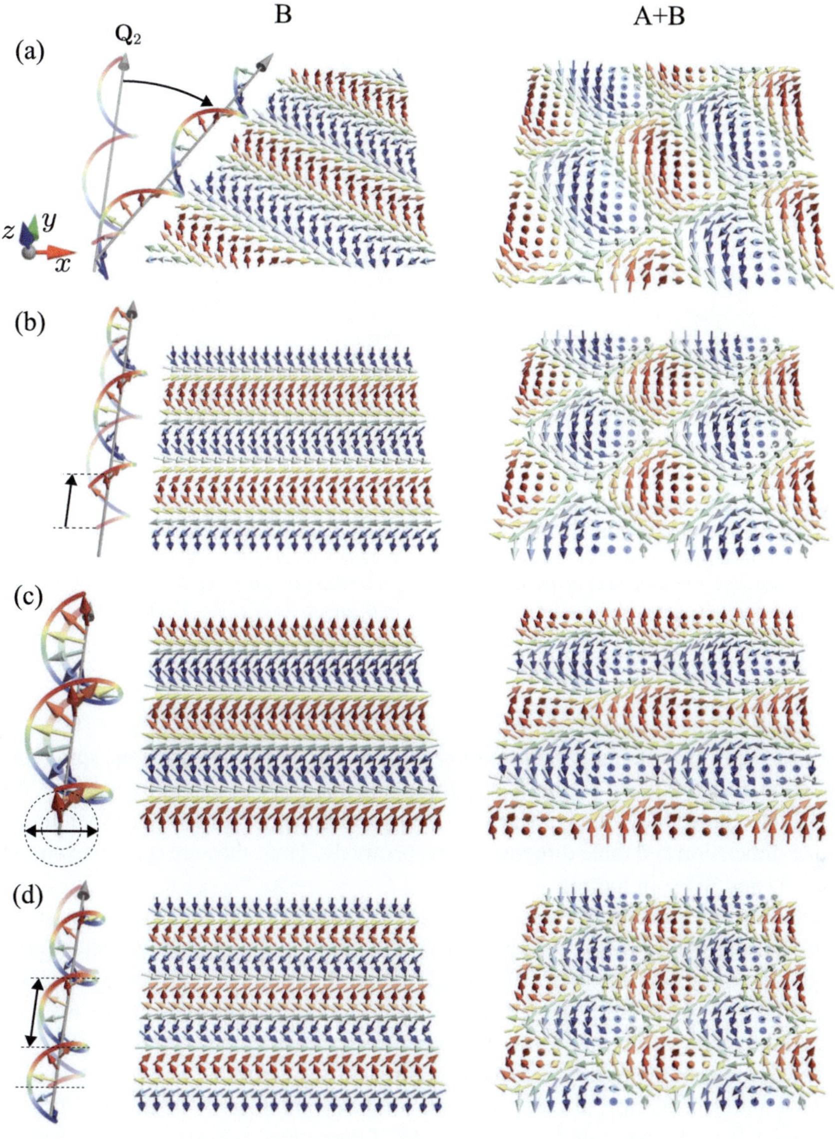

Fig. 2.4 Variations of spin moirés in Fig. 1.7b while changing the composed helix with the wave vector $\mathbf{Q}_2$ denoted by B. Each spin moiré A+B is obtained by changing B in **a** the propagating direction, **b** the phase, **c** the amplitude, and **d** the period. The pale and thick curves in the left panels represent the trajectories of the original helix in Fig. 1.7a and the modulated helix

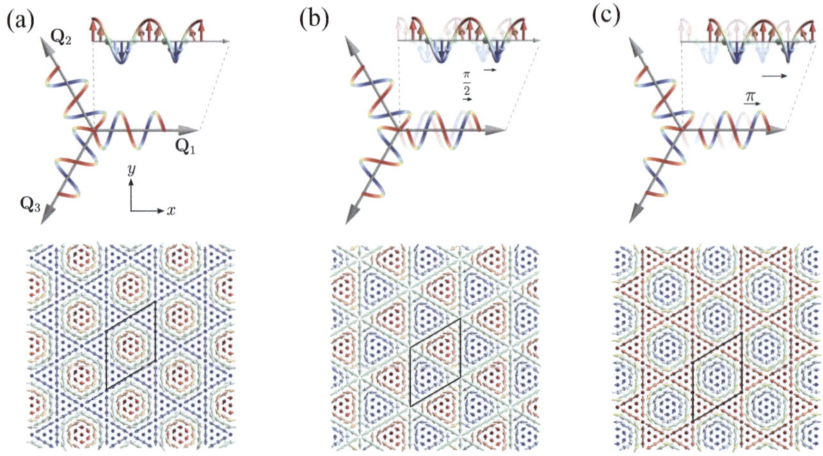

Fig. 2.5 Variations of spin moirés while changing the phase in the superpositions of three proper screws with the wave vectors $\mathbf{Q}_1$, $\mathbf{Q}_2$, and $\mathbf{Q}_3$ shown in Fig. 1.7c. **b** and **c** are obtained from (**a**) by the phase shift of $\frac{\pi}{2}$ and π, respectively. The top panels display the schematic pictures of the superposed waves and the bottom panels show the spin textures obtained by the superpositions. The skyrmion number changes from **a** $N_{\mathrm{sk}} = 1$ to **b** $N_{\mathrm{sk}} = 0$, and to (c) $N_{\mathrm{sk}} = -1$, and the symmetry of the spin texture changes from **a** sixfold, **b** threefold, and **c** sixfold rotational symmetry. The black rhombus represents the MUC

- First, while the structural moiré is mostly limited to two dimensions, the spin moiré can be easily extended from one to three dimensions. For example, the CSL and HL introduced in Sects. 1.2.1 and 1.2.3 are composed of spin helices spanning in one dimension and three dimensions, respectively. Thus, they are typical examples of 1D and 3D spin moirés.
- Second, the spin moirés are superpositions of vector (spinor) fields, providing diverse parameters to be controlled and accordingly, a richer superstructure compared to the scalar potential field in the structural moiré. For instance, spin helices manifest the ellipticity degree of freedom ϵ, whose control leads to the CSL introduced in Sect. 1.2.1, while such a parameter does not exist in the scalar field.
- Third, the spin moiré allows flexible control through external fields such as magnetic fields, electric currents, and temperature, although it is usually difficult to modulate the structural moiré after the fabrication, necessitating mechanical control.
- Fourth, the resultant spin structures can feature a topological nature characterized by the topological numbers, as represented in the CSL, SkL, and HL in Sect. 1.2, which makes them topologically protected against perturbations.
- Last but not least, they lead to interesting electronic, optical, and transport phenomena as well as the magnetic properties due to the Berry phase associated with their noncollinear and noncoplanar structures as introduced in Sect. 1.3.

In the following chapters, we investigate the spin moiré engineering of the magnetic and electronic properties through the control parameters partly exemplified above. Among many parameters, we specifically discuss the modulations by the number, amplitudes, ellipticities, twist angles, and relative phases of the superposed waves. We focus on controlling spin moirés through not only the static external fields for equilibrium states but also dynamically oscillating fields and electric currents for nonequilibrium states. Through these studies, we elucidate the unique features in spin moirés, i.e., wide dimensionality, the vector nature of the constituent waves, topological properties, and the Berry phase physics, and discuss how to control these properties by the parameters in spin moirés.

References

1. Oster, G., Wasserman, M., & Zwerling, C. (1964). *Journal of the Optical Society of America, 54*, 169.
2. Sciammarella, C. A. (1982). *Experimental Mechanics, 22*, 418.
3. Lopes dos Santos, J. M. B., Peres, N. M. R., & Castro Neto, A. H. (2007). *Physical Review Letters, 99*, 256802.
4. Trambly de Laissardière, G., Mayou, D., & Magaud, L. (2010). *Nano Letters, 10*, 804.
5. Suárez Morell, E., Correa, J. D., Vargas, P., Pacheco, M., & Barticevic, Z. (2010). *Physical Review B, 82*, 121407(R).
6. Kim, K., DaSilva, A., Huang, S., Fallahazad, B., Larentis, S., Taniguchi, T., Watanabe, K., LeRoy, B. J., MacDonald, A. H., & Tutuc, E. (2017). *Proceedings of the National Academy of Sciences, 114*, 3364.
7. Tarnopolsky, G., Kruchkov, A. J., & Vishwanath, A. (2019). *Physical Review Letters, 122*, 106405.
8. Cao, Y., Fatemi, V., Fang, S., Watanabe, K., Taniguchi, T., Kaxiras, E., & Jarillo-Herrero, P. (2018). *Nature, 556*, 43.
9. Cao, Y., Fatemi, V., Demir, A., Fang, S., Tomarken, S. L., Luo, J. Y., Sanchez-Yamagishi, J. D., Watanabe, K., Taniguchi, T., Kaxiras, E., Ashoori, R. C., & Jarillo-Herrero, P. (2018). *Nature, 556*, 80.

Chapter 3
Model

In this chapter, we introduce the models which we use in this thesis. In Sect. 3.1, we introduce a phenomenological model with classical spins in continuous space for chiral magnets based on the Ginzburg-Landau theory. In Sect. 3.2, we introduce a spin model defined on a discrete lattice for describing typical noncentrosymmetric magnets that exhibits the topological spin crystals. In Sect. 3.3, we introduce a spin-charge coupled model for a metallic magnet by taking into account itinerant electrons explicitly. In Sect. 3.4, we briefly review the derivation of the effective spin models for SkLs and HLs from the spin-charge coupled model.

3.1 Phenomenological Model

First, we introduce the Ginzburg-Landau free energy, which is a phenomenological model for describing the thermodynamic properties of chiral magnets. Specifically, following Refs. [1, 2], we consider the energy functional given by

$$F^{\mathrm{GL}} = F_2^{\mathrm{GL}} + F_4^{\mathrm{GL}}, \tag{3.1}$$

with

$$F_2^{\mathrm{GL}} = \sum_{\mathbf{q}} \left[(r_0 + Jq^2)|\mathbf{S}_{\mathbf{q}}|^2 + 2iD\mathbf{q} \cdot (\mathbf{S}_{\mathbf{q}} \times \mathbf{S}_{-\mathbf{q}}) \right], \tag{3.2}$$

$$F_4^{\mathrm{GL}} = \sum_{\mathbf{q}_1, \mathbf{q}_2, \mathbf{q}_3, \mathbf{q}_4} U(\mathbf{q}_1, \mathbf{q}_2, \mathbf{q}_3)(\mathbf{S}_{\mathbf{q}_1} \cdot \mathbf{S}_{\mathbf{q}_2})(\mathbf{S}_{\mathbf{q}_3} \cdot \mathbf{S}_{\mathbf{q}_4})\delta_{\mathbf{q}_1+\mathbf{q}_2+\mathbf{q}_3+\mathbf{q}_4,0}, \tag{3.3}$$

where $\mathbf{S}_{\mathbf{q}} = \frac{1}{V}\int d^3\mathbf{r}\mathbf{S}(\mathbf{r})e^{-i\mathbf{q}\cdot\mathbf{r}}$; $\mathbf{S}(\mathbf{r})$ is a classical spin at position $\mathbf{r}$ in continuous space, and V is the volume of the system. F_2^{GL} denotes the two-spin interactions in terms of $\mathbf{S}_{\mathbf{q}}$: r_0, J, and D represent the coefficients of the phenomenological term in the Ginzburg-Landau theory, Heisenberg interaction, and DM interaction,

K. Shimizu, *Spin Moiré Engineering and Emergent Electromagnetism in Topological Spin Crystals*, Springer Theses, https://doi.org/10.1007/978-981-96-5992-0_3

respectively. We take $J > 0$ and $D > 0$; the DM vector is assumed to be parallel to $\mathbf{q}$ so that it prefers a proper-screw spin structure. F_4^{GL} represents the four-spin interactions with the coupling constant $U(\mathbf{q}_1, \mathbf{q}_2, \mathbf{q}_3)$. Assuming that F_2^{GL} is the leading term with $r_0 - JQ_0^2 < 0$ where $Q_0 = D/J$, the system prefers a helix with the wave vector $\mathbf{Q}$ with $|\mathbf{Q}| = Q_0$ (see Sect. 1.5.1). The subleading F_4^{GL} tends to stabilize a noncollinear or noncoplanar spin texture composed of a superposition of the spin helices, as introduced in Sect. 1.5.3.

While F_4^{GL} includes various types of the four-spin interactions, one of the most standard and straightforward form is given by replacing $U(\mathbf{q}_1, \mathbf{q}_2, \mathbf{q}_3)$ with a constant U_0. Previous studies have successfully reproduced the phase diagram for MnSi, in which the SkL is stabilized at finite temperatures, by this simple form [3] [see Fig. 1.13a]. Other four-spin interactions can bring about various noncollinear and noncoplanar spin textures in addition to the SkL. In the previous studies, it was shown that this phenomenological model stabilizes superpositions of two, three, four, and six helices as well as the helical state at zero field by taking into account the four-spin interactions other than U_0 [1, 2]. This model will be used for investigating the spin moiré engineering through the twist angle for the superposition of three helices in 3D space in Sect. 7.2.4.

3.2 Spin Model

Next, we introduce a minimal spin model for describing the magnetic properties of chiral magnets. The Hamiltonian including the explicit time dependence is given by

$$\mathcal{H}(t) = \sum_{\langle ij \rangle} \left[-J\mathbf{S}_i(t) \cdot \mathbf{S}_j(t) - \mathbf{D}_{ij} \cdot \left(\mathbf{S}_i(t) \times \mathbf{S}_j(t) \right) \right] + \sum_i \mathbf{H}(t) \cdot \mathbf{S}_i(t), \quad (3.4)$$

where the summation is taken over the nearest-neighbor sites. The first and second terms in the square brackets denote the Heisenberg and DM interactions, respectively. The DM vector $\mathbf{D}_{ij}$ is determined from the crystalline symmetry. Hereafter, we consider the case where the DM vector is parallel to the bond direction. The last term in Eq. (3.4) describes the Zeeman coupling to a time-dependent magnetic field denoted by $\mathbf{H}(t)$ in the unit of $g\mu_{\mathrm{B}}$, where g is the electron g-factor and μ_{B} is the Bohr magneton. Note that we take the convention of a positive sign for this term. The Hamiltonian in Eq. (3.4) describes the fundamental magnetic properties of chiral magnets both in and out of equilibrium. In the following, we briefly review the thermodynamic properties of the model in Eq. (3.4) for 1D, 2D, and 3D cases. Since this model is used only for the spin moiré engineering in Chap. 5 in this thesis, we will review the 1D case in detail and only touch briefly on the 2D and 3D cases.

In the 1D case, under a static magnetic field, the model in Eq. (3.4) exhibits several spin structures with different magnetic modulation periods depending on the field strength and directions [5–9]. In this case, without loss of generality, we set the chain direction as the x axis and consider the static magnetic field in the xz plane, i.e.,

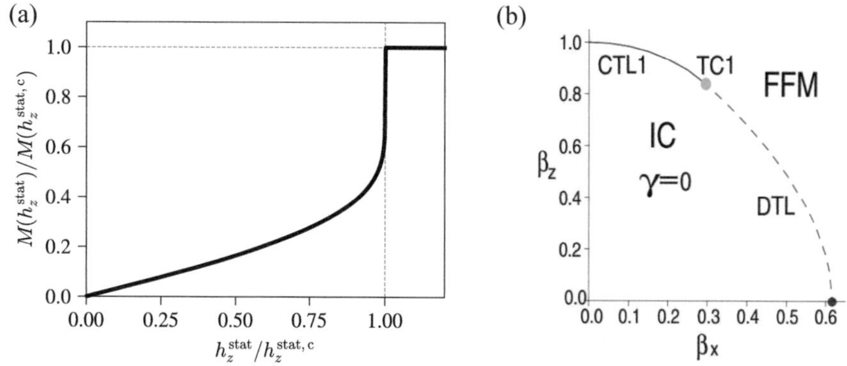

Fig. 3.1 **a** Field dependence of the magnetization $M(h_z^{\mathrm{stat}})$. **b** Phase diagram for the continuum version of the Hamiltonian in Eq. (3.4) [5]. β_x and β_z in the figure denote the normalized magnetic field perpendicular and parallel to the chiral axis, respectively. IC represents the incommensurate phase, which includes the helical, conical, CSL, and the mixture of the conical and CSL. The FFM transition denoted by the solid blue and dashed red lines represent the continuous and discontinuous transitions, respectively. The green circle shows the tricritical point. Reprinted figure with permission from Ref. [5]. Copyright (2016) by the American Physical Society

$\mathbf{H}(t) = \mathbf{h}^{\mathrm{stat}} = (h_x^{\mathrm{stat}}, 0, h_z^{\mathrm{stat}})$. When $h_z^{\mathrm{stat}} = 0$, a helical state with spins swirling in the yz plane is stabilized at $h_x^{\mathrm{stat}} = 0$, and it turns into a conical state for nonzero h_x^{stat}, exhibiting a net magnetization proportional to h_x^{stat}. Meanwhile, when $h_x^{\mathrm{stat}} = 0$, the model stabilizes a CSL, exhibiting divergent increase of the modulation period with h_z^{stat} (see Sect. 1.4.1) and the magnetization curve with nonlinear field dependence, as shown in Fig. 3.1a. The phase diagram is shown in Fig. 3.1b [5]. We note that β_x and β_z in the figure correspond to the normalized h_z^{stat} and h_x^{stat} in our notation, respectively. The ground state exhibits the incommensurate spin structures, including the helical, conical, CSL, and their mixture, below the critical field. Depending on the direction of the magnetic field, the system exhibits continuous and discontinuous FFM transitions denoted by the blue and red lines, respectively. Furthermore, the continuous transition can be further classified into nucleation-type accompanied by the divergence of the modulation period, as observed in the CSL, and instability-type keeping the magnetic modulation period finite, as observed in the conical state [9–11].

In the 2D and 3D cases, the SkL and HL can also be stabilized by the model in Eq. (3.4). The previous Monte Carlo simulation for the 2D case showed that the 3Q-SkL were stabilized down to zero temperature, similar to the experimental phase diagram for a thin film of $Fe_{1-x}Co_xSi$ shown in Fig. 1.13c [12]. In the 3D case, another Monte Carlo simulation based on the Hamiltonian including the third-neighbor interactions showed that around zero temperature, the SkL stabilized in the 2D case turns into the conical state. Furthermore, it was also demonstrated that the SkL only appears in the finite-temperature region, which is consistent with the experimental phase diagram for the bulk MnSi shown in Fig. 1.13a [13]. It was also shown that this model stabilizes the HL down to zero temperature by employing a large DM interaction [14].

3.3 Spin-Charge Coupled Model

Next, we introduce a model for a metallic magnet in which itinerant electrons are explicitly taken into account. Specifically, we consider a generalized version of the s-d model or the Kondo lattice model introduced in Sect. 1.5.3, where itinerant electrons are coupled with localized spins via Hund's coupling. The Hamiltonian is given by

$$\mathcal{H}(t) = -\sum_{ij\sigma\sigma'} t_{ij\sigma\sigma'} c^{\dagger}_{i\sigma} c_{j\sigma'} - J_{\mathrm{K}} \sum_{i\sigma\sigma'} c^{\dagger}_{i\sigma} \boldsymbol{\sigma}_{\sigma\sigma'} c_{i\sigma'} \cdot \mathbf{S}_i(t) - \mathbf{H}(t) \cdot \sum_i \mathbf{S}_i(t), \quad (3.5)$$

where $t_{ij\sigma\sigma'}$ is a spin-dependent hopping. Note that we take the convention of a negative sign for the Zeeman coupling in the last term. Here, we neglect the explicit interactions between the localized spins for simplicity. Due to the mechanism described in Sect. 1.5.3, various SkL phases emerge in this model. For instance, on the triangular lattice with spin-independent nearest- and third-neighbor hoppings, this model stabilizes the $3Q$-SkL and exhibits field-induced topological phase transitions accompanied by the two-step changes of the skyrmion number, as shown in Fig. 3.2a [15]. Real-space spin configurations for the SkL with a high topological number $|N_{\mathrm{sk}}| = 2$ at zero field and the conventional one with $|N_{\mathrm{sk}}| = 1$ at $B = 0.005$ are shown in Fig. 3.2b and c, respectively. We note that the skyrmions with opposite signs of N_{sk} are energetically degenerate.

For translationally symmetric systems, Eq. (3.5) is expressed in Fourier space as

$$\mathcal{H}(t) = -\sum_{\mathbf{k}\sigma\sigma'} t_{\mathbf{k}\sigma\sigma'} c^{\dagger}_{\mathbf{k}\sigma} c_{\mathbf{k}\sigma'} - J_{\mathrm{K}} \sum_{\mathbf{kq}\sigma\sigma'} c^{\dagger}_{\mathbf{k}\sigma} \boldsymbol{\sigma}_{\sigma\sigma'} c_{\mathbf{k}+\mathbf{q}\sigma'} \cdot \mathbf{S}_{\mathbf{q}}(t) - \mathbf{H}(t) \cdot \mathbf{S}_{\mathbf{q}=0}(t), \quad (3.6)$$

where $c_{\mathbf{k}\sigma}$ and $\mathbf{S}_{\mathbf{q}}(t)$ are the Fourier transformation of $c_{i\sigma}$ and $\mathbf{S}_i(t)$, respectively. The hopping $t_{\mathbf{k}\sigma\sigma'}$ can be spin-dependent due to the antisymmetric spin-orbit coupling (ASOC) arising from inversion symmetry breaking [16, 17]. In this case, $t_{\mathbf{k}\sigma\sigma'}$ is $\sigma\sigma'$

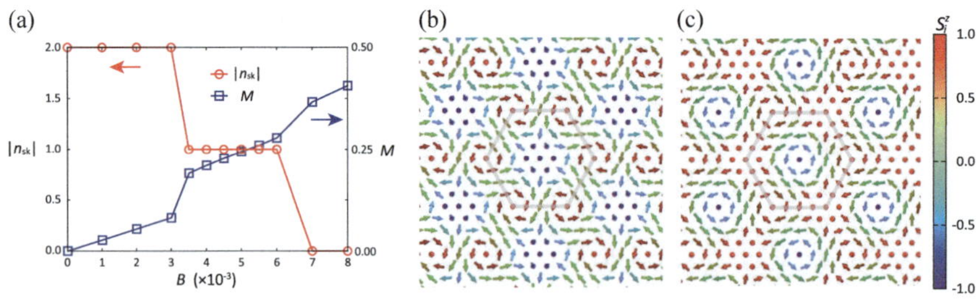

Fig. 3.2 **a** Magnetic field (B) dependence of the skyrmion number n_{sk} and the uniform magnetization M for the model in Eq. (3.5). Real-space spin configurations for the $3Q$-SkLs: **b** $n_{\mathrm{sk}} = 2$ and $B = 0$, and **c** $n_{\mathrm{sk}} = -1$ and $B = 0.005$. Reprinted figures with permission from Ref. [15]. Copyright (2017) by the American Physical Society

component of the matrix $\mathsf{t}_{\mathbf{k}}$ given by

$$\mathsf{t}_{\mathbf{k}} = \varepsilon_{\mathbf{k}}\sigma_0 + \mathbf{g}_{\mathbf{k}} \cdot \boldsymbol{\sigma}, \tag{3.7}$$

where $\mathbf{g}_{\mathbf{k}}$ is an odd-function of $\mathbf{k}$, reflecting symmetry of the crystal structure. For instance, the Rashba-type ASOC in the square lattice with the point group C_{4v} leads to $\mathbf{g}_{\mathbf{k}} \propto (\sin k_y, -\sin k_x, 0)$. The matrix $\mathsf{t}_{\mathbf{k}}$ can be diagonalized with the unitary matrix $\mathsf{U}_{\mathbf{k}}$ as

$$\mathsf{U}_{\mathbf{k}}^{\dagger}\mathsf{t}_{\mathbf{k}}\mathsf{U}_{\mathbf{k}} = \begin{pmatrix} E_{\mathbf{k}+} & 0 \\ 0 & E_{\mathbf{k}-} \end{pmatrix}, \tag{3.8}$$

where $E_{\mathbf{k}\sigma} = \varepsilon_{\mathbf{k}} + \sigma|\mathbf{g}_{\mathbf{k}}|$. The ASOC fixes the chirality of skyrmions and lifts the degeneracy between the skyrmions and antiskyrmions observed in Fig. 3.2. Furthermore, the spin-splitting of the energy band by the ASOC brings about the spin-dependent Fermi surfaces, leading to modulations of characteristic wave vectors for different spins. Hence, the relevant number of wave vectors in the mechanism in Sect. 1.5.3 is increased, and accordingly, further complicated multiple-Q spin textures are expected. For instance, $3Q$ and $6Q$ states were found on the square lattice at zero field with the Rashba and Dresselhaus spin-orbit couplings [16].

3.4 Effective Spin Model

Finally, to study 2D and 3D systems, we consider an effective spin model derived by integrating out the electron degree of freedom in Eq. (3.6). This is partly due to huge numerical costs for solving Eq. (3.6), making a large-scale simulation difficult. The computational cost for the brute-force diagonalization of the Hamiltonian in Eq. (3.6) is of order $\mathcal{O}(N^3)$, where N is the system size. This can be alleviated and reduced to $\mathcal{O}(N)$ by using a sophisticated method using the Chebyshev expansion based algorithms with the kernel polynomial method (KPM) [18]. Nevertheless, it is laborious to study a huge system such as 10^6-site system due to the additional numerical cost arising from the KPM calculation.

The effective spin model for Eq. (3.6) was derived by a perturbation expansion in the weak-coupling limit of $J_{\mathrm{K}}/W \ll 1$, where W is the bandwidth of electron dispersions [17, 19]. The free energy F can be expanded with respect to the order of J_{K} as

$$F - F^{(0)} = F^{(2)} + F^{(4)} + \cdots, \tag{3.9}$$

where $F^{(n)}$ denotes the n-th order contribution. Here, the odd-order terms are zero due to time-reversal symmetry [19]. In the following, we first introduce the effective Hamiltonian up to the second order in Sect. 3.4.1, and then, higher-order contributions in Sect. 3.4.2.

3.4.1 Bilinear Effective Spin Model

The lowest-order contribution can be represented as

$$F^{(2)} = \frac{J_{\mathrm{K}}^2}{2\beta} \sum_{\alpha\beta} \sum_{\mathbf{k}\mathbf{q}\sigma\sigma'\omega_n} G_{\mathbf{k}\sigma\omega_n} G_{\mathbf{k}+\mathbf{q}\sigma'\omega_n} V^{\alpha}_{\mathbf{k},\mathbf{q},\sigma,\sigma'} V^{\beta}_{\mathbf{k}+\mathbf{q},-\mathbf{q},\sigma',\sigma} S^{\alpha}_{\mathbf{q}} S^{\beta}_{-\mathbf{q}}, \tag{3.10}$$

where $G_{\mathbf{k}\sigma\omega_n} = (-i\omega_n + \xi_{\mathbf{k}\sigma})^{-1}$ is the non-interacting Green function, $\xi_{\mathbf{k}\sigma} = E_{\mathbf{k}\sigma} - \mu$ (μ is the chemical potential), $V^{\alpha}_{\mathbf{k},\mathbf{q},\sigma,\sigma'} = \left(\mathsf{U}^{\dagger}_{\mathbf{k}}\sigma_{\alpha}\mathsf{U}_{\mathbf{k}+\mathbf{q}}\right)_{\sigma\sigma'}$, $\omega_n = (2n+1)\pi/\beta$ (n is an integer) is the Matsubara frequency, and $\beta = \frac{1}{k_{\mathrm{B}}T}$ (k_{B} is the Boltzmann constant and T is the temperature). By taking the Matsubara summation ω_n, Eq. (3.10) leads to

$$F^{(2)} = -\frac{J_{\mathrm{K}}^2}{2} \sum_{\mathbf{q}\alpha\beta} \chi^{\alpha\beta}_{\mathbf{q}} S^{\alpha}_{\mathbf{q}} S^{\beta}_{-\mathbf{q}}, \tag{3.11}$$

with

$$\chi^{\alpha\beta}_{\mathbf{q}} = \sum_{\mathbf{k}\sigma\sigma'} \frac{f(\xi_{\mathbf{k}\sigma}) - f(\xi_{\mathbf{k}+\mathbf{q}\sigma'})}{\xi_{\mathbf{k}+\mathbf{q}\sigma'} - \xi_{\mathbf{k}\sigma}} V^{\alpha}_{\mathbf{k},\mathbf{q},\sigma,\sigma'} V^{\beta}_{\mathbf{k}+\mathbf{q},-\mathbf{q},\sigma',\sigma}, \tag{3.12}$$

where $f(\xi)$ is the Fermi-Dirac distribution function. This term is the generalized RKKY interaction [16, 17]. Thus, up to the second order with respect to J_{K}, the effective Hamiltonian including the Zeeman coupling reads

$$\mathcal{H}(t) = \sum_{\mathbf{q}} \mathbf{S}^{\mathsf{T}}_{-\mathbf{q}}(t) \mathsf{J}_{\mathbf{q}} \mathbf{S}_{\mathbf{q}}(t) - \mathbf{H}(t) \cdot \mathbf{S}_{\mathbf{q}=0}(t), \tag{3.13}$$

where T denotes the transpose of a vector and the interaction matrix element $J^{\alpha\beta}_{\mathbf{q}} = -\frac{J_{\mathrm{K}}^2}{2}\chi^{\alpha\beta}_{\mathbf{q}}$ depends on the electronic state. In general, the effective Hamiltonian consists of the DM interaction from the odd-functional part of the off-diagonal components in $\chi^{\alpha\beta}_{\mathbf{q}}$. Moreover, It also consists of the anisotropic terms such as $\Delta S^{x}_{\mathbf{q}} S^{x}_{-\mathbf{q}}$ and $\Gamma(S^{x}_{\mathbf{q}} S^{y}_{-\mathbf{q}} + S^{y}_{\mathbf{q}} S^{x}_{-\mathbf{q}})$, arising from the even-functional part of the diagonal and off-diagonal components, respectively.

In the effective Hamiltonian Eq. (3.13), the dominant contributions arise from the peaks of $\chi^{\alpha\beta}_{\mathbf{q}}$, which inherit from the Fermi surface nature of electrons. Usually, $\chi^{\alpha\beta}_{\mathbf{q}}$ exhibits multiple peaks reflecting crystalline symmetry, and the peak amplitude at $\mathbf{q} = \mathbf{Q}_{\eta}$ becomes significant when the Fermi surface is nested with the wave vector $\mathbf{Q}_{\eta}$. Let us consider the effective spin models for chiral magnets in 3D space by exploiting these properties. We consider the case where $\chi^{\alpha\beta}_{\mathbf{q}}$ exhibits the peaks at $\mathbf{Q}_1 = (Q, 0, 0)$, $\mathbf{Q}_2 = (0, Q, 0)$, and $\mathbf{Q}_3 = (0, 0, Q)$, which are related with each other by threefold rotational symmetry about the [111] axis, and proceed to construct a model incorporating the DM interaction and the anisotropic exchange interaction as well as the isotropic exchange interaction. The matrix $\mathsf{J}(\mathbf{q})$ in Eq. (3.13) is expressed as

$$\mathsf{J}_{\mathbf{q}} = \sum_{\eta=1}^{3} \left[\mathsf{B}_{\mathbf{Q}_\eta} \bar{M}_{\gamma,\mathbf{Q}_\eta}(\mathbf{q}) + \mathsf{B}^{*}_{\mathbf{Q}_\eta} \bar{M}_{\gamma,-\mathbf{Q}_\eta}(\mathbf{q}) \right], \tag{3.14}$$

with

$$\mathsf{B}_{\mathbf{Q}_1} = \begin{pmatrix} -J(1-\Delta) & 0 & 0 \\ 0 & -J(1+2\Delta) & iD \\ 0 & -iD & -J(1-\Delta) \end{pmatrix}, \tag{3.15}$$

$$\mathsf{B}_{\mathbf{Q}_2} = \begin{pmatrix} -J(1-\Delta) & 0 & -iD \\ 0 & -J(1-\Delta) & 0 \\ iD & 0 & -J(1+2\Delta) \end{pmatrix}, \tag{3.16}$$

$$\mathsf{B}_{\mathbf{Q}_3} = \begin{pmatrix} -J(1+2\Delta) & iD & 0 \\ -iD & -J(1-\Delta) & 0 \\ 0 & 0 & -J(1-\Delta) \end{pmatrix}, \tag{3.17}$$

$$\bar{M}_{\gamma,\mathbf{Q}}(\mathbf{q}) = \bar{L}_{\gamma}(\mathbf{q}-\mathbf{Q}) - \bar{L}_{\gamma}(\mathbf{Q})\bar{L}_{\gamma}(\mathbf{q}), \tag{3.18}$$

$$\bar{L}_{\gamma}(\mathbf{q}) = \frac{L_{\gamma}(q_x)}{L_{\gamma}(0)} \frac{L_{\gamma}(q_y)}{L_{\gamma}(0)} \frac{L_{\gamma}(q_z)}{L_{\gamma}(0)}, \tag{3.19}$$

$$L_{\gamma}(q) = \frac{\sinh\gamma}{\cosh\gamma - \cos q}, \tag{3.20}$$

where J, Δ, and D represent the coefficients for the isotropic exchange, anisotropic exchange, and DM interactions, respectively. We note that in Fourier space, the DM interaction is expressed as, e.g., $-i\mathbf{D}_{\mathbf{Q}_1} \cdot (\mathbf{S}_{-\mathbf{Q}_1} \times \mathbf{S}_{\mathbf{Q}_1})$ at $\mathbf{q} = \mathbf{Q}_1$. In Eq. (3.14), $\mathsf{J}_{\mathbf{q}}$ exhibits the peak around $\mathbf{q} = \pm\mathbf{Q}_1$, $\pm\mathbf{Q}_2$, and $\pm\mathbf{Q}_3$ as the Lorentzian-like function $\bar{L}_{\gamma}(\mathbf{q}-\mathbf{Q}_\eta)$ has a peak at $\mathbf{q} = \mathbf{Q}_\eta$ with the peak width of γ. By taking the limit of $\gamma \to 0$, the model consists of the delta-functional peaks at $\pm\mathbf{Q}_1$, $\pm\mathbf{Q}_2$, and $\pm\mathbf{Q}_3$, leading to the infinite-ranged interaction, which never decays in real space. Hence, γ plays a role in controlling the range of the interaction. The pictorial representation of the model is shown in Fig. 3.3b. It was shown that the similar model stabilized the $3Q$-HL as well as the trivial $2Q$ and $1Q$ states [20, 21].

In the actual calculations, we use the real-space representation by inverse Fourier transforming Eq. (3.14). The effective real-space Hamiltonian reads

$$\mathcal{H}(t) = \sum_{\substack{ij \\ g_3(\mathbf{r}_{ij}) \le r_c}} \mathbf{S}_i^{\mathsf{T}}(t)\mathsf{J}_{ij}\mathbf{S}_j(t) - \sum_i \mathbf{H}(t) \cdot \mathbf{S}_i(t), \tag{3.21}$$

with

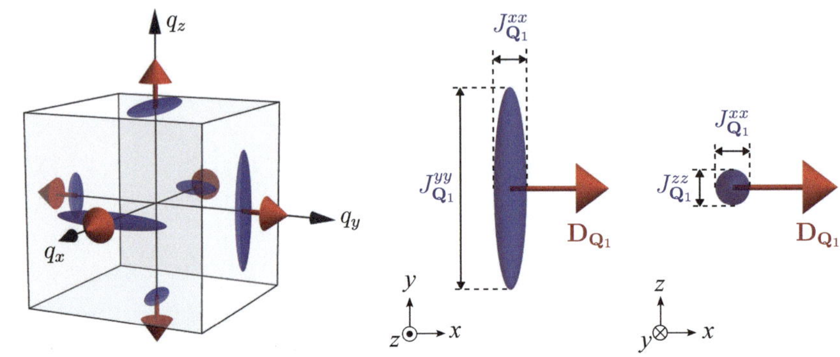

Fig. 3.3 Pictorial representations of the bilinear effective spin model for a three-dimensional chiral magnet given by Eq. (3.21) following the notations of Refs. [20, 21]. The blue ellipsoids placed at $\mathbf{q} = \mathbf{Q}_\eta$ represent $J^{\alpha\alpha}_{\mathbf{Q}_\eta}$: The length along the α-axis denotes the amplitude of $J^{\alpha\alpha}_{\mathbf{Q}_\eta}$. The red arrows represent the DM vector $\mathbf{D}_{\mathbf{Q}_\eta}$

$$\mathsf{J}_{ij} = \frac{2}{L_\gamma^3(0)} e^{-\gamma g_3(\mathbf{r}_{ij})} \sum_{\eta=1}^{3} \left[\mathrm{Re}\mathsf{B}_{\mathbf{Q}_\eta}\left(\cos(\mathbf{Q}_\eta \cdot \mathbf{r}_{ij}) - \bar{L}_\gamma(\mathbf{Q}_\eta)\right) - \mathrm{Im}\mathsf{B}_{\mathbf{Q}_\eta} \sin(\mathbf{Q}_\eta \cdot \mathbf{r}_{ij})\right], \tag{3.22}$$

$$g_3(\mathbf{r}) = |x| + |y| + |z|, \tag{3.23}$$

where $\mathbf{r}_i$ is the position of site i and $\mathbf{r}_{ij} = \mathbf{r}_j - \mathbf{r}_i$. Here, we reduce the numerical cost by limiting the summation in real space for $g_3(\mathbf{r}_{ij}) \leq r_{\mathrm{c}}$, where r_{c} is the truncation length, as the interaction decays exponentially. This model will be used to study the current-induced dynamics of the HL in Sect. 7.4.

3.4.2 Bilinear-Biquadratic Effective Spin Model

We introduce the effective spin model by taking into account the contributions from the fourth-order perturbation. In general, $F^{(4)}$ is much more complicated and consists of several different terms. To simplify the discussion, we consider the case without ASOC, i.e., $\mathbf{g}_{\mathbf{k}} = 0$ in Eq. (3.7). When $\mathbf{g}_{\mathbf{k}} = 0$, an important contribution from the fourth-order perturbation is given by [19]

$$F_1^{(4)} = 2J_{\mathrm{K}}^4 \sum_{\mathbf{q}} X_{\mathbf{q}} (\mathbf{S}_{\mathbf{q}} \cdot \mathbf{S}_{-\mathbf{q}})^2, \tag{3.24}$$

where

$$X_{\mathbf{q}} = \frac{1}{\beta}\sum_{\mathbf{k}\omega_n} G^2_{\mathbf{k}\omega_n} G^2_{\mathbf{k}+\mathbf{q}\omega_n}$$
$$= \sum_{\mathbf{k}} \frac{1}{(\xi_{\mathbf{k}+\mathbf{q}} - \xi_{\mathbf{k}})^2}\left(f'(\xi_{\mathbf{k}}) + f'(\xi_{\mathbf{k}+\mathbf{q}}) + 2\frac{f(\xi_{\mathbf{k}}) - f(\xi_{\mathbf{k}+\mathbf{q}})}{\xi_{\mathbf{k}+\mathbf{q}} - \xi_{\mathbf{k}}}\right), \quad (3.25)$$

and $f'(\xi) = df(\xi)/d\xi$. This is the biquadratic interaction. The importance of the biquadratic interaction and other four-spin interactions to stabilize noncoplanar spin textures was first pointed out to explain the noncoplanar $3Q$ state on the triangular lattice. Based on the first-principles calculation, the significant contribution from the four-spin exchange interactions in Mn/Cu(111) was revealed [22]. In addition, the perturbation analysis for the model in Eq. (3.5) on the triangular lattice also revealed that the four-spin interaction is enhanced when the Fermi surface is nested, leading to the $3Q$ state [23]. Later, this mechanism was extended to explain the multiple-Q states including the SkL in metallic magnets [19]. Notably, the effective spin model incorporating the biquadratic interaction into the Hamiltonian in Eq. (3.13) explains the $3Q$-SkL with $|N_{\mathrm{sk}}| = 2$ on the triangular lattice shown in Fig. 3.2b and the $2Q$ chiral stripe on the square lattice, both of which were initially found as the ground states for the model in Eq. (3.5) [15, 19, 24]. It was also shown that the biquadratic interaction as well as the anisotropic exchange interaction is also important to explain the $2Q$-SkL observed in $GdRu_2Si_2$ introduced in Sect. 1.4.2 [25].

A similar effective model including the biquadratic interactions was also studied for a 3D cubic chiral magnet. The Hamiltonian is given by

$$\mathcal{H}(t) = 2\sum_{\eta}\Big[-J\mathbf{S}_{\mathbf{Q}_\eta}(t)\mathbf{S}_{-\mathbf{Q}_\eta}(t) - iD\hat{\mathbf{Q}}_\eta \cdot (\mathbf{S}_{\mathbf{Q}_\eta}(t) \times \mathbf{S}_{-\mathbf{Q}_\eta}(t))$$
$$+K(\mathbf{S}_{\mathbf{Q}_\eta}(t)\cdot\mathbf{S}_{-\mathbf{Q}_\eta}(t))^2\Big] - \mathbf{H}(t)\cdot\mathbf{S}_{\mathbf{q}=0}(t), \quad (3.26)$$

where the first and third terms in the square bracket represent the bilinear and biquadratic interactions, respectively; the second term denotes the DM-type interaction assumed to stem from the ASOC. The model stabilizes $3Q$- and $4Q$-HLs introduced in Sect. 1.2.3, even at zero field, by taking the wave vectors in Eq. (3.26) as $\mathbf{Q}_1 = (Q, 0, 0)$, $\mathbf{Q}_2 = (0, Q, 0)$, and $\mathbf{Q}_3 = (0, 0, Q)$ for the $3Q$ case, and $\mathbf{Q}_1 = (Q, Q, Q)$, $\mathbf{Q}_2 = (-Q, -Q, Q)$, $\mathbf{Q}_3 = (-Q, Q, -Q)$, and $\mathbf{Q}_4 = (Q, -Q, -Q)$ for the $4Q$ case, respectively [26]. Furthermore, these HLs exhibit the topological phase transitions accompanied by the pair annihilation of the hedgehogs and antihedgehogs with changing the magnetic field. In Fig. 3.4a and b, we present the positions of the hedgehogs and antihedgehogs as well as the real-space spin configurations for the $3Q$- and $4Q$-HLs at zero field, respectively. Figure 3.4c–f display the evolution of the hedgehogs and antihedgehogs in the $4Q$-HL under the magnetic field $\mathbf{H}(t) = (0, 0, h)$ for the $4Q$ case. The eight hedgehogs and eight antihedgehogs at zero field reduce the distance between them, and half of them exhibit pair

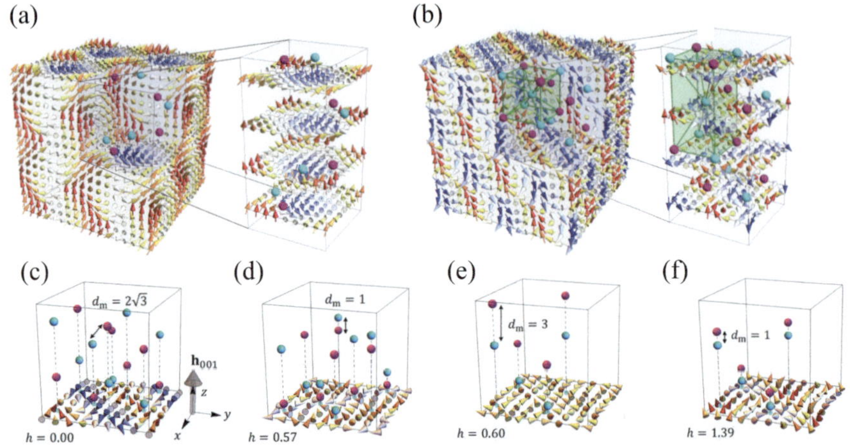

Fig. 3.4 Real-space spin configurations of **a** the 3Q-HL and **b** the 4Q-HL stabilized in the model in Eq. (3.26) at zero field for the 3Q and 4Q cases, respectively. Evolution of the positions of the hedgehogs and antihedgehogs in the 4Q-HL while changing the magnetic field in the [001] direction, h, denoted by the gray arrow: **c** $h = 0$, **d** $h = 0.57$, **e** $h = 0.60$, and **f** $h = 1.39$. d_{m} represents the minimum distance between the hedgehogs and antihedgehogs. Reprinted figures with permissions from Ref. [26]. Copyright (2020) by the American Physical Society

annihilation in the MUC by increasing h. The remaining half also shows the pair annihilation by further increasing h. Recently, a related model was studied to understand the transition between the 3Q- and 4Q-HLs introduced in Sect. 1.4.3 [27]. In Sect. 7.3, we will use the model in Eq. (3.26) for the 4Q case to analyze the phase degree of freedom in the 4Q-HL and discuss the topological phase transitions shown in Fig. 3.4.

In Eq. (3.26), due to cubic symmetry, the interactions J, D, and K for the distinct wave vectors $\mathbf{Q}_\eta$ connected by symmetry are equivalent. However, once this cubic symmetry is broken, for instance, through physical or chemical pressure, the three or four wave vectors are no longer related by symmetry, leading to variations in the magnitudes of the interactions among them and accordingly, constituting waves of spin moirés. In order to investigate the effect of such anisotropy in chiral magnet, we extend the model in Eq. (3.26) to the spatially anisotropic cases as

$$\mathcal{H} = 2\sum_\eta \left[-J_\eta \mathbf{S}_{\mathbf{Q}_\eta} \cdot \mathbf{S}_{-\mathbf{Q}_\eta} + K_\eta \left(\mathbf{S}_{\mathbf{Q}_\eta} \cdot \mathbf{S}_{-\mathbf{Q}_\eta} \right)^2 + i\mathbf{D}_\eta \cdot \left(\mathbf{S}_{\mathbf{Q}_\eta} \times \mathbf{S}_{-\mathbf{Q}_\eta} \right) \right], \quad (3.27)$$

with $\mathbf{Q}_1 = (Q, 0, 0)$, $\mathbf{Q}_2 = (0, Q, 0)$, and $\mathbf{Q}_3 = (0, 0, Q)$for the 3Q case and $\mathbf{Q}_1 = (Q, -Q, -Q)$, $\mathbf{Q}_2 = (-Q, Q, -Q)$, $\mathbf{Q}_3 = (-Q, -Q, Q)$, and $\mathbf{Q}_4 = (Q, Q, Q)$ for the 4Q case. We note that the order of $\mathbf{Q}_\eta$ is different from that in Eq. (3.26) for the 4Q case. Here, we introduce the anisotropy in the coupling constants J_η, K_η, and D_η, which can be different for different wave vectors $\mathbf{Q}_\eta$. For the 3Q case, we introduce the anisotropy along the [001] direction parallel to $\mathbf{Q}_3$ as

$$\begin{cases} J_1 = J_2 = J, \ D_1 = D_2 = D, \ K_1 = K_2 = K, \\ J_3 = J', \ D_3 = D', \ K_3 = K', \end{cases} \tag{3.28}$$

Meanwhile, for the $4Q$ case, we introduce the anisotropy along the [111] direction parallel to $\mathbf{Q}_4$ as

$$\begin{cases} J_1 = J_2 = J_3 = J, \ D_1 = D_2 = D_3 = D, \ K_1 = K_2 = K_3 = K, \\ J_4 = J', \ D_4 = D', \ K_4 = K', \end{cases} \tag{3.29}$$

In both cases, for simplicity, we assume the relations

$$\frac{D'}{D} = \frac{J'}{J} \quad \text{and} \quad \frac{K'}{K} = \left(\frac{J'}{J}\right)^2, \tag{3.30}$$

considering the order of the perturbation expansion. Thus, $J'/J > 1$ (< 1) corresponds to the easy-axis (hard-axis) anisotropy for constituting waves of spin moirés along the [001] and [111] directions for the $3Q$ and $4Q$ cases, respectively.

References

1. Binz, B., & Vishwanath, A. (2006). *Physical Review B, 74*, 214408.
2. Binz, B., Vishwanath, A., & Aji, V. (2006). *Physical Review Letters, 96*, 207202.
3. Mühlbauer, S., Binz, B., Jonietz, F., Pfleiderer, C., Rosch, A., Neubauer, A., Georgii, R., & Böni, P. (2009). *Science, 323*, 915.
4. Togawa, Y., Kousaka, Y., Inoue, K., & Kishine, J.-I. (2016). *Journal of the Physical Society of Japan, 85*, 112001.
5. Laliena, V., Campo, J., Kishine, J.-I., Ovchinnikov, A. S., Togawa, Y., Kousaka, Y., & Inoue, K. (2016). *Physical Review B, 93*, 134424.
6. Dzyaloshinskii, I. E. (1964). *Journal of Experimental and Theoretical Physics, 19*, 960.
7. Izyumov, Y. A. (1984). *Soviet Physics Uspekhi, 27*, 845.
8. Kishine, J.-I., Inoue, K., & Yoshida, Y. (2005). *Progress of Theoretical Physics Supplement, 159*, 82.
9. Masaki, Y., Aoki, R., Togawa, Y., & Kato, Y. (2018). *Physical Review B, 98*, 100402(R).
10. Laliena, V., Campo, J., & Kousaka, Y. (2017). *Physical Review B, 95*, 224410.
11. Masaki, Y. (2020). *Physical Review B, 101*, 214424.
12. Yu, X. Z., Onose, Y., Kanazawa, N., Park, J. H., Han, J. H., Matsui, Y., Nagaosa, N., & Tokura, Y. (2010). *Nature, 465*, 901.
13. Buhrandt, S., & Fritz, L. (2013). *Physical Review B, 88*, 195137.
14. Yang, S.-G., Liu, Y.-H., & Han, J. H. (2016). *Physical Review B, 94*, 054420.
15. Ozawa, R., Hayami, S., & Motome, Y. (2017). *Physical Review Letters, 118*, 147205.
16. Okada, K. N., Kato, Y., & Motome, Y. (2018). *Physical Review B, 98*, 224406.
17. Hayami, S., & Motome, Y. (2018). *Physical Review Letters, 121*, 137202.
18. Weiße, A., Wellein, G., Alvermann, A., & Fehske, H. (2006). *Reviews of Modern Physics, 78*, 275.
19. Hayami, S., Ozawa, R., & Motome, Y. (2017). *Physical Review B, 95*, 224424.

20. Kato, Y., Hayami, S., & Motome, Y. (2021). *Physical Review B, 104*, 224405.
21. Kato, Y., & Motome, Y. (2022). *Physical Review B, 105*, 174413.
22. Kurz, P., Bihlmayer, G., Hirai, K., & Blügel, S. (2001). *Physical Review Letters, 86*, 1106.
23. Akagi, Y., Udagawa, M., & Motome, Y. (2012). *Physical Review Letters, 108*, 096401.
24. Ozawa, R., Hayami, S., Barros, K., Chern, G. W., Motome, Y., & Batista, C. D. (2016). *Journal of the Physical Society of Japan, 85*, 103703.
25. Hayami, S., & Motome, Y. (2021). *Physical Review B, 103*, 024439.
26. Okumura, S., Hayami, S., Kato, Y., & Motome, Y. (2020). *Physical Review B, 101*, 144416.
27. Kato, Y., & Motome, Y. (2023). *Physical Review B, 107*, 094437.

Chapter 4
Methods

In this chapter, we introduce the methods which we use in this thesis. First, we introduce the variational method to obtain the ground state in Sect. 4.1. Next, we introduce a numerical method for energy optimization based on Adam in Sect. 4.2. In Sect. 4.3, we present an approach to extracting parameters characterizing the superposed waves in spin moirés from numerical data generated by real-space simulations. In Sect. 4.4, we introduce the LLG equation describing the time evolution of the classical spins. In Sect. 4.5, we present a numerical method based on the linear spin-wave theory to calculate the magnon spectrum. Finally, in Sect. 4.6 we present the hyperspace representation, which is employed to systematically analyze the effect of phase degrees of freedom in spin moirés.

4.1 Variational Calculation

To obtain the lowest-energy state for a given classical-spin Hamiltonian, we employ the variational calculation, in which we directly compare the energy for several candidates of the stable spin configurations. Based on the simulation results or physical intuition, we prepare N^{var} variational spin configurations, each of which is parametrized by a set of N_k^{param} variational parameters $\{w_j\}$ as $\{\mathbf{S}_i(\{w_j\})\}_k$. For instance, when a set of the N_Q wave vectors characterizing the system is determined, one can construct the variational states as superpositions of helices by using some or all of the wave vectors. Once the spin configuration with the optimal parameters $\{w_j^*\}$, $\{\mathbf{S}_i(\{w_j^*\})\}_k$, is given, the optimal energy for the variational state E_k^* can be calculated in a straightforward manner. By comparing the energies for different variational states, the lowest-energy state is uniquely determined. Better solutions can be, in general, obtained by larger number of variational states N^{var} and variational parameters N_k^{param}. Since the analytical form of the variational state is given, one can get the complete information of the superposition for the lowest-energy spin configuration, and accordingly, discuss how the spin moiré is controlled by changing the parameters in the Hamiltonian.

K. Shimizu, *Spin Moiré Engineering and Emergent Electromagnetism in Topological Spin Crystals*, Springer Theses, https://doi.org/10.1007/978-981-96-5992-0_4

4.2 Energy Minimization

To minimize the energy $E(\{w_j\})$ calculated from a given Hamiltonian with a set of parameters $\{w_j\}$, a gradient-based optimization method is employed, where the parameters are updated in the direction of the gradient to reduce the energy. In this thesis, we use this method to optimize the spin configuration. When exploring the ground state, we parametrize each spin configuration by θ_i and ϕ_i, which are the zenith and azimuthal angles defined in Eq. (1.2) for the spin at site i, respectively, and optimize the energy E with respect to $\{w_j\} = \{\theta_i, \phi_i\}$. To update the parameters, we employ Adam [1]. Starting from given initial parameters, the parameters at step n, $w_j^{(n)}$, are updated by

$$w_j^{(n+1)} = w_j^{(n)} - \kappa \frac{\hat{m}_j^{(n)}}{\sqrt{\hat{v}_j^{(n)}} + \epsilon}, \tag{4.1}$$

where

$$\hat{m}_j^{(n)} = \frac{m_j^{(n)}}{1 - \beta_1^{n+1}}, \quad \hat{v}_j^{(n)} = \frac{v_j^{(n)}}{1 - \beta_2^{n+1}}, \tag{4.2}$$

with

$$m_j^{(n)} = \beta_1 m_j^{(n-1)} + (1 - \beta_1) g_j^{(n)}, \quad v_j^{(n)} = \beta_2 v_j^{(n-1)} + (1 - \beta_2) \left(g_j^{(n)} \right)^2, \tag{4.3}$$

$$g_j^{(n)} = \left. \frac{\partial E}{\partial w_j} \right|_{w_j = w_j^{(n)}}. \tag{4.4}$$

The initial conditions are set as $m_j^{(0)} = 0$ and $v_j^{(0)} = 0$. Here, we introduce four hyperparameters: κ is the learning rate, β_1 and β_2 are the exponential decay rates for the moment estimates, and ϵ is the small constant to improve the numerical stability. We use $\kappa = 0.1$–0.001 and set the other hyperparameters to the default values suggested in Ref. [1] as $\beta_1 = 0.9$, $\beta_2 = 0.999$, and $\epsilon = 10^{-8}$. The calculations are employed by using the libraries Optax [2] and JAX [3]. Although the convergence to a stable state is very fast with this method, the spin configuration can be trapped at the local minimum. Thus, to explore the global minimum for the ground state of the problem, we prepare typically 20–50 different initial spin configurations and select the optimal spin configuration with the lowest energy.[1] This method will be used to prepare the ground-state spin configurations in Chap. 5. In the actual calculations, however, we mainly employ the LLG equation to prepare the ground state, and use the above optimization technique as supplemental (see Sect. 4.4).

---

[1] A sufficient number of initial configurations and a better way of the preparation strongly depend on the problem.

4.3 Extracting Moiré Parameters

Spin moirés are represented by superpositions of multiple waves, each of which is characterized by several parameters, such as amplitudes and phases, as introduced in Chap. 2. To get detailed profiles of spin configurations generated by numerical simulations, $\{\mathbf{S}_i^{\mathrm{sim}}\}$, we find the optimal parameters by estimating the cost function

$$U(\{w_j\}) = \frac{1}{N}\sum_{i=1}^{N}\left(1 - \mathbf{S}_i^{\mathrm{sim}} \cdot \mathbf{S}_i^{\mathrm{fit}}(\{w_j\})\right), \tag{4.5}$$

where N is the number of spins and $\{\mathbf{S}_i^{\mathrm{fit}}(\{w_j\})\}$ is the spin configuration parameterized by a set of parameters $\{w_j\}$. For the general form of $\mathbf{S}_i^{\mathrm{fit}}$ given by Eq. (1.11), the wave vectors can be determined from the Fourier transform of $\mathbf{S}_i^{\mathrm{sim}}$; it is usually sufficient to select the wave vectors whose Fourier component is predominantly large. Once the set of N_Q wave vectors is determined, the general form of the superposition in Eq. (1.11) requires $5N_Q + 3$ parameters.[2] The number of parameters can be reduced by symmetry and physical intuition. For instance, when the DM interaction is present, it is natural to set the helical plane normal to the DM vector (see Sect. 7.3.4). We follow this procedure to get the phases from the numerically obtained spin configuration in Sects. 6.1.4 and 7.3.4.

4.4 Landau-Lifshitz-Gilbert Equation

To elucidate the spin moiré engineering through the time-dependent external fields, we study the real-time dynamics of spins by using the LLG equation given by [4, 5]

$$\frac{\partial \mathbf{S}_i(t)}{\partial t} = \frac{1}{1+\alpha^2}\left[-\mathbf{S}_i(t) \times \mathbf{h}_i^{\mathrm{eff}}(t) + \alpha \mathbf{S}_i(t) \times \left(\mathbf{S}_i(t) \times \mathbf{h}_i^{\mathrm{eff}}(t)\right)\right], \tag{4.6}$$

where α is the Gilbert damping and $\mathbf{h}_i^{\mathrm{eff}}(t)$ is the mean magnetic field at time t, defined as

$$\mathbf{h}_i^{\mathrm{eff}}(t) = \frac{\partial \mathcal{H}(t)}{\partial \mathbf{S}_i(t)} \tag{4.7}$$

Note that the length constraint of $|\mathbf{S}_i(t)| = 1$ is deferred only when taking the derivative of the Hamiltonian with respect to $\mathbf{S}_i(t)$. As an example, we specifically consider

[2] Each superposed wave requires 5 parameters: the amplitudes ψ_η^{c} and ψ_η^{s}, the phase φ_η, and the direction of $\mathbf{e}_\eta^1 \times \mathbf{e}_\eta^2$. The remaining 3 parameters are for the magnetization $\mathbf{m}$.

the model Hamiltonian for a 1D chiral magnet in Eq. (3.4). In the 1D case, we use l for the site index and L for the number of spins; namely, $l = 0, 1, \ldots, L-1$. Then, $\mathbf{h}_l^{\mathrm{eff}}(t)$ reads

$$\mathbf{h}_l^{\mathrm{eff}}(t) = -J\big(\mathbf{S}_{l+1}(t) + \mathbf{S}_{l-1}(t)\big) + D\hat{\mathbf{x}} \times \big(\mathbf{S}_{l+1}(t) - \mathbf{S}_{l-1}(t)\big) + \mathbf{h}(t). \quad (4.8)$$

In this case, we impose $\mathbf{S}_L(t) = \mathbf{S}_0(t)$ and $\mathbf{S}_{-1}(t) = \mathbf{S}_{L-1}(t)$ under the periodic boundary condition (PBC), and $\mathbf{S}_L(t) = \mathbf{S}_{-1}(t) = 0$ under the open boundary condition (OBC).

In addition to investigating the real-time spin dynamics, the LLG equation in Eq. (4.6) can be employed to find the ground state, similar to the minimization method introduced in Sect. 4.2. In both methods, the update of the spin structure is halted when the gradients of the Hamiltonian with respect to the spin directions are zero, i.e., when $\mathbf{S}_i \times \frac{\partial \mathcal{H}(t)}{\partial \mathbf{S}_i(t)} = 0$. Hence, these methods essentially give the same ground state. When α is sufficiently large, the LLG equation can be regarded as the gradient descent.[3] Thus, this method can also be trapped to the local minimum, similar to the energy optimization method introduced in Sect. 4.2. We will mainly use the LLG equation to get the ground state in Chaps. 5–7.

In this thesis, we also investigate the real-time dynamics of spins under an electric current. When the current flows, the spin-transfer torque drives the spin structure, and its effect can be incorporated into the LLG equation as [6, 7]

$$\frac{\partial \mathbf{S}_i(t)}{\partial t} = \frac{1}{1+\alpha^2}\Big[-\mathbf{S}_i(t) \times \mathbf{h}_i^{\mathrm{eff}}(t) + \alpha \mathbf{S}_i(t) \times \big(\mathbf{S}_i(t) \times \mathbf{h}_i^{\mathrm{eff}}(t)\big) - \tilde{j}_e\Big((1+\alpha\beta)(\hat{\mathbf{j}}_e \cdot \hat{\boldsymbol{\delta}})\mathbf{S}_i(t) + (\beta - \alpha)\mathbf{S}_i(t) \times \Big((\hat{\mathbf{j}}_e \cdot \hat{\boldsymbol{\delta}})\mathbf{S}_i(t)\Big)\Big)\Big], \quad (4.9)$$

where $\tilde{j}_e = \frac{p}{2} j_e$, p is the spin polarization of the current, j_e and $\hat{\mathbf{j}}_e$ are the amplitude and the direction of the electron flow, respectively, and β is the nonadiabatic coefficient. Here, $\hat{\boldsymbol{\delta}} = (\hat{\delta}_x, \hat{\delta}_y, \hat{\delta}_z)$ stands for the discrete difference defined as

$$\hat{\delta}_\mu \mathbf{S}_{\mathbf{r}_i}(t) = \frac{1}{2}\left(\mathbf{S}_{\mathbf{r}_i+\hat{\mu}} - \mathbf{S}_{\mathbf{r}_i-\hat{\mu}}\right), \quad (4.10)$$

where $\mathbf{r}_i$ is the position of the site i. This is an approximation for the spatial derivative in the discretized lattice system.

In the following calculations, we numerically solve Eqs. (4.6) and (4.9) by using the fourth-order Runge-Kutta method. We mainly take $\alpha = 0.04$ and $\alpha = 0.1$, which are typical values for ferromagnetic metals [8–11].

[3] The second term in the square bracket in Eq. (4.6) represents $-\mathbf{h}_i^{\mathrm{eff}}(t)$ projected onto the plane perpendicular to $\mathbf{S}_i(t)$, hence can be written as $-\alpha \left.\frac{\partial \mathcal{H}(t)}{\partial \mathbf{S}_i(t)}\right|_{|\mathbf{S}_i(t)|=1}$. Here, the derivative of $\mathcal{H}(t)$ is taken by considering the length constraint for $\mathbf{S}_i(t)$.

4.5 Linear Spin-Wave Theory

In investing the spin moiré engineering through the oscillating magnetic field, it is also important to clarify the fundamental excitation properties of the system. To clarify the spin-wave excitation for a 1D chiral magnet, in Sect. 5.1.2, we study the magnon spectrum and the spatial profile of the excitation modes by using the linear spin-wave theory; technical details can be found, e.g., in Refs. [12–14]. For the stable spin configuration in the ground state, we introduce a new local spin frame where the spin configuration is regarded as the collinear ferromagnetic state. The spin in the new frame is denoted by $\tilde{\mathbf{S}}_l$ and the stable spin configuration is represented by $\tilde{\mathbf{S}}_l = (0, 0, 1)^\mathsf{T}$. By using the Holstein-Primakoff transformation, $\tilde{\mathbf{S}}_l$ is represented by the boson operators up to the leading order as

$$\begin{pmatrix} \tilde{S}_{l,x} \\ \tilde{S}_{l,y} \\ \tilde{S}_{l,z} \end{pmatrix} \simeq \begin{pmatrix} \sqrt{\frac{1}{2}}(\hat{a}_l + \hat{a}_l^\dagger) \\ -\mathrm{i}\sqrt{\frac{1}{2}}(\hat{a}_l - \hat{a}_l^\dagger) \\ 1 - \hat{a}_l^\dagger \hat{a}_l \end{pmatrix}, \tag{4.11}$$

where i is the imaginary unit, and $\hat{a}_l$ ($\hat{a}_l^\dagger$) represents the annihilation (creation) operator of the magnon at site l. By substituting Eq. (4.11) into the model Hamiltonian for a 1D chiral magnet in Eq. (3.4) with $\mathbf{h}(t) = \mathbf{h}^{\mathrm{stat}}$, we obtain the magnon Hamiltonian expressed as

$$\mathcal{H}_{\mathrm{SW}} = \sum_{lm} \hat{\mathbf{a}}_l^\dagger \mathsf{K}_{lm} \hat{\mathbf{a}}_m, \tag{4.12}$$

where $\hat{\mathbf{a}}_l = \left(\hat{a}_l, \hat{a}_l^\dagger\right)^\mathsf{T}$ and K_{lm} is a 2×2 matrix being the lm block of the $2L \times 2L$ Hamiltonian matrix H. The eigenenergy of the magnon, ω_i $(i = 1, 2, \ldots, L)$, appears twice in $2L$ eigenenergies obtained by numerically diagonalizing the matrix H.

In general, the Hamiltonian in Eq. (3.4) exhibits the incommensurate ground state. To numerically obtain the magnon band structure for such a case, we calculate the magnon band structure for the spin texture with the magnetic modulation period of an integer Λ, approximating the true modulation period, in the system with $L = M\Lambda$ sites under the PBC, where M is also an integer. The spin structure with the period of Λ which gives the lowest energy can be regarded as the commensurate approximant of the incommensurate ground-state spin texture in the thermodynamic limit. By regarding this spin structure as a MUC, we introduce the Fourier transformation as

$$\hat{\mathbf{a}}_{\alpha,q} = \frac{1}{M} \sum_{m=0}^{M-1} \hat{\mathbf{a}}_{\alpha+m\Lambda} e^{-iqm\Lambda} = \begin{pmatrix} \hat{a}_{\alpha,q} \\ \hat{a}_{\alpha,-q}^\dagger \end{pmatrix}, \tag{4.13}$$

where the subscript $\alpha = 0, 1, \ldots, \Lambda - 1$ denotes the coordinate in the MUC and q is the wave number. By Fourier transforming Eq. (4.12), we obtain

$$\mathcal{H}_{\rm SW} = \sum_q \sum_{\alpha\beta} \hat{\mathbf{a}}^{\dagger}_{\alpha,q} \mathsf{K}_{\alpha\beta,q} \hat{\mathbf{a}}_{\beta,q}, \tag{4.14}$$

where $\mathsf{K}_{\alpha\beta,q}$ is a 2×2 matrix being the $\alpha\beta$ block of the $2\Lambda \times 2\Lambda$ Hamiltonian matrix H_q. Finally, by diagonalizing H_q, the p-th magnon band dispersion with the wave number q, $\omega_{p,q}$ $(p = 1, 2, \ldots, \Lambda)$, is obtained.

4.6 Phase Degree of Freedom and Hyperspace Representation

In this section, we introduce a theoretical framework that we developed to systematically analyze the effect of phase shifts in multiple-Q spin structures. In Sect. 4.6.1, we introduce the phase degree of freedom for a general multiple-Q spin texture and briefly review the effect of the phase shifts. In Sect. 4.6.2, we present a generic framework to systematically deal with a phase shift by introducing a hyperspace with additional dimensions representing the phase degree of freedom, inspired by the description of the phason degree of freedom in quasicrystals [15–18].

4.6.1 Phase Degree of Freedom

Here, we consider a generic form of the multiple-Q spin structures in d-dimensional continuous space, which is given by Eq. (1.11) with $\mathbf{r} = (x_1, \ldots, x_d)$, $\mathcal{Q}_\eta = \mathbf{q}_\eta \cdot \mathbf{r} + \varphi_\eta$, and $\mathbf{q}_\eta = (q^1_\eta, \ldots, q^d_\eta)$. Here, we use $\mathbf{q}_\eta$ instead of $\mathbf{Q}_\eta$ for the later use in Sect. 4.6.2.

The spin length is normalized as $|\mathbf{S}(\mathbf{r})| = 1$ for any $\mathbf{r}$ unless the spin length vanishes. Note that $\mathbf{m}$ is not the net magnetization because of the normalization.

Let us consider how the spin structures are modulated by changing the phases φ_η. When $N_Q \le d$ and $\mathbf{q}_\eta$ are linearly independent, the periodic spin textures are described by the set of N_Q linearly-independent magnetic translation vectors $\mathbf{a}_\eta$, which satisfy

$$\mathbf{a}_\eta \cdot \mathbf{q}_{\eta'} = 2\pi \delta_{\eta\eta'}, \tag{4.15}$$

where $\delta_{\eta\eta'}$ is the Kronecker delta. In this case, a phase shift from φ_η to $\varphi_\eta + \Delta\varphi_\eta$ is reduced to a spatial translation from $\mathbf{r}$ to $\mathbf{r} + \Delta\mathbf{r}$ with

$$\Delta\mathbf{r} = \sum_{\eta=1}^{N_Q} \frac{\Delta\varphi_\eta}{2\pi} \mathbf{a}_\eta, \tag{4.16}$$

since the following relation holds:

$$\mathbf{q}_\eta \cdot \Delta\mathbf{r} = \sum_{\eta'} \frac{\Delta\varphi_{\eta'}}{2\pi}\mathbf{q}_\eta \cdot \mathbf{a}_{\eta'} = \Delta\varphi_\eta. \quad (4.17)$$

Hence, the phase degree of freedom is irrelevant when $N_Q \leq d$.[4]

The situation is, however, different for $N_Q > d$. In this case, the wave vectors $\mathbf{q}_\eta$ are not linearly independent of each other. This means that a phase shift cannot be reduced to a spatial translation, as $\mathbf{a}_\eta$ defined by Eq. (4.15) with d out of N_Q wave vectors $\mathbf{q}_\eta$ do not satisfy the relation in Eq. (4.16). In this case, phase shifts bring about not only different spin textures but also different topological properties and symmetry. An illustrative example for the 3Q-SkL is given in Fig. 2.5.

4.6.2 Hyperspace Representation

To discuss the effect of the phase shift in the case of $N_Q > d$ systematically, it is convenient to introduce N_Q-dimensional (N_QD) hyperspace. In the hyperspace, we can introduce a position vector $\mathbf{R} = (X_1, X_2, \ldots, X_{N_Q})$ and N_Q linearly-independent wave vectors $\mathbf{Q}_\eta = (Q_\eta^1, Q_\eta^2, \ldots, Q_\eta^{N_Q})$ ($\eta = 1, 2, \ldots, N_Q$) so that they satisfy the relations[5]

$$\mathcal{Q}_\eta = \mathbf{q}_\eta \cdot \mathbf{r} + \varphi_\eta = \mathbf{Q}_\eta \cdot \mathbf{R}. \quad (4.18)$$

In this hyperspace, we can define the N_Q linearly-independent magnetic translation vectors $\mathbf{A}_\eta$ ($\eta = 1, 2, \ldots, N_Q$) satisfying

$$\mathbf{A}_\eta \cdot \mathbf{Q}_{\eta'} = 2\pi\delta_{\eta\eta'}. \quad (4.19)$$

By using Eqs. (4.18) and (4.19), the hyperspace position $\mathbf{R}$ and the real-space one $\mathbf{r}$ are related with each other as

$$\begin{pmatrix} X_1 \\ X_2 \\ \vdots \\ X_{N_Q} \end{pmatrix} = M_Q^{-1} M_q \begin{pmatrix} x_1 \\ x_2 \\ \vdots \\ x_d \\ 1 \end{pmatrix}, \quad (4.20)$$

---

[4] In discrete lattice systems, the phase degree of freedom becomes relevant even when $N_Q \leq d$.

[5] While one can freely choose $\mathbf{Q}_\eta$ as long as they are linearly independent, we assume that their projections onto the original space are $\mathbf{q}_\eta$ in the following analyses.

where

$$M_Q = \begin{pmatrix} \mathbf{Q}_1 \\ \mathbf{Q}_2 \\ \vdots \\ \mathbf{Q}_{N_Q} \end{pmatrix} = \begin{pmatrix} Q_1^1 & Q_1^2 & \dots & Q_1^{N_Q} \\ Q_2^1 & Q_2^2 & \dots & Q_2^{N_Q} \\ \vdots & \vdots & \ddots & \vdots \\ Q_{N_Q}^1 & Q_{N_Q}^2 & \dots & Q_{N_Q}^{N_Q} \end{pmatrix}, \tag{4.21}$$

$$M_q = \left(\begin{array}{c|c} \mathbf{q}_1 & \varphi_1 \\ \mathbf{q}_2 & \varphi_2 \\ \vdots & \vdots \\ \mathbf{q}_{N_Q} & \varphi_{N_Q} \end{array}\right) = \begin{pmatrix} q_1^1 & \dots & q_1^d & \varphi_1 \\ q_2^1 & \dots & q_2^d & \varphi_2 \\ \vdots & \ddots & \vdots & \vdots \\ q_{N_Q}^1 & \dots & q_{N_Q}^d & \varphi_{N_Q} \end{pmatrix}. \tag{4.22}$$

Note that M_Q^{-1} is given as

$$\begin{aligned} M_Q^{-1} &= \frac{1}{2\pi} \left(\mathbf{A}_1 \; \mathbf{A}_2 \dots \mathbf{A}_{N_Q} \right) \\ &= \frac{1}{2\pi} \begin{pmatrix} A_1^1 & A_2^1 & \dots & A_{N_Q}^1 \\ A_1^2 & A_2^2 & \dots & A_{N_Q}^2 \\ \vdots & \vdots & \ddots & \vdots \\ A_1^{N_Q} & A_2^{N_Q} & \dots & A_{N_Q}^{N_Q} \end{pmatrix}. \end{aligned} \tag{4.23}$$

The relation in Eq. (4.20) can be regarded as a surjective mapping from the set of the real-space position $\mathbf{r}$ and the phases φ_η onto the hyperspace position $\mathbf{R}$.[6] In this setting, Eq. (4.18) gives one-to-one correspondence between a multiple-Q spin configuration with the phase variables φ_η in the original d-dimensional real space and that in the N_QD hyperspace with fixed phases (in other words, without the phase degrees of freedom). In this representation, a phase shift by $\Delta\varphi_\eta$ in the original real space corresponds to a translation in the hyperspace by

$$\Delta\mathbf{R} = \sum_{\eta=1}^{N_Q} \frac{\Delta\varphi_\eta}{2\pi} \mathbf{A}_\eta. \tag{4.24}$$

Consequently, the hyperspace representation enables us to treat the phase degree of freedom as additional coordinates in the hyperspace. We note that the situation is analogous to the hyperspace introduced to understand the structures of quasiperiodic crystals, where the number of translation vectors is in general larger than the system dimension and the quasicrystals are obtained by a "slice" of a periodic structure in the hyperspace with additional dimensions spanned by the same number of vectors [15–18]. In the quasicrystals, the excitations associated with the additional variables in

[6] One can obtain the hyperspace position $\mathbf{R}$ for any real-space position $\mathbf{r}$ and phases φ_η from Eq. (4.20), but the inverse mapping is not unique, as different sets of $\mathbf{r}$ and φ_η can give the same $\mathbf{R}$.

the hyperspace are called phasons [19–22], which also supports the analogy. The examples of the hyperspace representation for 2D $3Q$ and 3D $4Q$ states will be given in Sects. 6.1.1 and 7.3.1, respectively.

References

1. Kingma, D. P., & Ba, J. (2014). arXiv:1412.6980
2. Babuschkin, I., Baumli, K., Bell, A., Bhupatiraju, S., Bruce, J., Buchlovsky, P., Budden, D., Cai, T., Clark, A., Danihelka, I., Dedieu, A., Fantacci, C., Godwin, J., Jones, C., Hemsley, R., Hennigan, T., Hessel, M., Hou, S., Kapturowski, S., ... Viola, F. (2020). *The DeepMind JAX Ecosystem*.
3. Bradbury, J., Frostig, R., Hawkins, P., Johnson, M. J., Leary, C., Maclaurin, D., Necula, G., Paszke, A., VanderPlas, J., Wanderman-Milne, S., & Zhang, Q. (2018). JAX: Composable transformations of Python+NumPy programs.
4. Landau, L. D., & Lifshitz, E. M. (1935). *Physikalische Zeitschrift der Sowjetunion, 8*, 153.
5. Gilbert, T. L. (1955). *Physical Review, 100*, 1243.
6. Zhang, S., & Li, Z. (2004). *Physical Review Letters, 93*, 127204.
7. Vansteenkiste, A., Leliaert, J., Dvornik, M., Helsen, M., Garcia-Sanchez, F., & Van Waeyenberge, B. (2014). *AIP Advances, 4*, 107133.
8. Mizukami, S., Ando, Y., & Miyazaki, T. (2001). *Japanese Journal of Applied Physics, 40*, 580.
9. Mizukami, S., Sajitha, E. P., Watanabe, D., Wu, F., Miyazaki, T., Naganuma, H., Oogane, M., & Ando, Y. (2010). *Applied Physics Letters, 96*, 152502.
10. Oogane, M., Wakitani, T., Yakata, S., Yilgin, R., Ando, Y., Sakuma, A., & Miyazaki, T. (2006). *Japanese Journal of Applied Physics, 45*, 3889.
11. Oogane, M., Kubota, T., Kota, Y., Mizukami, S., Naganuma, H., Sakuma, A., & Ando, Y. (2010). *Applied Physics Letters, 96*, 252501.
12. Colpa, J. (1978). *Physica A: Statistical Mechanics and its Applications, 93*, 327.
13. Toth, S., & Lake, B. (2015). *Journal of Physics: Condensed Matter, 27*, 166002.
14. Kato, Y., Hayami, S., & Motome, Y. (2021). *Physical Review B, 104*, 224405.
15. Levine, D., & Steinhardt, P. J. (1984). *Physical Review Letters, 53*, 2477.
16. Levine, D., & Steinhardt, P. J. (1986). *Physical Review B, 34*, 596.
17. Socolar, J. E. S., & Steinhardt, P. J. (1986). *Physical Review B, 34*, 617.
18. Steinhardt, P. J., & Ostlund, S. (1987). *The physics of quasicrystals*. World Scientific.
19. Levine, D., Lubensky, T. C., Ostlund, S., Ramaswamy, S., Steinhardt, P. J., & Toner, J. (1985). *Physical Review Letters, 54*, 1520.
20. Bak, P. (1985). *Physical Review B, 32*, 5764.
21. Kalugin, P., Kitayev, A. Y., & Levitov, L. (1985). *Journal de Physique Letttres, 46*, 601.
22. Hu, C., Wang, R., & Ding, D.-H. (2000). *Reports on Progress in Physics, 63*, 1.

Chapter 5
Spin Moiré Engineering in One Dimension

In this chapter, we investigate the spin moiré engineering in one dimension. Focusing on the model for a 1D chiral magnet, whose Hamiltonian is given by Eq. (3.4), we discuss how the topological properties and the emergent phenomena can be controlled through the static and the time-dependent magnetic fields. In Sect. 5.1, we perform the systematic analysis of the fundamental spin-wave excitation properties by using the linear spin-wave theory. Furthermore, we also investigate the dynamical control of the number of chiral solitons by selectively activating the excitation modes localized near the edges, through numerical simulations based on the LLG equation. In Sect. 5.2, we perform a comprehensive study on the momentum-frequency profile of the EEF using the LLG equation while changing the strength and direction of the static and linearly-polarized AC magnetic fields for both bulk and finite-size chains with edges. Section 5.3 is devoted to the concluding remarks. Hereafter, we set the energy scale as $J = 1$ and take $D = \tan\left(\frac{\pi}{10}\right)$ in Eq. (3.4).

5.1 Magnetic Resonance and Soliton Penetration

In this section, we investigate the dynamical control of soliton number in a 1D chiral magnet, through the edge modes, which are magnetic excitation modes localized near the edges [1–4]. In the topological spin crystals, the creation and annihilation of magnetic solitons in the bulk require considerable energy cost, thereby underscoring the importance of edges and surfaces of the system to control the number of solitons. Particularly, in the CSL, chiral solitons experience a potential barrier at the edges [5], leading to a large hysteresis in the field dependence of the magnetization and the magnetoresistance [5–11]. Thus far, the soliton generation has been mainly discussed for skyrmions in two dimensions and investigated for the process of creating a single or a few solitons in the FFM state, which is a vacuum of solitons [12–14].

K. Shimizu, *Spin Moiré Engineering and Emergent Electromagnetism in Topological Spin Crystals*, Springer Theses, https://doi.org/10.1007/978-981-96-5992-0_5

Nevertheless, generation processes in the CSL are more intriguing since they affect the magnetic and electronic properties over a wider range of the magnetic field.

To clarify the dynamical control of the number of solitons, we investigate the nonequilibrium process accompanied by the soliton penetration from the edges of the system. For the dynamical control of the spin texture, we selectively activate the edge modes by a circularly-polarized AC magnetic field, which brings about the spin dynamics associated with the spin-wave excitation. In Sect. 5.1.1, we briefly review the ground-state phase diagram. To clarify the fundamental spin-wave excitation properties, in Sect. 5.1.2, we present the bulk magnon band structures that exhibit multiple band gaps opened by the solitonic feature of the CSL, obtained by using the linear spin-wave theory. We also show that, in the presence of edges, additional edge modes appear within the magnon band gaps. In Sect. 5.1.3, we show the results of the dynamical process that exhibits the soliton penetration from the edges of the system obtained by numerical simulation with the LLG equation.

5.1.1 Ground-State Phase Diagram

Before discussing the dynamics, we first present the bulk phase diagram for the Hamiltonian in Eq. (3.4) under a static magnetic field $\mathbf{h}(t) = \mathbf{h}^{\text{stat}} = (h_x^{\text{stat}}, 0, h_z^{\text{stat}})$. In Fig. 5.1a, we illustrate the ordering wave number of the ground state, Q^*, as a function of h_x^{stat} and h_z^{stat}. This result is obtained by minimizing the system's energy for $L = 1000$ sites with periodic boundary conditions (PBC). The energy minimization is performed either through the time evolution via the LLG equation in Eq. (4.6) or by applying the energy minimization process described in Sect. 4.2, both of which yield essentially identical results. The obtained phase diagram closely resembles that of the continuum model shown in Fig. 3.1b.

In the absence of an external magnetic field, the ground state is a helical state, where the spins rotate within the yz plane with a periodicity of 20 lattice sites, i.e., $Q^* = Q_0 = \frac{\pi}{10}$ [see Eq. (1.37)]. The real-space spin configuration of this helical state is depicted in Fig. 5.1b. As h_x^{stat} increases with $h_z^{\text{stat}} = 0$, the spins in the yz plane of the helical state tilt uniformly in the $-x$ direction, forming a conical state. Note that the sign of the Zeeman coupling in the Hamiltonian in Eq. (3.4). The real-space spin configuration for $h_x^{\text{stat}} = 0.06$ is displayed in Fig. 5.1c. The spin configuration for the helical and conical states is expressed as

$$\mathbf{S}_l = \left(m, \sqrt{1-m^2}\cos(Q_0 l), \sqrt{1-m^2}\sin(Q_0 l)\right), \tag{5.1}$$

where the uniform magnetization is given by $m = -\frac{h_x^{\text{stat}}}{2}\frac{1}{\sqrt{J^2+D^2}-J}$. In this case, the periodicity remains unchanged until the system undergoes a phase transition to the FFM phase at $h_x^{\text{stat,c}} = 2\left(\sqrt{J^2+D^2} - J\right) \simeq 0.103$.

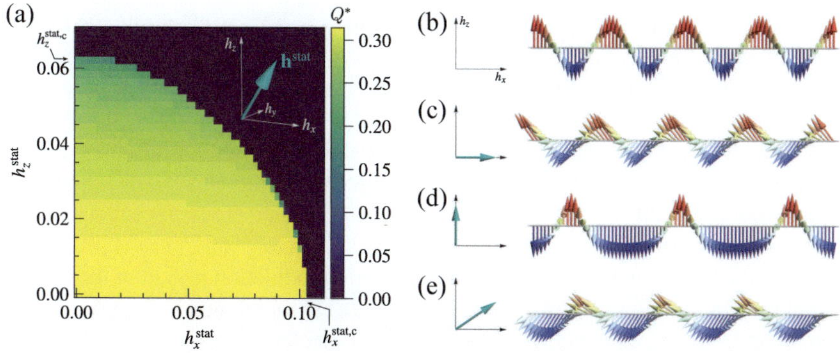

Fig. 5.1 **a** Phase diagram for the Hamiltonian in Eq. (3.4) on the plane of $h_x^{\rm stat}$ and $h_z^{\rm stat}$. The color represents the ordering wave number Q^* calculated for the $L = 1000$ system; $Q^* = \frac{\pi}{10}$ for the helical state at $h_x^{\rm stat} = h_z^{\rm stat} = 0$ and the conical states for $0 < h_x^{\rm stat} < h_x^{\rm stat,c} \simeq 0.103$ at $h_z^{\rm stat} = 0$, $Q^* = 0$ for the FFM state above the critical field curve connecting $h_x^{\rm stat,c}$ and $h_z^{\rm stat,c} \simeq 0.063$, and $0 < Q^* \le \frac{\pi}{10}$ otherwise. Real-space spin configurations of the ground state at **b** $(h_x^{\rm stat}, h_z^{\rm stat}) = (0, 0)$, **c** $(h_x^{\rm stat}, h_z^{\rm stat}) = (0.06, 0)$, **d** $(h_x^{\rm stat}, h_z^{\rm stat}) = (0, 0.06)$, and **e** $(h_x^{\rm stat}, h_z^{\rm stat}) = (0.06, 0.04)$. The results are shown for 80 sites of the 1000-site system

For $h_x^{\rm stat} = 0$, the presence of $h_z^{\rm stat}$ introduces a solitonic nature into the spin configuration, leading to the CSL, as illustrated in Fig. 5.1d. In this case, Q^* gradually decreases with increasing $h_z^{\rm stat}$ until reaching the critical field $h_z^{\rm stat,c} \simeq 0.063$ (see Sect. 1.4.1). The real-space spin configuration of the CSL state follows the lattice discretized form of Eq. (1.12), with parameters given by Eqs. (1.33)–(1.35). For $Q_0 = \frac{\pi}{10}$, the critical field $H_{\rm c}$ is evaluated as $H_{\rm c} \simeq 0.061$ using Eq. (1.35).

When both $h_x^{\rm stat}$ and $h_z^{\rm stat}$ are finite, the system stabilizes a complex spin texture that combines features of both the conical state and the CSL. An example of the real-space spin configuration for $(h_x^{\rm stat}, h_z^{\rm stat}) = (0.06, 0.04)$ is provided in Fig. 5.1e.

5.1.2 Linear Spin-Wave Analysis

In this section, we present the bulk and edge properties of the spin-wave excitation based on the linear spin-wave theory. First, we show the magnon band structure for the helical and CSL states by changing $h_z^{\rm stat}$ with $h_x^{\rm stat} = 0$. Next, we show the spin-wave spectrum under the PBC and OBC, and also present the real-space distribution of the magnon wave function of the edge mode.

Magnon Band Structure

To compute the magnon band structure, we employ the linear spin-wave theory, following the procedure outlined in Sect. 4.5. First, we optimize the spin configuration to minimize the model energy in Eq. (3.4) with $\mathbf{h}(t) = \mathbf{h}^{\text{stat}}$ by changing the integer approximant of the magnetic modulation period Λ from 20 to 40. When $h_x^{\text{stat}} = 0$, Λ increases monotonically with h_z^{stat}, as illustrated in Fig. 5.2a, which is in agreement with the h_z^{stat} dependence of Q^* shown in Fig. 5.1a. Next, we determine the magnon band structure by numerically diagonalizing the magnon Hamiltonian in Eq. (4.14). The magnon band dispersions $\omega_{p,q}$ in the repeated Brillouin zone (BZ) scheme are displayed in Fig. 5.2b.

Let us first examine the results for the helical state at $(h_x^{\text{stat}}, h_z^{\text{stat}}) = (0, 0)$, represented by the black lines. In this case, the magnetic modulation period of the ground state in the thermodynamic limit is exactly $\Lambda = 20$, as mentioned in Sect. 5.1.1. The band structure in the extended BZ scheme is analytically given by

$$\omega_q = 2\sqrt{\tilde{J}\,(1 - \cos q)\left(\tilde{J} - \cos q\right)}, \tag{5.2}$$

where $\tilde{J} = \sqrt{J^2 + D^2}$ [15–17]. This dispersion remains continuous without a gap, resembling the electron band dispersion under the empty lattice approximation. Furthermore, the helical state exhibits a Goldstone mode with a linear dispersion around $q = 0$, which results from the continuous rotational symmetry of the helical spin structure about the chiral axis.

Next, we discuss the results for the CSL at $(h_x^{\text{stat}}, h_z^{\text{stat}}) = (0, 0.06)$, represented by the blue lines in Fig. 5.2b. In this case, the magnetic modulation period extends to $\Lambda = 29$, as shown in Fig. 5.2a. We observe gap openings in the magnon dispersion: the gaps between the $(2n-1)$-th and $2n$-th bands, and between the $2n$-th and $(2n+1)$-th bands, appear at $q = \pm\frac{\pi}{\Lambda}$ and $q = 0$, respectively, where n is an integer. This is

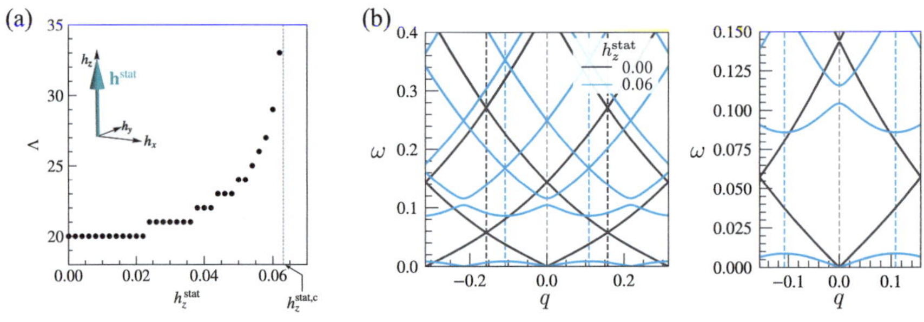

Fig. 5.2 **a** The magnetic modulation period for the commensurate approximant of the stable spin configuration, Λ, for $h_x^{\text{stat}} = 0$. **b** The magnon band dispersion for $h_z^{\text{stat}} = 0$ and $h_z^{\text{stat}} = 0.06$ represented by the black and blue curves, respectively, at $h_x^{\text{stat}} = 0$. The results are shown in the repeated BZ scheme up to the second and first magnetic BZs of $h_z^{\text{stat}} = 0$ in the left and right panels, respectively

attributed to the solitonic nature of the spin texture induced by h_z^{stat}, which acts as a periodic potential and scatters the magnons. As h_z^{stat} increases, the gap magnitude becomes larger, leading to a reduction in the width of each magnon band. We also note that the CSL exhibits a linear Goldstone mode near $q = 0$ in the lowest branch [18–20].

Edge Mode

To examine the impact of edge effects on spin-wave excitations, we compare the magnon spectra for systems with PBC and OBC. Figure 5.3a presents the magnon band structure alongside the eigenfrequencies of the real-space magnon Hamiltonian in Eq. (4.12) for an $L = 1000$-site system at $(h_x^{\text{stat}}, h_z^{\text{stat}}) = (0, 0.02)$. The left panel illustrates the magnon band dispersion for the PBC case, which features a gap in the range of $0.45 \lesssim \omega \lesssim 0.07$ due to the finite h_z^{stat}. The right panel displays the eigenfrequencies in ascending order.

For the OBC case, the results largely coincide with those of the PBC case, except for two additional modes that emerge within the band gap. These mid-gap states correspond to edge modes. Indeed, their magnon wave functions are localized near the system's edges, as depicted in Fig. 5.3b, which illustrates one of the two modes localized at the left edge (the other is localized at the right edge). As shown in Fig. 5.3b, the localization of these edge modes becomes more pronounced as h_z^{stat} increases. This behavior aligns with the increase of the bulk band gap in the bulk band spectrum. We note that for $h_x^{\text{stat}} = 0$ two eigenfrequencies are degenerate due to twofold rotational symmetry about the z axis; in our numerically optimized solutions, the degeneracy is slightly lifted, allowing the two modes localized at the left and right edges to be resolved.

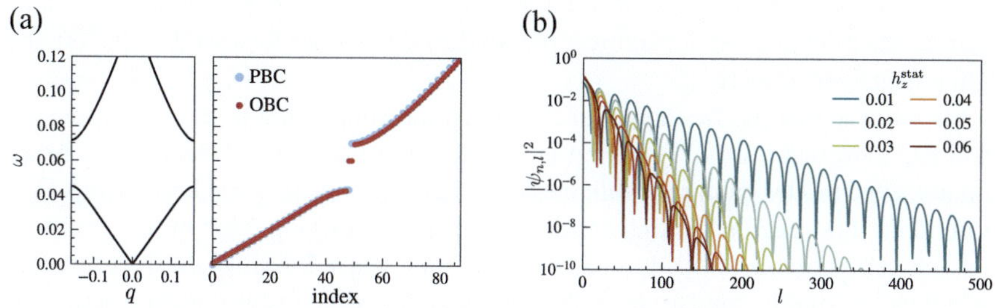

Fig. 5.3 **a** Magnon band dispersion for the first BZ (left) and eigenfrequencies of the magnon Hamiltonian for the 1000-site system under the PBC and OBC shown by the blue and red points, respectively (right). The results are shown for $(h_x^{\text{stat}}, h_z^{\text{stat}}) = (0, 0.02)$. In the right panel, the eigenfrequencies are shown in ascending order. **b** Real-space distribution of the square amplitude of the wavefunctions $|\psi_{n,l}|^2$ at $h_x^{\text{stat}} = 0$ for several values of h_z^{stat}. For each h_z^{stat}, the eigenfrequency is set at the edge mode localized at the left edge of the system

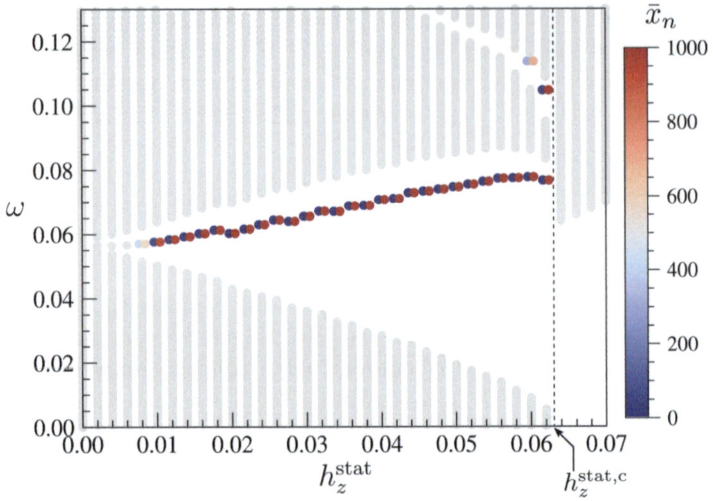

Fig. 5.4 Eigenfrequencies of the magnon Hamiltonian with varying h_z^{stat} at $h_x^{\mathrm{stat}} = 0$. The color represents the center of mass for the wavefunction [Eq. (5.3)]. For a better visibility, each edge mode pair is plotted slightly offset to the left and right

In Fig. 5.4, we present the h_z^{stat} dependence of the eigenspectra for $h_x^{\mathrm{stat}} = 0$ under the OBC. Each dot represents an eigenfrequency ω_n, with the color indicating the center of mass of the corresponding magnon wave function $\psi_{n,l}$, defined by

$$\bar{x}_n = \frac{\sum_{l=0}^{L-1} l \left|\psi_{n,l}\right|^2}{\sum_{l=0}^{L-1} \left|\psi_{n,l}\right|^2}. \tag{5.3}$$

The bulk modes, which extend throughout the system, are represented by gray dots, whereas the blue and red dots correspond to the edge modes localized at the left and right edges, respectively.

We observe that the frequency of the edge mode appearing in the lowest-energy band gap increases with h_z^{stat}, except in the vicinity of the critical field $h_z^{\mathrm{stat,c}}$. Additionally, edge modes emerge within the second gap, though they are less distinct than those in the first gap. In the higher-energy region, identifying edge modes becomes more challenging due to the diminishing size of the gaps. It is also worth noting that these localized modes appear not only at the system's edges but also at the boundaries of domains with different chirality (see Appendix A).

5.1.3 Edge Excitation and Soliton Penetration

In the preceding section, we elucidated the characteristics of the magnetic excitation spectrum under a static magnetic field, with a particular focus on edge modes. In this section, we explore a fascinating dynamical phenomenon wherein edge modes are strongly excited, resulting in a progressive increase in the number of solitons over time through successive penetration from the edges.

Here, the spin dynamics is driven by a rotating magnetic field with frequency ω, given by

$$\mathbf{h}^{\mathrm{rot}}(t) = h^{\mathrm{rot}}\left(\hat{\mathbf{y}}\cos(\omega t) + \epsilon\hat{\mathbf{z}}\sin(\omega t)\right)\left(1 - e^{-(t/t_0)^2}\right), \qquad (5.4)$$

where h^{rot} and $\epsilon = \pm 1$ denote the amplitude and handedness of the rotating magnetic field, respectively. Consequently, the total magnetic field in Eq. (3.4) is given by $\mathbf{h}(t) = \mathbf{h}^{\mathrm{stat}} + \mathbf{h}^{\mathrm{rot}}(t)$. We refer to $\mathbf{h}^{\mathrm{rot}}(t)$ with $\epsilon = 1$ (-1) as the counterclockwise (clockwise) rotating magnetic field. In Eq. (5.4), an exponential damping factor with $t_0 = 200$ is introduced to mitigate the initial disturbance caused by switching on the field. In the following analysis, we compute the spin dynamics up to $t_f = 2 \times 10^5$ by numerically solving the LLG equation in Eq. (4.6) with a time step of $\Delta t = 0.1$ and a Gilbert damping coefficient of $\alpha = 0.04$ for an $L = 1000$-site system under OBC.

In the subsequent discussion, we first present results for $h_x^{\mathrm{stat}} = 0$, examining the spatiotemporal evolution of the spin configuration and the correlation between the amplitude of the rotating magnetic field and the number of solitons infiltrating the system. We then extend our analysis to the case of $h_x^{\mathrm{stat}} \neq 0$, elucidating the impact of a static magnetic field aligned with the chiral axis.

$h_x^{\mathrm{stat}} = 0$ Case

First, we present the results in the absence of a static magnetic field parallel to the chiral axis, h_x^{stat}. Figure 5.5 illustrates the spatiotemporal evolution of $S_{l,z}(t)$ at $(h_x^{\mathrm{stat}}, h_z^{\mathrm{stat}}) = (0, 0.05)$ under the rotating magnetic field given by Eq. (5.4), with $\epsilon = 1$ and $\omega = 0.075$. The frequency ω is chosen to resonate with the edge modes within the first band gap (see Fig. 5.4). In each panel of Fig. 5.5, the left panel presents the evolution up to $t = 1.2 \times 10^5$ across all sites of the 1000-site system, while the right subplot provides an enlarged plot of the rightmost 100 sites over a shorter duration, up to $t = 4 \times 10^4$.

Figure 5.5a depicts the results for $h^{\mathrm{rot}} = 0.01$. Applying the rotating magnetic field at the edge mode frequency induces substantial modulations of the spins at the left and right edges, which exhibit the precessional motion. Simultaneously, bulk solitons start moving leftward, as shown in the left panel of Fig. 5.5a. This directed motion arises due to the breaking of twofold rotational symmetry about the z axis by the rotating field, thereby facilitating unidirectional spin-wave propagation. As a consequence,

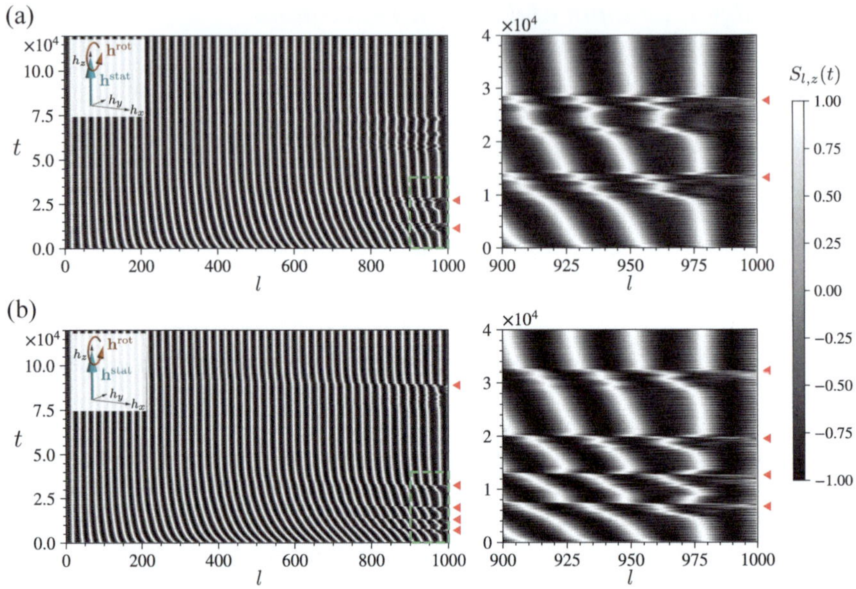

Fig. 5.5 Real-space and real-time distributions of the z component of spins $S_{l,z}(t)$ on the plane of l and t for the counterclockwise rotating magnetic field ($\epsilon = 1$) at $(h_x^{\rm stat}, h_z^{\rm stat}) = (0, 0.05)$ and $\omega = 0.075$: **a** $h^{\rm rot} = 0.01$ and **b** $h^{\rm rot} = 0.012$. The frequency ω is set at the edge modes in the first gap (see Fig. 5.4). The red triangles represent the time when solitons penetrate from the right edge of the system. In each panel, the left figure shows the whole system, while the right one is an enlarged plot near the right edge in a shorter time window, denoted by the dashed green square in the left one

the soliton lattice becomes compressed towards the left edge, creating soliton-sparse regions near the right edge where spins predominantly point downward. Then, a pronounced precursor effect characterized by the rapid displacement of the rightmost soliton is observed at $t \simeq 0.9 \times 10^4$, which is followed by the penetration of a new soliton from the right edge at $t \simeq 1.4 \times 10^4$, as indicated by the lower red triangle in the right panel of Fig. 5.5a. A subsequent soliton penetration occurs at $t \simeq 2.8 \times 10^4$, preceded by a precursor shift at $t \simeq 2.2 \times 10^4$. After the insertion of these two solitons, further penetration ceases under the given field parameters, and the system attains a nonequilibrium steady state.

Through detailed numerical simulations, the mechanism of soliton penetration can be understood as follows. The leftward motion of solitons extends the domain of downward-pointing spins near the right boundary, effectively creating a local FFM state. Since the energy of the FFM state is higher that that of the CSL in the static case, this configuration is energetically unstable. To restore stability, an additional soliton is introduced via the large spin precession at the system's edges (see also the discussion on the energy barrier for a single soliton in Appendix B). Once a sufficient number of solitons accumulate, the system reaches a nonequilibrium steady state, wherein the driving force promoting soliton injection is counterbalanced by the repulsive interactions between neighboring solitons. In the present case, a counterclockwise rotating magnetic field induces soliton penetration from the right edge. Conversely, a clockwise rotating field results in soliton penetration from the left edge, ultimately leading to a left-right inverted nonequilibrium steady state. We note that similar soliton penetration also occurs at boundaries of domains with different chirality. In this case, a soliton pair is created at the boundary, with each soliton propagating into the different domains (see Appendix A).

Figure 5.5b presents the results for a stronger rotating magnetic field, $h^{\mathrm{rot}} = 0.012$. Here, soliton penetration is observed to commence earlier, the time interval between successive penetration events decreases, and the total number of penetrating solitons increases. This behavior can be understood based on the above discussion. As the amplitude of the rotating magnetic field grows, the exerted force on solitons intensifies, driving them more rapidly toward the left. The acceleration of this precursor motion shortens both the onset time of soliton penetration and the intervals between subsequent events. Moreover, to counterbalance the stronger driving force with the repulsive interactions among solitons, a larger number of soliton penetrations is required.

While changing the values of h_z^{stat} and h^{rot}, we systematically investigate the number of solitons penetrating into the system. We calculate the difference of the soliton number between the initial state at $t = 0$ and the final state at $t = t_f$ as

$$\Delta N_{\mathrm{soliton}} = \frac{1}{50T} \int_{t_f - 50T}^{t_f} N_{\mathrm{soliton}}(t) dt - N_{\mathrm{soliton}}(0), \tag{5.5}$$

where $T = \frac{2\pi}{\omega}$ is the time period of the rotating magnetic field, and $N_{\mathrm{soliton}}(t)$ is the number of solitons given by [see also Eq. (1.3)]

$$N_{\mathrm{soliton}}(t) = \frac{1}{2\pi} \sum_l \mathrm{Arg}\left(\frac{S_{l+1}^+(t)}{S_l^+(t)} \right), \tag{5.6}$$

with

$$S_l^+(t) = S_{l,x}(t) + \mathrm{i} S_{l,y}(t). \tag{5.7}$$

Here, by averaging the values of $N_{\mathrm{soliton}}(t)$ during the final 50 periods of the rotating magnetic field, we eliminate the fine time dependence of $N_{\mathrm{soliton}}(t)$ and obtain the value close to an integer.

Figure 5.6a shows $\Delta N_{\mathrm{soliton}}$ for the counterclockwise rotating magnetic field with $\epsilon = 1$ on the plane of h_z^{stat} and h^{rot}. The results are common to the clockwise field with $\epsilon = -1$ due to the symmetry for $h_x^{\mathrm{stat}} = 0$. The frequency of the rotating field is set at the edge modes in the first magnon gap for each h_z^{stat}. We find that for all h_z^{stat} the soliton penetration occurs above a certain threshold value of h^{rot}. Above the threshold, $\Delta N_{\mathrm{soliton}}$ starts to incrementally grow in a sequential manner with h^{rot}. This trend is explicitly shown for $h_z^{\mathrm{stat}} = 0.034$, 0.042, and 0.05 in Fig. 5.6b.

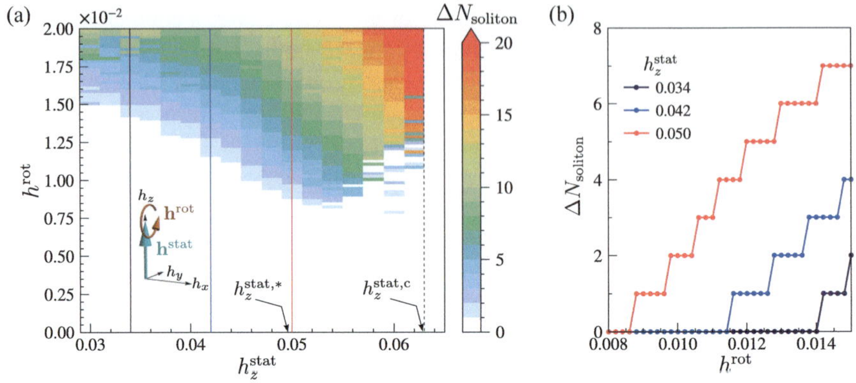

Fig. 5.6 **a** The total number of solitons penetrating into the system, $\Delta N_{\mathrm{soliton}}$ [Eq. (5.5)], on the plane of h_z^{stat} and h^{rot} with the counterclockwise rotating magnetic field at $h_x^{\mathrm{stat}} = 0$. The frequency of the magnetic field is set as the edge mode in the lowest-energy magnon gap for each h_z^{stat}. The dotted line denotes the critical magnetic field $h_z^{\mathrm{stat,c}}$ for the transition from the CSL to the FFM state. **b** h^{rot} dependence of $\Delta N_{\mathrm{soliton}}$ for several values of h_z^{stat}: $h_z^{\mathrm{stat}} = 0.034$, 0.042, and 0.05, corresponding to the black, blue, and red lines in (**a**), respectively

For $h^{\mathrm{rot}} \gtrsim 0.015$, the overall trend of $\Delta N_{\mathrm{soliton}}$ increasing is maintained, but the sequential increment is no longer observed due to strong fluctuations under large h^{rot}, as represented by the nonmonotonic color changes in Fig. 5.6a. The threshold value of h^{rot} for the soliton penetration steadily diminishes with increasing h_z^{stat}, but it becomes ambiguous and seemingly increases with h_z^{stat} above $h_z^{\mathrm{stat},*} \simeq 0.052$, where the increment of $\Delta N_{\mathrm{soliton}}$ with h^{rot} is also fluctuating.

The reduction of the threshold for $h_z^{\mathrm{stat}} \lesssim h_z^{\mathrm{stat},*}$ can be attributed to the enhancement of the precession motion of spins around the edges with h_z^{stat}. In fact, the dynamical spin susceptibility of the edge mode monotonically increases with h_z^{stat}, as discussed in see Sect. 5.2.3. Thus, larger h_z^{stat} facilitates a stronger flow of spin waves in the system, decreasing the threshold value. Meanwhile, the increase of the threshold for $h_z^{\mathrm{stat}} \gtrsim h_z^{\mathrm{stat},*}$ can be understood from the energy experienced by a chiral soliton around the edges that can be locally regarded as the FFM state. Considering an addition of a single chiral soliton to the FFM state, for $h_z^{\mathrm{stat}} \lesssim h_z^{\mathrm{stat},*}$, the energy is reduced by the addition of the soliton inside the bulk, while for $h_z^{\mathrm{stat}} \gtrsim h_z^{\mathrm{stat},*}$, the system prefers maintaining the bulk in the FFM state by putting the soliton outside and limiting its influence to modulations of spins around the edges (see also Appendix B for the discussion on the energy gain accompanied by the soliton penetration). Consequently, the penetration of solitons from the edge becomes unstable for $h_z^{\mathrm{stat}} \gtrsim h_z^{\mathrm{stat},*}$, whereas the significant spin precession at the edge with large h^{rot} can force a chiral soliton to be pushed deeply inside the bulk, leading to the increase of the threshold value of h^{rot}.

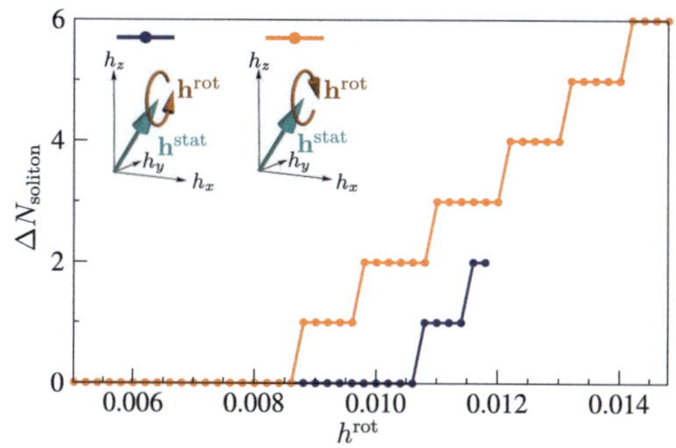

Fig. 5.7 h^{rot} dependence of $\Delta N_{\text{soliton}}$ at $(h_x^{\text{stat}}, h_z^{\text{stat}}) = (0.04, 0.042)$ and $\omega = 0.068$. The blue and orange lines represent the results for the counterclockwise ($\epsilon = 1$) and clockwise ($\epsilon = -1$) rotating magnetic fields, respectively. The results for $h^{\text{rot}} \geq 0.012$ with $\epsilon = 1$ are not shown as the spin texture is strongly disturbed

$h_x^{\text{stat}} \neq 0$ Case

Next, we examine soliton penetration in the presence of a static magnetic field parallel to the chiral axis, h_x^{stat}. Even for $h_x^{\text{stat}} \neq 0$, the magnon band gap remains open due to h_z^{stat}, and edge modes emerge within the gap, similar to the case with $h_x^{\text{stat}} = 0$. However, the introduction of h_x^{stat} breaks the twofold rotational symmetry about the z axis, leading to asymmetric responses under clockwise and counterclockwise rotating magnetic fields.

Figure 5.7 presents $\Delta N_{\text{soliton}}$ as a function of h^{rot} at $(h_x^{\text{stat}}, h_z^{\text{stat}}) = (0.04, 0.042)$ for both rotating directions. The blue and orange lines correspond to counterclockwise ($\epsilon = 1$) and clockwise ($\epsilon = -1$) rotating fields, respectively. A key observation is that the threshold values of h^{rot}, at which $\Delta N_{\text{soliton}}$ begins to increase, differ between the two cases. Specifically, the threshold values are 0.0087 for the clockwise field and 0.0107 for the counterclockwise field, both of which are lower than the threshold of 0.0115 observed at $h_x^{\text{stat}} = 0$ in Fig. 5.6b. This reduction can be attributed to the conical spin canting induced by h_x^{stat}, which effectively lowers the energy barrier for a chiral soliton at the system's edge.

Beyond these threshold values, $\Delta N_{\text{soliton}}$ exhibits a progressive increase, reflecting the trend observed in the $h_x^{\text{stat}} = 0$ case. For the counterclockwise rotating magnetic field, however, when $h^{\text{rot}} \gtrsim 0.0119$, the system undergoes substantial disturbances, breaking the CSL structure; thus, these data points are omitted from Fig. 5.7.

5.2 Emergent Electric Field from Magnetic Resonance

In this section, we explore the emergence of the EEF driven by the time evolution of noncollinear spin textures and examine its control via external magnetic fields. While the EMF vanishes in 1D space, spin dynamics can induce a finite EEF, thereby enabling emergent electric phenomena across a broader class of magnetic materials compared to emergent magnetic phenomena (see Sect. 1.3). In a 1D chiral magnet, diverse spin textures, including helical, conical, and CSL states, give rise to distinct spin dynamics, leading to rich emergent electric phenomena. Previous studies have investigated the EEF in 1D chiral magnets within the continuum limit, primarily employing the collective coordinate method [21–23]. However, the interplay between the EEF and magnetic excitations, including the spin-wave modes analyzed in Sect. 5.1, remains largely unexplored. Furthermore, contributions to the EEF from both bulk and edge regions are particularly relevant for experiments on nano- and microscale samples.

To address these issues, we systematically investigate the dependence of the EEF on static and AC magnetic fields in both bulk and finite-size systems. We numerically solve the LLG equation in Eq. (4.6) with a time step $\Delta t = 0.02$ and a Gilbert damping $\alpha = 0.04$ for a system of $L = 1000$ sites under the PBC and OBC. In Sect. 5.2.1, we introduce the physical quantities relevant to the EEF. In Sect. 5.2.2, we present the results for the bulk contribution under PBC, analyzing the complex admittance of the EEF for various static and AC magnetic field configurations and comparing these findings with the dynamical spin susceptibility. We also examine the spatiotemporal evolution of spin textures and the EEF in resonance modes. In Sect. 5.2.3, we extend our analysis to edge contributions under OBC, following an approach similar to the bulk case. Finally, in Sect. 5.2.4, we estimate the expected magnitude of the EEF in real materials.

5.2.1 Physical Quantities

The EEF is generated by the time evolution of noncollinear spin textures and is computed as [24]

$$\bar{E}^{\mathrm{em}}(t) = \frac{1}{L} \sum_{l} E_l^{\mathrm{em}}(t), \tag{5.8}$$

where

$$E_l^{\mathrm{em}}(t) = \mathbf{S}_l(t) \cdot \left(\hat{\delta}_l \mathbf{S}_l(t) \times \frac{\partial \mathbf{S}_l(t)}{\partial t} \right). \tag{5.9}$$

Here, the EEF is obtained from the spin configurations computed by numerically solving the LLG equation in Eq. (4.6). This formulation represents a lattice-discretized version of Eq. (1.28). In Eq. (5.9), the spatial derivative $\frac{\partial \mathbf{S}(x)}{\partial x}$ in the continuum formulation [24–28] is approximated by the discrete difference $\hat{\delta}_l \mathbf{S}_l(t) = \frac{1}{2}\left(\mathbf{S}_{l+1}(t) - \mathbf{S}_{l-1}(t)\right)$. For the edge spins at $l = 0$ and $l = L - 1$ in the OBC case, we use $\hat{\delta}_0 \mathbf{S}_0(t) = -\frac{3}{2}\mathbf{S}_0(t) + 2\mathbf{S}_1(t) - \frac{1}{2}\mathbf{S}_2(t)$ and $\hat{\delta}_{L-1}\mathbf{S}_{L-1}(t) = \frac{3}{2}\mathbf{S}_{L-1}(t) - 2\mathbf{S}_{L-2}(t) + \frac{1}{2}\mathbf{S}_{L-3}(t)$, respectively, to reduce the discretization errors [29].

The linear response of the EEF to a time-dependent magnetic field is characterized by the complex admittance, which quantifies the response to a linearly polarized AC magnetic field $\mathbf{h}^{\mathrm{AC}}(t)$,

$$\chi_{E,\mu}(\omega) = \frac{\bar{E}^{\mathrm{em}}(\omega)}{h_\mu^{\mathrm{AC}}(\omega)}, \tag{5.10}$$

where $\mu = x, y, z$ and the Fourier component of a time-dependent quantity $\mathcal{O}(t)$ is obtained by

$$\mathcal{O}(\omega) = \frac{1}{N_t}\sum_{n=0}^{N_t-1} \mathcal{O}(t_n) e^{-i\omega t_n}. \tag{5.11}$$

In our numerical simulations, measurements are taken every 50 time steps, with the n-th measurement occurring at $t_n = 50n\Delta t = n$, where N_t denotes the total number of measurements. We also compute the dynamical spin susceptibility defined as

$$\chi_{S,\nu\mu}(\omega) = \frac{\Delta \bar{S}_\nu(\omega)}{h_\mu^{\mathrm{AC}}(\omega)}, \tag{5.12}$$

where $\Delta\bar{\mathbf{S}}(t) = \frac{1}{L}\sum_l [\mathbf{S}_l(t) - \mathbf{S}_l(0)]$ represents the magnetization variation over time.

To compute $\chi_{E,\mu}(\omega)$ and $\chi_{S,\nu\mu}(\omega)$ efficiently, we apply a pulse magnetic field $\mathbf{h}^{\mathrm{pulse}}(t)$ instead of a continuous AC field $\mathbf{h}^{\mathrm{AC}}(t)$, such that

$$\mathbf{h}(t) = \mathbf{h}^{\mathrm{stat}} + \mathbf{h}^{\mathrm{pulse}}(t). \tag{5.13}$$

This approach allows for simultaneous extraction of the full frequency spectrum without requiring steady-state simulations at each frequency. For the pulse field, we use

$$\mathbf{h}^{\mathrm{pulse}}(t) = \begin{cases} \Delta h \hat{\boldsymbol{\mu}} & (0 \le t < 1), \\ 0 & (1 \le t \le 6000), \end{cases} \tag{5.14}$$

where $\Delta h = 0.002$, and $\hat{\boldsymbol{\mu}}$ denotes the unit vector along $\mu = x, y, z$. By applying the Fourier transform to the response, the spectra of $\chi_{E,\mu}(\omega)$ and $\chi_{S,\nu\mu}(\omega)$ are obtained from Eqs. (5.10) and (5.12) with the substitution of $h_\mu^{\mathrm{AC}}(\omega)$ by $h_\mu^{\mathrm{pulse}}(\omega)$.

For analyzing the spatiotemporal evolution of spin configurations and the EEF associated with magnetic resonance, we instead apply a linearly polarized AC magnetic field with frequency ω in the direction of $\hat{\boldsymbol{\mu}}$, i.e., $\mathbf{h}(t) = \mathbf{h}^{\mathrm{stat}} + \mathbf{h}^{\mathrm{AC}}(t)$. The AC field is given by

$$\mathbf{h}^{\mathrm{AC}}(t) = \Delta h \left(1 - e^{-t/t_0}\right) \sin(\omega t) \hat{\boldsymbol{\mu}} \quad (0 \le t \le 10000), \tag{5.15}$$

where $t_0 = 50$ is introduced to suppress transient disturbances from the initial switching of the field.

5.2.2 *Bulk Properties*

In this section, we show the results of the bulk EEF obtained from the calculations for the system with the PBC. First, we present the maximum values of $|\chi_{E,\mu}(\omega)|$ on the plane of h_x^{stat} and h_z^{stat}, and thc typical frequency spectra. Next, we discuss the relation between the EEF and the magnetic excitations by comparing $\chi_{E,\mu}(\omega)$ and $\chi_{S,\nu\mu}(\omega)$. Finally, we show the spatiotemporal profiles of the spin textures and the EEF of the resonance modes.

Complex Admittance of the EEF

We present the results for the complex admittance $\chi_{E,\mu}(\omega)$ in Eq. (5.10) for three distinct orientations of the AC magnetic field: $\mathbf{h}^{\mathrm{AC}} \parallel \hat{\mathbf{x}}$ ($\mu = x$), $\mathbf{h}^{\mathrm{AC}} \parallel \hat{\mathbf{y}}$ ($\mu = y$), and $\mathbf{h}^{\mathrm{AC}} \parallel \hat{\mathbf{z}}$ ($\mu = z$).

Let us first consider the results for $\mathbf{h}^{\mathrm{AC}} \parallel \hat{\mathbf{x}}$. In Fig. 5.8a, we display the maximum values of $|\chi_{E,x}(\omega)|$, denoted as $\chi_{E,x}^{\max}$, on the h_x^{stat}-h_z^{stat} plane. We observe that $\chi_{E,x}^{\max}$ takes nonzero values within the region $0 \le h_x^{\mathrm{stat}} \lesssim 0.103$ and $0 \le h_z^{\mathrm{stat}} \lesssim 0.063$, where the system exhibits a noncollinear spin texture in the static magnetic field, specifically the CSL for $h_x^{\mathrm{stat}} = 0$, conical for $h_z^{\mathrm{stat}} = 0$, and a mixture of both otherwise. In this region, $\chi_{E,x}^{\max}$ increases with increasing h_z^{stat}, but decreases with increasing h_x^{stat}; the EEF is maximally generated at $(h_x^{\mathrm{stat}}, h_z^{\mathrm{stat}}) \approx (0, 0.06)$, which is near the phase transition from the CSL to the FFM. Outside this region, the spins become fully polarized by the magnetic field, and $\chi_{E,x}(\omega)$ vanishes, indicating that the FFM phase does not contribute to the generation of the EEF.

We present the spectrum of $|\chi_{E,x}(\omega)|$ in Fig. 5.8d by varying h_z^{stat} at $h_x^{\mathrm{stat}} = 0$, along the yellow dashed line in Fig. 5.8a. $|\chi_{E,x}(\omega)|$ exhibits several peaks superimposed on an almost ω-independent background (the origin of which will be discussed in the next subsection). As h_z^{stat} approaches the critical value for the FFM transition, $h_z^{\mathrm{stat,c}} \approx 0.063$, the peak frequencies decrease while their intensities increase. The maximum value of $|\chi_{E,x}(\omega)|$ occurs at the lowest-frequency peak, denoted by $\omega_1^{\mathrm{bulk}} \approx 0.1$. A second peak is observed at $\omega_2^{\mathrm{bulk}} \gtrsim 0.2$, although higher peaks are difficult to discern in the contour plot [see the ω profile in Fig. 5.9a]. These peak structures are attributed to magnetic resonance, as discussed in the following subsection.

Next, we turn to the analysis of $\chi_{E,y}^{\max}$ for $\mathbf{h}^{\mathrm{AC}} \parallel \hat{\mathbf{y}}$, as shown in Fig. 5.8b. The EEF is again maximally generated in the CSL phase for $h_x^{\mathrm{stat}} = 0$, but at a slightly lower $h_z^{\mathrm{stat}} \approx 0.05$, just below the critical value for the FFM transition. The inten-

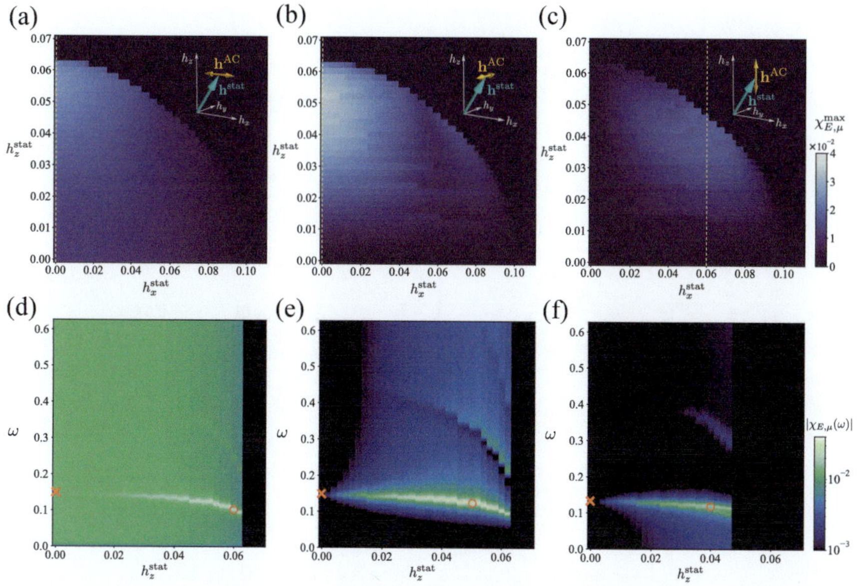

Fig. 5.8 Maximum values of $|\chi_{E,\mu}(\omega)|$, $\chi^{\max}_{E,\mu}$, calculated for the model in Eq. (3.4) with the PBC for **a** $\mathbf{h}^{\rm AC} \parallel \hat{\mathbf{x}}$ ($\mu = x$), **b** $\mathbf{h}^{\rm AC} \parallel \hat{\mathbf{y}}$ ($\mu = y$), and **c** $\mathbf{h}^{\rm AC} \parallel \hat{\mathbf{z}}$ ($\mu = z$), as represented in the insets. The frequency dependence of $|\chi_{E,\mu}(\omega)|$ along the yellow dashed lines in (**a**), (**b**), and (**c**) are shown in (**d**), (**e**), and (**f**), respectively: **d** $\mathbf{h}^{\rm AC} \parallel \hat{\mathbf{x}}$ with $h_x^{\rm stat} = 0$, **e** $\mathbf{h}^{\rm AC} \parallel \hat{\mathbf{y}}$ with $h_x^{\rm stat} = 0$, and **f** $\mathbf{h}^{\rm AC} \parallel \hat{\mathbf{z}}$ with $h_x^{\rm stat} = 0.06$. The orange circles and crosses in (**d**), (**e**), and (**f**) indicate the parameters for which the real-space behaviors are presented in Fig. 5.10. Reproduced from Ref. [4]. Copyright (2023) by the American Physical Society

sity is greater than that observed in Fig. 5.8a. A weaker peak is identified at $(h_x^{\rm stat}, h_z^{\rm stat}) \approx (0.07, 0.04)$, which is a remnant of the peak observed for $\mathbf{h}^{\rm AC} \parallel \hat{\mathbf{z}}$, discussed further below in Fig. 5.8c. It is important to note that $\chi^{\max}_{E,y}$ is significantly suppressed in the conical state at $h_z^{\rm stat} = 0$, in contrast to the case of $\mathbf{h}^{\rm AC} \parallel \hat{\mathbf{x}}$ in Fig. 5.8a. This suppression arises due to the twofold rotational symmetry about the x axis accompanied by time translation by half the period of $\mathbf{h}^{\rm AC}$, $\frac{\pi}{\omega}$.[1] The spectrum of $|\chi_{E,y}(\omega)|$ for $h_x^{\rm stat} = 0$ [the yellow dashed line in Fig. 5.8b] is shown in Fig. 5.8e. Three distinct peaks are clearly identifiable within this ω range, more so than in Fig. 5.8d. In contrast to the case in Fig. 5.8d, no ω-independent contribution is observed, and the peak intensities are maximized at $h_z^{\rm stat} \approx 0.05$, diminishing for $h_z^{\rm stat} \gtrsim 0.05$.

Lastly, we present the results for $\mathbf{h}^{\rm AC} \parallel \hat{\mathbf{z}}$ in Fig. 5.8c and f. In this case, the EEF is maximally generated at $(h_x^{\rm stat}, h_z^{\rm stat}) \approx (0.06, 0.045)$, which is close to the subdominant maximum observed in Fig. 5.8b. Similar to the case in Fig. 5.8b, $\chi^{\max}_{E,z}$ is strongly suppressed for $h_z^{\rm stat} = 0$ due to symmetry. Moreover, in contrast to the results in Fig. 5.8a and b, $\chi^{\max}_{E,z}$ vanishes for $h_x^{\rm stat} = 0$ due to the twofold rotational symmetry

[1] We note that weak responses persist for $\mathbf{h}^{\rm AC} \perp \mathbf{h}^{\rm stat}$ due to nonlinear contributions.

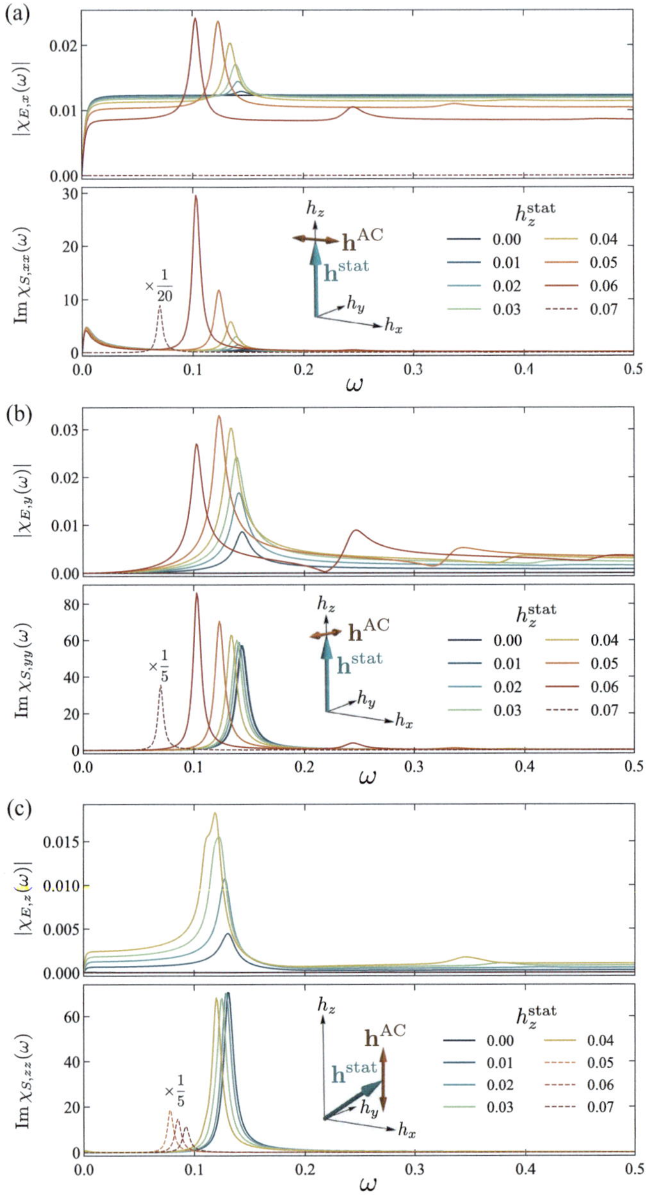

Fig. 5.9 Comparison between $|\chi_{E,\mu}(\omega)|$ and the imaginary part of the dynamical spin susceptibility, Im $\chi_{S,\mu\mu}(\omega)$, for several values of h_z^{stat}: **a** $\mathbf{h}^{\mathrm{AC}} \parallel \hat{\mathbf{x}}$ with $h_x^{\mathrm{stat}} = 0$, **b** $\mathbf{h}^{\mathrm{AC}} \parallel \hat{\mathbf{y}}$ with $h_x^{\mathrm{stat}} = 0$, and **c** $\mathbf{h}^{\mathrm{AC}} \parallel \hat{\mathbf{z}}$ with $h_x^{\mathrm{stat}} = 0.06$ for the model in Eq. (3.4) with the PBC. The data for the FFM phases are denoted by the dotted lines and scaled for better visibility. Reproduced from Ref. [4]. Copyright (2023) by the American Physical Society

about the z axis. Under these symmetric constraints, the peak for $\mathbf{h}^{\mathrm{AC}} \parallel \hat{\mathbf{z}}$ appears in the intermediate region near the phase boundary to the FFM state. Figure 5.8f shows $|\chi_{E,z}(\omega)|$ for $h_x^{\mathrm{stat}} = 0.06$, along the yellow dashed line passing through the peak of $\chi_{E,z}^{\mathrm{max}}$ in Fig. 5.8c. Several peaks are observed without an ω-independent background, as in Fig. 5.8e. Notably, each peak splits into two, most clearly visible at the lowest-frequency peak, ω_1^{bulk}. This phenomenon can be understood in terms of the nonreciprocal magnon band dispersion for $h_x^{\mathrm{stat}} > 0$.

Comparison with Magnetic Excitations

In this section, we discuss the results of $|\chi_{E,\mu}(\omega)|$ in comparison with the magnetic excitation spectrum. Figure 5.9 displays the comparisons between $|\chi_{E,\mu}(\omega)|$ and the imaginary part of the dynamical spin susceptibility, Im $\chi_{S,\mu\mu}(\omega)$, calculated by Eq. (5.12) for different field configurations: $\mathbf{h}^{\mathrm{AC}} \parallel \hat{\mathbf{x}}$ ($\mu = x$) and $h_x^{\mathrm{stat}} = 0$ in Fig. 5.9a, $\mathbf{h}^{\mathrm{AC}} \parallel \hat{\mathbf{y}}$ ($\mu = y$) and $h_x^{\mathrm{stat}} = 0$ in Fig. 5.9b, and $\mathbf{h}^{\mathrm{AC}} \parallel \hat{\mathbf{z}}$ ($\mu = z$) and $h_x^{\mathrm{stat}} = 0.06$ in Fig. 5.9c [on the yellow dashed lines in Fig. 5.8a, b, and c, respectively]. In each figure, the different colors represent the different values of h_z^{stat}. The solid lines denote the results in the noncollinear spin phase where $|\chi_{E,\mu}(\omega)|$ is nonzero in Fig. 5.8a–c, while the dashed lines denote the results in the outside, i.e., the FFM phase where $|\chi_{E,\mu}(\omega)| = 0$.

Let us begin by discussing the results for $\mathbf{h}^{\mathrm{AC}} \parallel \hat{\mathbf{x}}$ and $h_x^{\mathrm{stat}} = 0$. As illustrated in Fig. 5.9a, both $|\chi_{E,x}(\omega)|$ and Im $\chi_{S,xx}(\omega)$ exhibit peaks at identical frequencies, with the exception of $h_z^{\mathrm{stat}} = 0.07$, where the system enters the FFM phase and $|\chi_{E,x}(\omega)|$ vanishes. This observation suggests that the EEF is enhanced at the magnetic resonance frequencies in the noncollinear spin phase. The resonance frequencies decrease as h_z^{stat} increases, which aligns with the bulk magnon band gap studied in Sect. 5.1.2 and previous studies on magnetic resonance in the CSL phase [15, 30]. Notably, the higher-frequency peaks are more pronounced in $|\chi_{E,x}(\omega)|$ than in Im $\chi_{S,xx}(\omega)$, indicating that the EEF response is more sensitive to higher-frequency excitations compared to the magnetic response. This enhancement is likely due to an additional ω-linear factor originating from $\frac{\partial \mathbf{S}_l(t)}{\partial t}$ in Eq. (5.9); see also Eq. (C.6) in Appendix C.1.

While $|\chi_{E,x}(\omega)|$ includes an almost ω-independent contribution (discussed below), the peak intensities increase from zero as h_z^{stat} increases, which is in good agreement with the growth of the peaks in Im $\chi_{S,xx}(\omega)$. However, the growth rate of the peaks in $|\chi_{E,x}(\omega)|$ slows down as h_z^{stat} nears the critical field $h_z^{\mathrm{stat,c}}$, in contrast to the rapid increase in Im $\chi_{S,xx}(\omega)$. This behavior can be qualitatively understood in terms of the decreasing frequency ω_1^{bulk} and the vector spin chirality that connects $\chi_{E,x}(\omega)$ and $\chi_{S,xx}$. The vector chirality in the ground state is defined as

$$\bar{\mathbf{C}}^{\mathrm{vc}} = \frac{1}{L} \sum_{l=0}^{L-1} \mathbf{S}_l(0) \times \mathbf{S}_{l+1}(0), \tag{5.16}$$

which measures the overall spin noncollinearity. As h_z^{stat} approaches $h_z^{\text{stat,c}}$, where the spin texture adopts a prominent solitonic feature, this quantity diminishes rapidly, thereby suppressing the growth rate of the peaks in $|\chi_{E,x}(\omega)|$; refer to Eq. (C.6) in Appendix C.1.

The ω-independent component in $|\chi_{E,x}(\omega)|$ can be explained by considering the large-ω behavior. For larger ω, we derive the relation

$$\chi_{E,\mu}(\omega) \simeq -\frac{\alpha}{1+\alpha^2}\bar{\mathbf{C}}^{\text{vc}} \cdot \hat{\boldsymbol{\mu}}. \tag{5.17}$$

For the derivation, see Appendix C.2. Thus, the large-ω behavior is independent of ω and is governed by the overall spin noncollinearity $\bar{\mathbf{C}}^{\text{vc}} \cdot \hat{\mathbf{x}}$. As $\bar{\mathbf{C}}^{\text{vc}} \cdot \hat{\mathbf{x}}$ decreases with increasing h_z^{stat}, the ω-independent component of $|\chi_{E,x}(\omega)|$ diminishes, as depicted in the upper panel of Fig. 5.9a. It is important to note that this contribution corresponds to the collective dynamics of spin textures, referred to as coherent sliding dynamics [21]. This will be discussed further in the next subsubsection.

Next, we examine the results for $\mathbf{h}^{\text{AC}} \parallel \hat{\mathbf{y}}$ and $h_x^{\text{stat}} = 0$. In this case, $|\chi_{E,y}(\omega)|$ also shows peaks corresponding to the magnetic resonances in Im $\chi_{S,yy}(\omega)$, and the higher-frequency modes are more clearly visible in $|\chi_{E,y}(\omega)|$ than in Im $\chi_{S,yy}(\omega)$, as shown in Fig. 5.9b. However, in contrast to the case of $\mathbf{h}^{\text{AC}} \parallel \hat{\mathbf{x}}$, the peak heights in $|\chi_{E,y}(\omega)|$ increase for $0 \leq h_z^{\text{stat}} \lesssim 0.055$ but decrease for $h_z^{\text{stat}} \gtrsim 0.055$, whereas Im $\chi_{S,yy}(\omega)$ exhibits resonance peaks even for $h_z^{\text{stat}} = 0$ and grows monotonically with h_z^{stat}. This nonmonotonic behavior of $|\chi_{E,y}(\omega)|$ was previously observed in Fig. 5.8b. The decrease in $|\chi_{E,y}(\omega)|$ can be attributed to the gradual increase of Im $\chi_{S,yy}(\omega)$ in the lower panel of Fig. 5.9b, in addition to the reductions in $\bar{C}_x^{\text{vc}}$ and ω_1^{bulk} discussed above. Furthermore, there are no ω-independent contributions, as $\bar{C}_y^{\text{vc}} = 0$ in Eq. (5.16), leading to $\chi_{E,y}(\omega) \simeq 0$ for large ω in Eq. (5.17).[2]

Finally, in Fig. 5.9c, we present the results for $\mathbf{h}^{\text{AC}} \parallel \hat{\mathbf{z}}$ and $h_x^{\text{stat}} = 0.06$. Similar to the previous cases, the higher-frequency peaks are more clearly visible for $|\chi_{E,z}(\omega)|$ than in Im $\chi_{S,zz}(\omega)$. The lowest-energy peak in $|\chi_{E,z}(\omega)|$, which displays a shoulder-like feature indicative of the splitting discussed in the previous subsubsection, increases with h_z^{stat}, as observed in the case of $\mathbf{h}^{\text{AC}} \parallel \hat{\mathbf{x}}$. This behavior can be explained by the increase of $\bar{C}_z^{\text{vc}}$ with h_z^{stat}; see Eq. (C.6) in Appendix C.1. In contrast, the peak heights in Im $\chi_{S,zz}(\omega)$ remain nearly unchanged for h_z^{stat} since the solitonic feature in the spin texture is not well developed within this range of h_z^{stat}.

Real-Space Behavior

Here, we present the spatiotemporal profiles of spins $\mathbf{S}_l(t)$ and EEF $E_l^{\text{em}}(t)$ in Eq. (5.9). Figure 5.10 displays the results for the lowest-ω resonance mode: The spin dynamics are obtained for the static magnetic field and the frequency of the

[2] The long tails of $|\chi_{E,y}(\omega)|$ in the large-ω region are a result of discretization errors in the numerical computation of the spatial derivative in Eq. (5.9).

AC magnetic field indicated by the circles and crosses in Fig. 5.8d–f. In each figure of Fig. 5.10, the top panel shows the real-space spin configuration and trajectories of spin precession motions, the middle panel shows the time evolution of each spin component, and the bottom panel shows the time evolution of $E_l^{\mathrm{em}}(t)$ in Eq. (5.9) during a single period of $\mathbf{h}^{\mathrm{AC}}$ for 60 sites at the center of the 1000-site system with the PBC.

First, we examine the results for $\mathbf{h}^{\mathrm{AC}} \parallel \hat{\mathbf{x}}$ and $h_x^{\mathrm{stat}} = 0$, as depicted in Fig. 5.8d. Figure 5.10a corresponds to the helical state at $(h_x^{\mathrm{stat}}, h_z^{\mathrm{stat}}) = (0, 0)$ and $\omega = 0.143$, indicated by the cross in Fig. 5.8d. In this case, the spin dynamics is significantly suppressed, leaving only tiny oscillations, as evident in the enlarged plot of $S_{l,x}(t)$ in the middle panel of Fig. 5.10a. We note that such oscillations arise not only at the resonance frequency but also for general ω, and they are referred to as coherent sliding dynamics [21]. In this case, we observe a small but spatially uniform $E_l^{\mathrm{em}}(t)$, as shown in the bottom panel of Fig. 5.10a. The spatial average in Eq. (5.8), however, can be comparable to those for the other resonance modes, which gives rise to the ω-independent contribution in Fig. 5.9a.

Figure 5.10b is also for $\mathbf{h}^{\mathrm{AC}} \parallel \hat{\mathbf{x}}$, but at $(h_x^{\mathrm{stat}}, h_z^{\mathrm{stat}}) = (0, 0.06)$ and $\omega = 0.103$, indicated by the circle in Fig. 5.8d. In contrast to the helical case mentioned above, the spin dynamics in this CSL state is resonantly activated, particularly in the regions between the solitons with $S_{l,z}(t) \simeq +1$, as shown in the top and middle panels of Fig. 5.10b. In contrast, the EEF is predominantly generated around the solitons, as shown in the bottom panel. This trend can be attributed to the large vector spin chirality near the solitons. The spatially averaged EEF reaches a large value, leading to the resonance peak of $|\chi_{E,x}(\omega)|$ in Figs. 5.8d and 5.9a. We note that in this resonance with $h_x^{\mathrm{stat}} = 0$, the spin texture merely oscillates around its original position in real space, and the EEF behaves like a standing wave, as shown in Fig. 5.11a. This behavior arises due to the twofold rotational symmetry about the z axis combined with time translation by $\frac{\pi}{\omega}$.

Next, we consider the results for $\mathbf{h}^{\mathrm{AC}} \parallel \hat{\mathbf{y}}$ and $h_x^{\mathrm{stat}} = 0$, as shown in Fig. 5.8e. Figure 5.10c corresponds to the helical state at $h_z^{\mathrm{stat}} = 0$ and $\omega = 0.143$, indicated by the cross in Fig. 5.8e. In this case, the spins pointing in the $\pm y$ directions do not precess and form the nodes of oscillating $\mathbf{S}_l(t)$, while those pointing in the $\pm z$ directions are the antinodes, where the spins exhibit maximal precession. The EEF vanishes at the nodes, while it is maximally generated at the antinodes, reflecting the large spin precessions. It is worth noting, however, that the average of the EEF is strongly suppressed due to the twofold rotational symmetry about the x axis, accompanied by time translation by $\frac{\pi}{\omega}$ of $\mathbf{h}^{\mathrm{AC}}$ (see footnote 1 of this chapter). Meanwhile, Fig. 5.10d depicts a CSL state at $h_z^{\mathrm{stat}} = 0.05$ with $\omega = 0.124$, corresponding to the circle in Fig. 5.8e. In this case, the amplitude of the spin precessions is enhanced (suppressed) in the regions between (near) the solitons, while $E_l^{\mathrm{em}}(t)$ is suppressed (slightly enhanced). The spatially averaged $E_l^{\mathrm{em}}(t)$ gives rise to the sharp resonance peak in Figs. 5.8e and 5.9b. In these cases with $h_x^{\mathrm{stat}} = 0$, the spin texture and the EEF behave like standing waves, similar to those in Fig. 5.11a due to the symmetry.

Finally, we examine the results for $\mathbf{h}^{\mathrm{AC}} \parallel \hat{\mathbf{z}}$, where we take $h_x^{\mathrm{stat}} = 0.06$, since $\bar{E}^{\mathrm{em}}(t)$ vanishes for $h_x^{\mathrm{stat}} = 0$, as explained in the previous subsubsections.

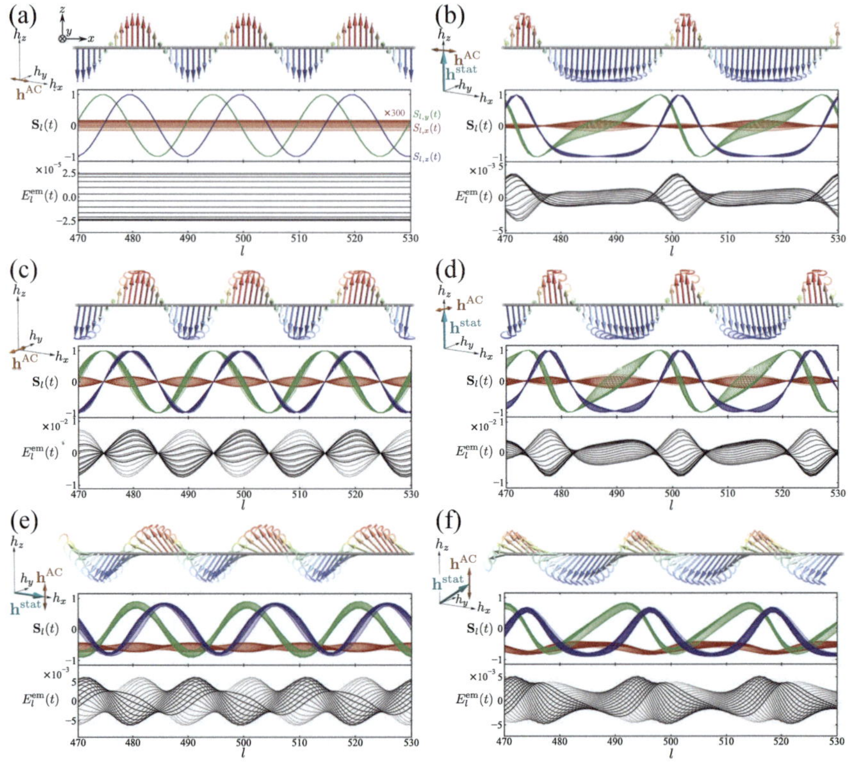

Fig. 5.10 Real-space spin configurations (top and middle) and real-space distributions of the EEF (bottom) in the steady states under the AC magnetic field $\mathbf{h}^{\mathrm{AC}}$. **a** and **b**, **c** and **d**, and **e** and **f** are for $\mathbf{h}^{\mathrm{AC}} \parallel \hat{\mathbf{x}}$, $\mathbf{h}^{\mathrm{AC}} \parallel \hat{\mathbf{y}}$, and $\mathbf{h}^{\mathrm{AC}} \parallel \hat{\mathbf{z}}$, respectively, at **a** $(h_x^{\mathrm{stat}}, h_z^{\mathrm{stat}}) = (0, 0)$ and $\omega = 0.143$, **b** $(h_x^{\mathrm{stat}}, h_z^{\mathrm{stat}}) = (0, 0.06)$ and $\omega = 0.103$, **c** $(h_x^{\mathrm{stat}}, h_z^{\mathrm{stat}}) = (0, 0)$ and $\omega = 0.143$, **d** $(h_x^{\mathrm{stat}}, h_z^{\mathrm{stat}}) = (0, 0.05)$ and $\omega = 0.124$, **e** $(h_x^{\mathrm{stat}}, h_z^{\mathrm{stat}}) = (0.06, 0)$ and $\omega = 0.131$, and **f** $(h_x^{\mathrm{stat}}, h_z^{\mathrm{stat}}) = (0.06, 0.04)$ and $\omega = 0.123$. All the frequencies are set at the values for the lowest-ω resonance mode; the values are denoted by the orange crosses and circles in Fig. 5.8d–f. The results are shown for 60 sites at the center of the 1000-site system with the PBC. In each figure, the top panel displays the spin configurations by the arrows with the lines representing the trajectories of the arrowheads during a single period of $\mathbf{h}^{\mathrm{AC}}$. Meanwhile, the middle panel displays the spin components $S_{l,x}(t)$, $S_{l,y}(t)$, and $S_{l,z}(t)$ by the red, green, and blue lines, respectively, and the bottom panel displays the EEF at each site, $E_l^{\mathrm{em}}(t)$, during the corresponding period of $\mathbf{h}^{\mathrm{AC}}$. The color and gray-scale intensities in the middle and bottom panels, respectively, increase with time evolution in the single period of $\mathbf{h}^{\mathrm{AC}}$. In the middle panel of (**a**), $S_{l,x}(t)$ is multiplied by a factor of 300 for better visibility. Reproduced from Ref. [4].

Figure 5.10e corresponds to the conical state at $h_z^{\mathrm{stat}} = 0$ and $\omega = 0.131$, indicated by the cross in Fig. 5.8f. In this case, the spin precessions become larger (smaller) for $S_{l,z}(t) \simeq 0$ ($S_{l,y}(t) \simeq 0$), since the AC magnetic field is applied along the z direction. In contrast to the previous cases with $h_x^{\mathrm{stat}} = 0$, we find that $E_l^{\mathrm{em}}(t)$ no longer behaves like a standing wave, and the EEF propagates from left to right, as shown more clearly

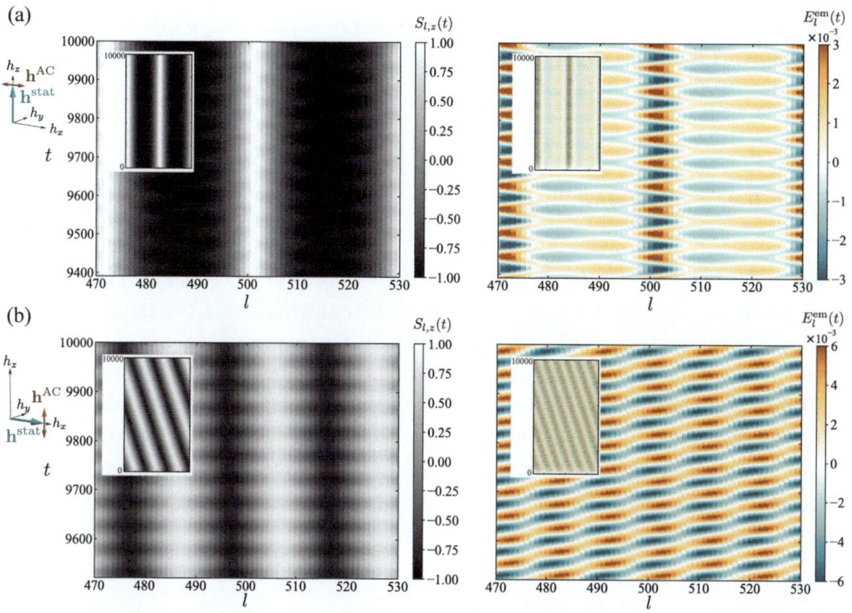

Fig. 5.11 Real-space and real-time distribution of the z component of spins $S_{l,z}(t)$ (left) and $E_l^{\mathrm{em}}(t)$ (right) within 10 periods of $\mathbf{h}^{\mathrm{AC}}$ on the plane of l and t: **a** and **b** correspond to Fig. 5.10b and e, respectively. The insets show the results for a longer time window. Reproduced from Ref. [4].

in the main panel of Fig. 5.11b. In this mode, the EEF propagates from one soliton to the next during a time cycle, and the velocity is estimated as $v \simeq \frac{\omega}{Q_0} \simeq 0.42$. Such motion is scarcely observed in $S_{l,y}(t)$ and $S_{l,z}(t)$, but is discernible in $S_{l,x}(t)$ in the middle panel of Fig. 5.10e. Since the motion in $S_{l,x}(t)$ corresponds to spin precession, the rapid propagation of the EEF in the field direction is associated with magnon propagation in the conical state. Additionally, the spin texture is driven from right to left with a much slower velocity, as shown in the inset of the left panel of Fig. 5.11b; the velocity is estimated as $v' \simeq -0.0035$. This is a drift motion known as Archimedean screw dynamics [31], induced as a counteraction to magnon propagation. Similar phenomena have been observed for skyrmions [32, 33]. Reflecting this drift motion, the EEF is also driven from right to left with the same velocity, as shown in the inset of the right panel of Fig. 5.11b. Thus, the parallel component of the static magnetic field, h_x^{stat}, activates two types of propagating motions of the EEF: the fast motion towards the field direction due to magnon propagation and the slow drift in the opposite direction due to the Archimedean screw dynamics of the spin texture. Although the spatially averaged EEF $\bar{E}^{\mathrm{em}}(t)$ vanishes for $h_z^{\mathrm{stat}} = 0$, it becomes nonzero for nonzero h_z^{stat}, as a solitonic feature is induced in the spin texture. Such results are shown in Fig. 5.10f for $h_z^{\mathrm{stat}} = 0.04$ and $\omega = 0.123$, corresponding to the circle in Fig. 5.8f. In this case as well, the spin texture and the EEF are driven to the left in association with Archimedean screw dynamics, while the EEF also propagates to the right with much faster velocity related to magnon propagation, similar to Fig. 5.11b.

5.2.3 Edge Contributions

Thus far, we have shown the results for the bulk contributions in the system with the PBC. In this section, we turn our attention to the contributions from the edges of the system by employing the OBC. First, we present the results of $|\chi_{E,\mu}(\omega)|$ to show the edge contributions to the EEF through additional resonance modes. Then, we discuss the difference between the bulk resonance modes and the additional modes based on the spectra of $|\chi_{E,\mu}(\omega)|$ and Im $\chi_{S,\mu\mu}(\omega)$, including the system size dependence. Finally, we show the spatiotemporal profiles of the spin textures and the EEF for the additional resonance modes to explicitly show that these modes are localized at the edges of the system.

Complex Admittance of the EEF

As outlined in the first subsubsection of Sect. 5.2.2, we present the results of the complex admittance $\chi_{E,\mu}(\omega)$ for the system with the OBC. First, we discuss the results for $\mathbf{h}^{\mathrm{AC}} \parallel \hat{\mathbf{x}}$, as depicted in Fig. 5.12a and d. Similarly to the periodic boundary conditions (PBC) case, in the noncollinear spin phase, $\chi_{E,x}^{\max}$ increases as h_x^{stat} decreases and h_z^{stat} increases, attaining a maximum at $(h_x^{\mathrm{stat}}, h_z^{\mathrm{stat}}) \simeq (0, 0.06)$. Notably, however, $\chi_{E,x}^{\max}$ becomes nonzero also in the FFM phase outside of this region, with a relatively large value near the phase boundary at $(h_x^{\mathrm{stat}}, h_z^{\mathrm{stat}}) \simeq (0.04, 0.055)$. This contribution arises exclusively from the edges of the system under OBC, as it was absent in the PBC case, as shown in Fig. 5.8a. Furthermore, we observe an additional resonance mode in the noncollinear spin phase, as illustrated in the spectrum of $|\chi_{E,x}(\omega)|$ in Fig. 5.12d along the yellow dashed line in Fig. 5.12a. In addition to several peaks superimposed on an almost ω-independent contribution present in the PBC case (Fig. 5.8d), $|\chi_{E,x}(\omega)|$ exhibits a sharp peak at a lower frequency. The resonance frequency of this additional mode, denoted as ω^{edge}, increases as h_z^{stat} increases, in contrast to the bulk resonance frequencies ω_n^{bulk}. This corresponds to the edge mode reported in previous studies [3]. The field dependence of ω^{edge} aligns with the result obtained from linear spin-wave theory, as shown in Fig. 5.4. In the following subsubsections, we will demonstrate that the intensity at ω^{edge} depends on the system size, and that the spin dynamics and the EEF associated with this resonance mode are localized near the system's edges.

A similar situation arises for the case of $\mathbf{h}^{\mathrm{AC}} \parallel \hat{\mathbf{y}}$, shown in Fig. 5.12b and e, where we again obtain an additional resonance mode at ω^{edge}, along with nonzero contributions even in the FFM phase. In contrast, for $\mathbf{h}^{\mathrm{AC}} \parallel \hat{\mathbf{z}}$, shown in Fig. 5.12c and f, instead of an additional resonance peak, we observe a sharp dip at ω^{edge},

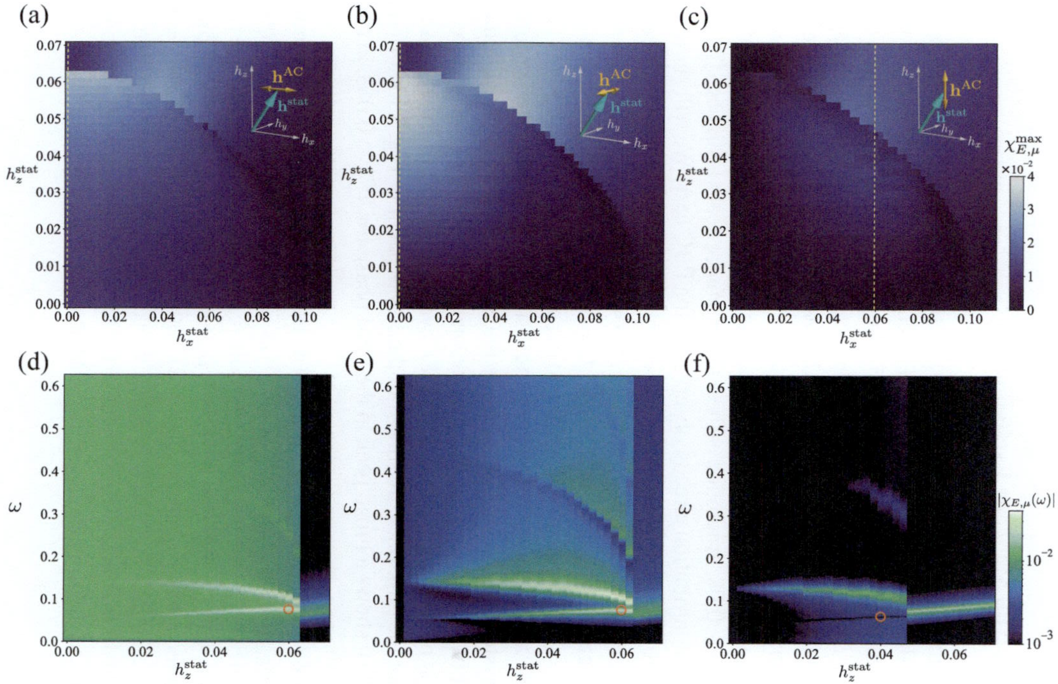

Fig. 5.12 Maximum value of $|\chi_{E,\mu}(\omega)|$, $\chi_{E,\mu}^{max}$, calculated for the model in Eq. (3.4) with the OBC for **a** $\mathbf{h}^{AC} \parallel \hat{\mathbf{x}}$ ($\mu = x$), **b** $\mathbf{h}^{AC} \parallel \hat{\mathbf{y}}$ ($\mu = y$), and **c** $\mathbf{h}^{AC} \parallel \hat{\mathbf{z}}$ ($\mu = z$). The system size is $L = 1000$. The frequency dependence of $|\chi_{E,\mu}(\omega)|$ along the yellow dashed lines in (**a**), (**b**), and (**c**) are shown in (**d**), (**e**), and (**f**), respectively: **d** $\mathbf{h}^{AC} \parallel \hat{\mathbf{x}}$ with $h_x^{stat} = 0$, **e** $\mathbf{h}^{AC} \parallel \hat{\mathbf{y}}$ with $h_x^{stat} = 0$, and **f** $\mathbf{h}^{AC} \parallel \hat{\mathbf{z}}$ with $h_x^{stat} = 0.06$. The orange circles in (**d**), (**e**), and (**f**) indicate the parameters for which the real-space behaviors are presented in Fig. 5.14. Reproduced from Ref. [4]. Copyright (2023) by the American Physical Society

superimposed on the long tail of ω_1^{bulk}. This behavior is attributed to the fact that $\chi_{E,z}(\omega)$ from the edges has the opposite sign compared to that of the bulk, as discussed in the subsequent subsubsection. Additionally, we note that $|\chi_{E,z}(\omega)|$ exhibits a sharper peak in the FFM phase, as compared to the other cases shown in Fig. 5.12d and e.

Comparison with Magnetic Excitations

Next, we compare the results of $|\chi_{E,\mu}(\omega)|$ with Im $\chi_{S,\mu\mu}(\omega)$ for the system with the OBC, as in the second subsubsection of Sect. 5.2.2. Let us first discuss the results for $\mathbf{h}^{\text{AC}} \parallel \hat{\mathbf{x}}$ and $h_x^{\text{stat}} = 0$ shown in Fig. 5.13a; the solid and dashed lines denote the results with $L = 1000$ and $L = 500$, respectively, in the noncollinear spin phase for $h_z^{\text{stat}} \lesssim 0.063$, while the dotted and dashed-dotted lines denote the results with $L = 1000$ and $L = 500$, respectively, in the FFM phase for $h_z^{\text{stat}} \gtrsim 0.063$. In the noncollinear spin phase, both $|\chi_{E,x}(\omega)|$ and Im $\chi_{S,xx}(\omega)$ exhibit a resonance peak at $\omega^{\text{edge}} \simeq 0.06 - 0.08$, in addition to the bulk resonance peaks at higher ω_n^{bulk} and the ω-independent contribution already observed in the PBC case. We will show explicitly that this additional contribution originates from the localized modes at the edges in the next subsubsection. Accordingly, the edge contribution shows conspicuous system size dependences: The peak heights for $L = 1000$ become about twice as small as those for $L = 500$, while the bulk ones do not change largely. As $|\chi_{E,x}(\omega)|$ describes the response averaged over the system, the result indicates that the edge contribution is almost system size independent, while the bulk one is proportional to the system size. Notably, the intensity of the edge resonance peak of $|\chi_{E,x}(\omega)|$ increases with h_z^{stat} and can be stronger than the bulk ones, even though that of Im $\chi_{S,xx}(\omega)$ is much weaker than the bulk ones, as shown in the lower panel of Fig. 5.13a. We also find that $|\chi_{E,x}(\omega)|$ shows a broad peak in the FFM phase for $h_z^{\text{stat}} \gtrsim 0.063$, in contrast to the PBC case.

Similar results are obtained for $\mathbf{h}^{\text{AC}} \parallel \hat{\mathbf{y}}$, as shown in Fig. 5.13b. Meanwhile, for $\mathbf{h}^{\text{AC}} \parallel \hat{\mathbf{z}}$, $|\chi_{E,z}(\omega)|$ shows a dip at ω^{edge} as shown in Fig. 5.13c, as a result of a cancellation between the long tail of the bulk response at ω_1^{bulk} and the edge resonance contribution with the opposite sign. This is confirmed by the observation that the dip becomes almost twice shallower for $L = 1000$ compared to that for $L = 500$. We note that the edge contribution is much weaker in both $|\chi_{E,z}(\omega)|$ and Im $\chi_{S,zz}(\omega)$ compared to the previous two cases due to the less solitonic feature of the spin texture. In contrast, the additional peaks in the FFM phase are much sharper than the previous ones.

Real-Space Behavior

The spatiotemporal profiles of $\mathbf{S}_l(t)$ and $E_l^{\text{em}}(t)$ for the additional mode at ω^{edge} are presented in Fig. 5.14. Here, we display the spin textures and the EEF for each of the 60 sites at the left and right edges of the system with $L = 1000$. Figure 5.14a

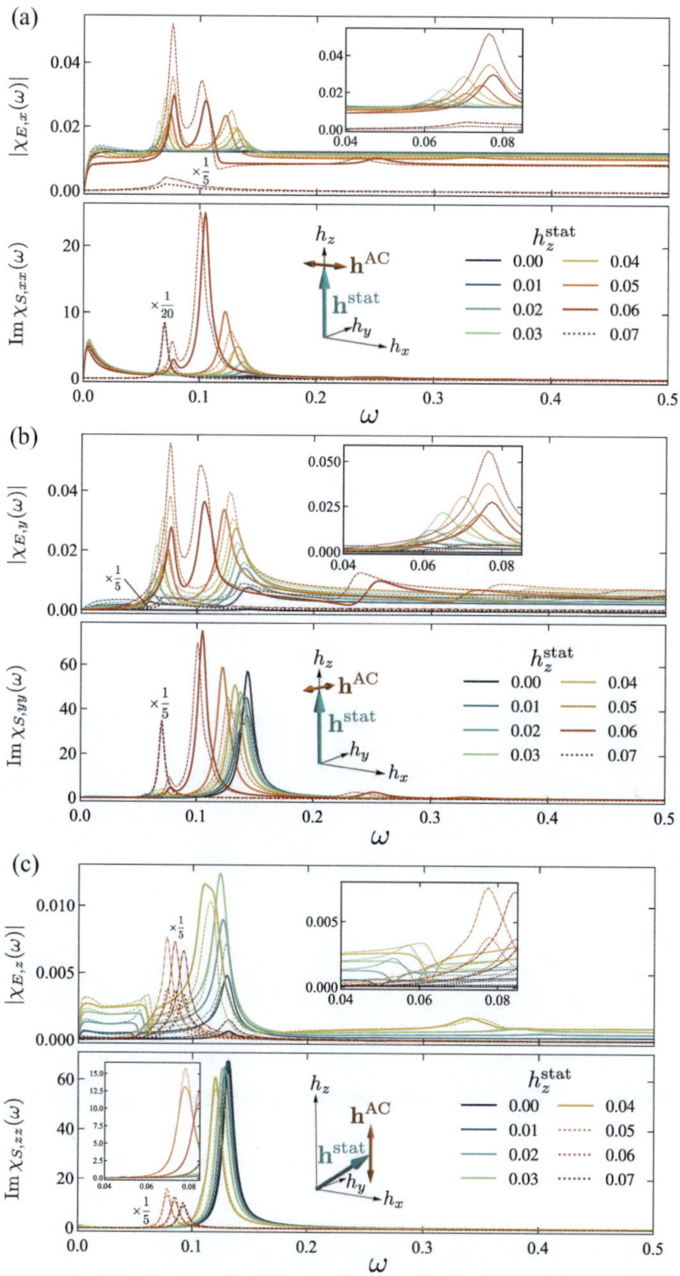

Fig. 5.13 Comparison between $|\chi_{E,\mu}(\omega)|$ and the imaginary part of the dynamical spin susceptibility, Im $\chi_{S,\mu\mu}(\omega)$, for several values of h_z^{stat}: **a** $\mathbf{h}^{\mathrm{AC}} \parallel \hat{\mathbf{x}}$ with $h_x^{\mathrm{stat}} = 0$, **b** $\mathbf{h}^{\mathrm{AC}} \parallel \hat{\mathbf{y}}$ with $h_x^{\mathrm{stat}} = 0$, and **c** $\mathbf{h}^{\mathrm{AC}} \parallel \hat{\mathbf{z}}$ with $h_x^{\mathrm{stat}} = 0.06$ for the model in Eq. (3.4) with the OBC. The solid and dashed (dotted and dashed-dotted) lines denote the data in the noncollinear spin (FFM) phase with $L = 1000$ and $L = 500$, respectively. The data for the FFM phases are scaled for better visibility. Reproduced from Ref. [4]. Copyright (2023) by the American Physical Society

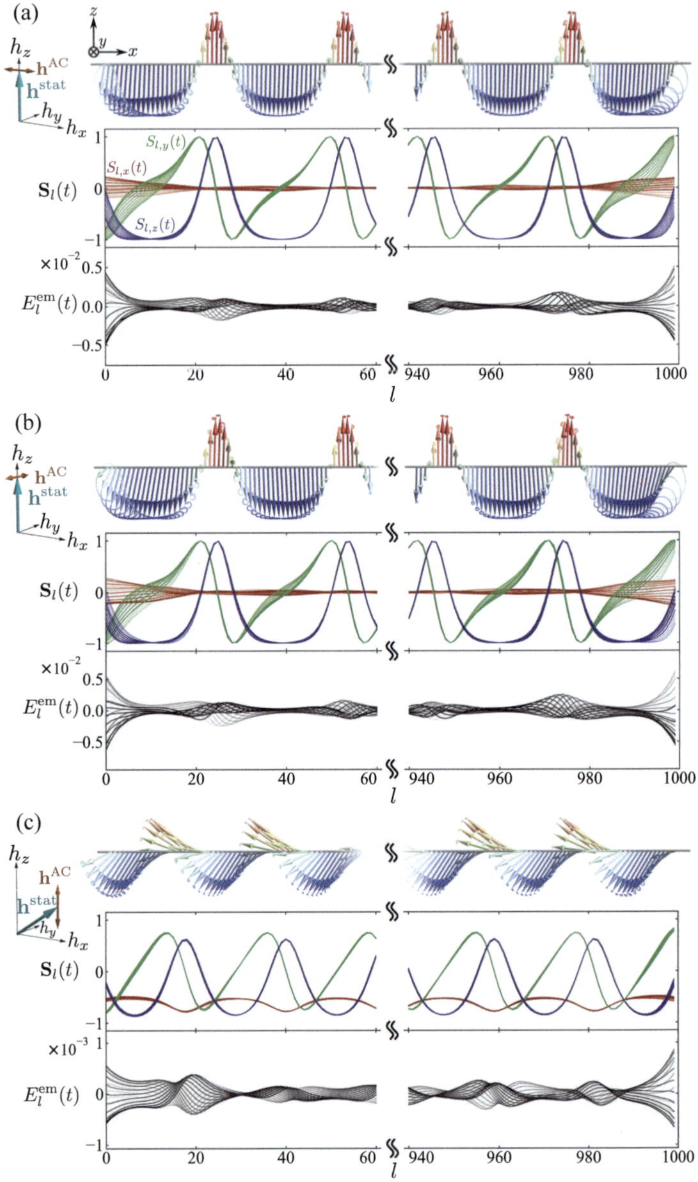

Fig. 5.14 Real-space spin configurations (top and middle) and real-space distributions of the EEF (bottom) in the steady states under the AC magnetic field $\mathbf{h}^{\rm AC}$. **a**, **b**, and **c** are for $\mathbf{h}^{\rm AC} \parallel \hat{\mathbf{x}}$, $\mathbf{h}^{\rm AC} \parallel \hat{\mathbf{y}}$, and $\mathbf{h}^{\rm AC} \parallel \hat{\mathbf{z}}$, respectively, at **a** $(h_x^{\rm stat}, h_z^{\rm stat}) = (0, 0.06)$ and $\omega = 0.077$, **b** $(h_x^{\rm stat}, h_z^{\rm stat}) = (0, 0.06)$ and $\omega = 0.077$, and **c** $(h_x^{\rm stat}, h_z^{\rm stat}) = (0.06, 0.04)$ and $\omega = 0.061$. All the frequencies are set at the values for the edge mode at $\omega^{\rm edge}$; the values are denoted by the orange circles in Fig. 5.12d–f. The results are shown for each of the 60 sites at the left and right edges of the 1000-site system with the OBC. The notations are common to those in Fig. 5.10. Reproduced from Ref. [4]. Copyright (2023) by the American Physical Society

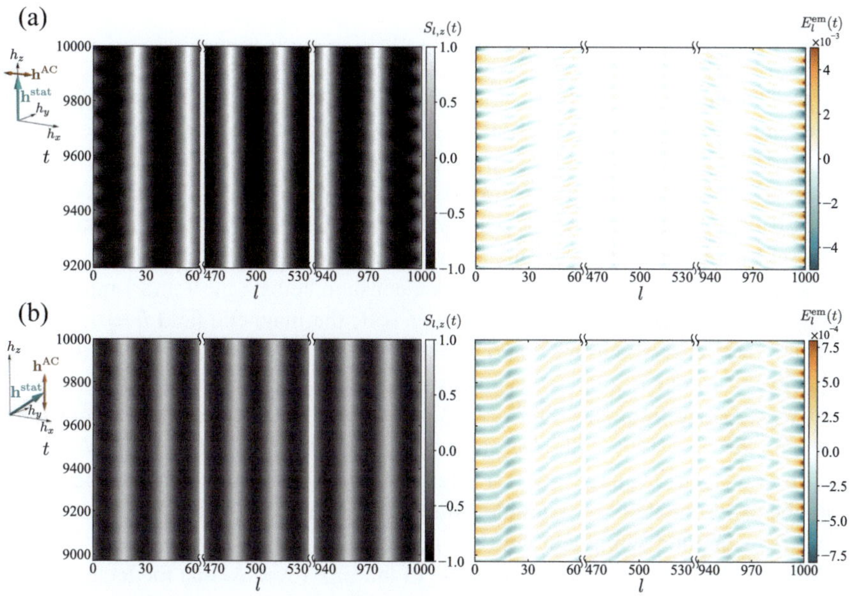

Fig. 5.15 Real-space and real-time distribution of the z component of spins $S_{l,z}(t)$ (left) and $E_l^{em}(t)$ (right) within 10 periods of $\mathbf{h}^{AC}$ on the plane of l and t: **a** and **b** correspond to Fig. 5.14a and c, respectively. The results are shown for 60 sites at the left edge, center, and right edge of the 1000-site system. Reproduced from Ref. [4]. Copyright (2023) by the American Physical Society

illustrates the results for $\mathbf{h}^{AC} \parallel \hat{\mathbf{x}}$ in the CSL state at $(h_x^{stat}, h_z^{stat}) = (0, 0.06)$ and $\omega = 0.077$, as indicated by the orange circle in Fig. 5.12d. In this case, as shown in the top and middle panels of Fig. 5.14a, we observe that the spin dynamics with ω^{edge} are localized at the edges, where the spins exhibit a twisted configuration in the ground state, known as the chiral surface twist [34]. Therefore, the additional mode identified in the previous subsubsections corresponds to the edge mode. It is noteworthy that the EEF $E_l^{em}(t)$ at both edges exhibits the same sign, and its amplitude is approximately ten times greater than that in the bulk, contributing significantly to the bulk responses, as shown in Fig. 5.13a. Figure 5.15a shows the time evolution of $S_{l,z}(t)$ and $E_l^{em}(t)$. Both patterns exhibit twofold rotational symmetry about the z-axis and time translation symmetry of $\frac{\pi}{\omega}$ with respect to the center of the system. However, $E_l^{em}(t)$ appears to propagate from the edges toward the interior, as illustrated in the right panel, while $S_{l,z}(t)$ on the left does not exhibit a similar behavior.

Similar spatiotemporal profiles of the edge mode are observed for $\mathbf{h}^{AC} \parallel \hat{\mathbf{y}}$, as shown in Fig. 5.14b. Meanwhile, as previously discussed in Sect. 5.2.3, the edge mode for $\mathbf{h}^{AC} \parallel \hat{\mathbf{z}}$ is less pronounced, as shown in Fig. 5.14c. In this case, the EEF propagates from left to right, similar to the PBC case, even near the edges, as shown in the right panel of Fig. 5.15b, while the pattern near the right edge becomes more complex.

5.2.4 *Order Estimate*

Let us estimate the amplitude of the EEF expected in real materials based on the results explored thus far. Experimentally, the noncollinear spin phase including the CSL has been observed, e.g., in $CrNb_3S_6$ [35], $MnNb_3S_6$ [36], strained Cu_2OSeO_3 [37], and $Yb(Ni_{1-x}Cu_x)_3Al_9$ [38] (see Sect. 1.4.1). The actual value of the EEF depends on the energy scale of the magnetic interactions and the length scale set by the lattice constant in each material. By assuming the typical values of the Heisenberg interaction J and the lattice constant being 1 meV and 1 nm, respectively, the time unit $t = 1$, the frequency $\omega = 1$, the magnetic field $h = 1$, and the EEF $\bar{E}^{\mathrm{em}} = 1$ correspond to ~0.66 psec, ~241 GHz, ~8.6 T, and $\sim 5 \times 10^5$ V/m, respectively. In our results, we typically obtained $|\chi_{E,\mu}(\omega)| \sim 0.01$ at the lowest-energy resonance frequency $\omega_1^{\mathrm{bulk}} \sim 0.1 \sim 24.1$ GHz, and hence the AC magnetic field with $|\mathbf{h}^{\mathrm{AC}}| \sim 0.01 \sim 860$ Oe can generate the total EEF of $\bar{E}^{\mathrm{em}} \sim 50$ V/m. This leads to the electric voltage ~500 μV for a sample with 10 μm length scale, which could be experimentally measurable. Note that the resonance frequency observed in $CrNb_3S_6$ is in good agreement with our results [39].

Meanwhile, our spatiotemporal profile of the EEF revealed that the local value of $E_l^{\mathrm{em}}(t)$ reaches $\sim 10^{-2}$, which corresponds to the electric field of ~5 KV/m. Thus, even though the total EEF is relatively small, electrons coupled to the noncollinear spin texture would be largely influenced by such a strong local electric field. This may bring about intriguing electric responses, such as the AC and DC electric currents. Note that the local EEF takes such a large value even in the helical and the conical states where the solitonic feature is absent.

We note that the generated EEF depends on the value of the Gilbert damping α in Eq. (4.6). The EEF associated with the resonance dynamics is roughly proportional to $1/\alpha$. This is because the peak intensity of $\chi_{E,\mu}(\omega)$ is approximately proportional to $\chi_{S,\mu\mu}(\omega)$ [see Eq. (C.6) in Appendix C.1] and $\mathrm{Im}\,\chi_{S,\mu\mu}(\omega)$ is proportional to $1/\alpha$. Meanwhile, the EEF by the coherent sliding dynamics, which gives the ω-independent contribution, is proportional to α, as implied by Eq. (5.17) for small α. The value of α depends on materials and the realistic estimate is not an easy task, while we used $\alpha = 0.04$ as a typical value for ferromagnetic metals [40–43].

5.3 Concluding Remarks of This Chapter

To conclude this chapter, we discuss some implications of our results. In Sect. 5.1, we demonstrated the soliton penetration dynamics associated with the edge mode. Interestingly, edge modes have also been found in a 2D chiral magnet exhibiting the SkL [44], and hence similar soliton penetration dynamics is expected also in other magnetic solitons including skyrmions. The number of magnetic solitons is intimately linked to the electrical properties such as magnetoresistance, particularly pronounced in micro-sized samples [5–11]. Our results in Fig. 5.7 showed that maximally about 20 solitons can be introduced to the CSL with 30–40 solitons, suggesting more than 50% change of the resistivity.

In Sect. 5.2, we clarified that the resonance peaks of the total EEF are more clearly visible compared to the magnetic ones for higher frequency modes in the bulk. Furthermore, the edge resonances of the EEF are also more conspicuous than the magnetic ones. Thus, the measurement of the EEF could be a good probe of the high-frequency magnetic resonances and the edge modes. Besides the total EEF, the spatiotemporal profile of the EEF is also important for electronic transport phenomena due to its large amplitude estimated in Sect. 5.2.4. To reveal such transport phenomena explicitly, further studies including electrons coupled to the chiral magnetic textures are desired.

References

1. Masaki, Y., Aoki, R., Togawa, Y., & Kato, Y. (2018). *Physical Review B, 98*, 100402(R).
2. Masaki, Y. (2020). *Physical Review B, 101*, 214424.
3. Hoshi, K., Kishine, J.-I., & Ohe, J.-I. (2020). *Physical Review B, 102*, 134414.
4. Shimizu, K., Okumura, S., Kato, Y., & Motome, Y. (2023). *Physical Review B, 108*, 134436.
5. Shinozaki, M., Masaki, Y., Aoki, R., Togawa, Y., & Kato, Y. (2018). *Physical Review B, 97*, 214413.
6. Wilson, M. N., Karhu, E. A., Lake, D. P., Quigley, A. S., Meynell, S., Bogdanov, A. N., Fritzsche, H., Rößler, U. K., & Monchesky, T. L. (2013). *Physical Review B, 88*, 214420.
7. Kishine, J.-I., Bostrem, I. G., Ovchinnikov, A. S., & Sinitsyn, V. I. E. (2014). *Physical Review B, 89*, 014419.
8. Togawa, Y., Koyama, T., Nishimori, Y., Matsumoto, Y., McVitie, S., McGrouther, D., Stamps, R. L., Kousaka, Y., Akimitsu, J., Nishihara, S., Inoue, K., Bostrem, I. G., Sinitsyn, V. I. E., Ovchinnikov, A. S., & Kishine, J. (2015). *Physical Review B, 92*, 220412(R) .
9. Wang, L., Chepiga, N., Ki, D.-K., Li, L., Li, F., Zhu, W., Kato, Y., Ovchinnikova, O. S., Mila, F., Martin, I., Mandrus, D., & Morpurgo, A. F. (2017). *Physical Review Letters, 118*, 257203.
10. Ohkuma, M., Mito, M., Nakamura, N., Tsuruta, K., Ohe, J., Shinozaki, M., Kato, Y., Kishine, J., Kousaka, Y., Akimitsu, J., & Inoue, K. (2019). *AIP Advances, 9*, 075212.
11. Paterson, G. W., Koyama, T., Shinozaki, M., Masaki, Y., Goncalves, F. J. T., Shimamoto, Y., Sogo, T., Nord, M., Kousaka, Y., Kato, Y., McVitie, S., & Togawa, Y. (2019). *Physical Review B, 99*, 224429.
12. Du, H., et al. (2015). *Nature Communications, 6*, 8504.
13. Müller, J., Rosch, A., & Garst, M. (2016). *New Journal of Physics, 18*, 065006.
14. Schäffer, A. F., Siegl, P., Stier, M., Posske, T., Berakdar, J., Thorwart, M., Wiesendanger, R., & Vedmedenko, E. Y. (2020). *Science and Reports, 10*, 20400.
15. Kishine, J.-I., & Ovchinnikov, A. S. (2009). *Physical Review B, 79*, 220405(R).
16. Kataoka, M. (1987). *Journal of the Physical Society of Japan, 56*, 3635.
17. Maleyev, S. V. (2006). *Physical Review B, 73*, 174402.
18. Izyumov, Y. A., & Laptev, V. M. (1985). *Soviet Physics JETP, 62*, 755.
19. Bostrem, I. G., Kishine, J.-I., & Ovchinnikov, A. S. (2008). *Physical Review B, 77*, 132405.
20. Bostrem, I. G., Kishine, J.-I., & Ovchinnikov, A. S. (2008). *Physical Review B, 78*, 064425.
21. Kishine, J.-I., Bostrem, I. G., Ovchinnikov, A. S., & Sinitsyn, V. I. E. (2012). *Phys. Rev. B, 86*, 214426.
22. Ovchinnikov, A. S., Sinitsyn, V. E., Bostrem, I. G., & Kishine, J. (2013). *Journal of Experimental and Theoretical Physics, 116*, 791.

23. Kishine, J.-I., Proskurin, I., Bostrem, I. G., Ovchinnikov, A. S., & Sinitsyn, V. I. E. (2016). *Physical Review B, 93*, 054403.
24. Nagaosa, N., & Tokura, Y. (2013). *Nature Nanotechnology, 8*, 899.
25. Volovik, G. E. (1987). *Journal of Physics C: Solid State Physics, 20*, L83.
26. Xiao, D., Chang, M.-C., & Niu, Q. (2010). *Reviews of Modern Physics, 82*, 1959.
27. Nagaosa, N., & Tokura, Y. (2012). *Physica Scripta, T146*, 014020.
28. Nagaosa, N., Yu, X. Z., & Tokura, Y. (2012). *Philosophical Transactions of the Royal Society A: Mathematical, Physical and Engineering Sciences, 370*, 5806.
29. Fornberg, B. (1988). *Mathematics of Computation, 51*, 699.
30. Kiselev, V. V., & Raskovalov, A. A. (2013). *Journal of Experimental and Theoretical Physics, 116*, 272.
31. del Ser, N., Heinen, L., & Rosch, A. (2021). *SciPost Physics, 11*, 009.
32. Iwasaki, J., Beekman, A. J., & Nagaosa, N. (2014). *Physical Review B, 89*, 064412.
33. del Ser, N., & Lohani, V. (2023). *SciPost Physics, 15*, 065.
34. Meynell, S. A., Wilson, M. N., Fritzsche, H., Bogdanov, A. N., & Monchesky, T. L. (2014). *Physical Review B, 90*, 014406.
35. Togawa, Y., Koyama, T., Takayanagi, K., Mori, S., Kousaka, Y., Akimitsu, J., Nishihara, S., Inoue, K., Ovchinnikov, A. S., & Kishine, J. (2012). *Physical Review Letters, 108*, 107202.
36. Karna, S. K., Womack, F. N., Chapai, R., Young, D. P., Marshall, M., Xie, W., Graf, D., Wu, Y., Cao, H., DeBeer-Schmitt, L., Adams, P. W., Jin, R., & DiTusa, J. F. (2019). *Physical Review B, 100*, 184413.
37. Okamura, Y., Yamasaki, Y., Morikawa, D., Honda, T., Ukleev, V., Nakao, H., Murakami, Y., Shibata, K., Kagawa, F., Seki, S., Arima, T., & Tokura, Y. (2017). *Physical Review B, 96*, 174417.
38. Matsumura, T., Kita, Y., Kubo, K., Yoshikawa, Y., Michimura, S., Inami, T., Kousaka, Y., Inoue, K., & Ohara, S. (2017). *Journal of the Physical Society of Japan, 86*, 124702.
39. Shimamoto, Y., Matsushima, Y., Hasegawa, T., Kousaka, Y., Proskurin, I., Kishine, J., Ovchinnikov, A. S., Goncalves, F. J. T., & Togawa, Y. (2022). *Physical Review Letters, 128*, 247203.
40. Mizukami, S., Ando, Y., & Miyazaki, T. (2001). *Japanese Journal of Applied Physics, 40*, 580.
41. Mizukami, S., Sajitha, E. P., Watanabe, D., Wu, F., Miyazaki, T., Naganuma, H., Oogane, M., & Ando, Y. (2010). *Applied Physics Letters, 96*, 152502.
42. Oogane, M., Wakitani, T., Yakata, S., Yilgin, R., Ando, Y., Sakuma, A., & Miyazaki, T. (2006). *Japanese Journal of Applied Physics, 45*, 3889.
43. Oogane, M., Kubota, T., Kota, Y., Mizukami, S., Naganuma, H., Sakuma, A., & Ando, Y. (2010). *Applied Physics Letters, 96*, 252501.
44. Díaz, S. A., Hirosawa, T., Klinovaja, J., & Loss, D. (2020). *Physical Review Research, 2*, 013231.

Chapter 6
Spin Moiré Engineering in Two Dimensions

In this chapter, we investigate the spin moiré engineering in two dimensions. Among many parameters in spin moirés, we focus on the phase degree of freedom, denoted by φ in Eq. (1.11). In Sect. 6.1, we systematically clarify the effect of phase shifts on the two types of $3Q$-SkLs composed of the helices and sinusoidal waves by using the hyperspace representation given in Sect. 4.6. Furthermore, we also analyze the previous numerical data for the model Hamiltonian in Eq. (3.5) by using the technique in Sect. 4.3 and study the evolution of the phases under the magnetic field. Section 6.2 is devoted to the concluding remarks.

6.1 Phase Shift in $3Q$-SkL

In this section, we discuss the effect of phase shifts on the spin structures composed of three waves in two dimensions. The effect of the phase shift in the $3Q$-SkL has already been exemplified in Fig. 2.5. In the $3Q$-SkL, phase shifts bring about spin textures with different symmetries and topological properties, highlighting a phase degree of freedom as an important parameter for the spin moiré engineering [1, 2]. Nevertheless, their impact on the SkL has not been fully elucidated thus far.

To reveal the effect of phase shifts on the spin structures composed of three waves, we employ the hyperspace representation introduced in Sect. 4.6 and comprehensively clarify how the spin texture, symmetry, and the topological properties are modulated through the phase shifts. Specifically, we consider the spin configuration in Eq. (1.11) with $\mathcal{Q}_\eta = \mathbf{q}_\eta \cdot \mathbf{r} + \varphi_\eta$ and three $\mathbf{q}_\eta$ given by

$$\mathbf{q}_1 = (q, 0)\,,\ \mathbf{q}_2 = \left(-\frac{q}{2}, \frac{\sqrt{3}q}{2}\right),\ \mathbf{q}_3 = \left(-\frac{q}{2}, -\frac{\sqrt{3}q}{2}\right), \tag{6.1}$$

K. Shimizu, *Spin Moiré Engineering and Emergent Electromagnetism in Topological Spin Crystals*, Springer Theses, https://doi.org/10.1007/978-981-96-5992-0_6

where $|\mathbf{q}_\eta| = q$. For this $3Q$ spin structure, the 2D magnetic translation vectors are defined as

$$\mathbf{a}_1 = \frac{2\pi}{q}\left(1, \frac{1}{\sqrt{3}}\right), \quad \mathbf{a}_2 = \frac{2\pi}{q}\left(0, \frac{2}{\sqrt{3}}\right). \tag{6.2}$$

In the following, we first construct the hyperspace representation of the 2D $3Q$ states in Sect. 6.1.1. Next, by using the hyperspace representation, we elucidate the effect of phase shifts on the two types of 2D $3Q$-SkLs: $3Q$-SkLs composed of three helices in Sect. 6.1.2 and those composed of three sinusoidal waves in Sect. 6.1.3. Finally, in Sect. 6.1.4, we study how the actual phases of the superposed waves evolve while increasing an external magnetic field for the 2D $3Q$-SkL by analyzing the actual numerical data of the real-space spin configurations obtained for the Kondo lattice model in Ref. [3].

6.1.1 Hyperspace Representation of the $3Q$ States

Using the framework introduced in Sect. 4.6, we construct the hyperspace representation of the $3Q$ spin structures with $N_Q = 3$ and $d = 2$. First, we set the 3D wave vectors $\mathbf{Q}_\eta = (Q^X_\eta, Q^Y_\eta, Q^Z_\eta)$ in Eq. (4.18) as

$$\mathbf{Q}_1 = q\left(1, 0, \frac{1}{\sqrt{2}}\right), \tag{6.3}$$

$$\mathbf{Q}_2 = q\left(-\frac{1}{2}, \frac{\sqrt{3}}{2}, \frac{1}{\sqrt{2}}\right), \tag{6.4}$$

$$\mathbf{Q}_3 = q\left(-\frac{1}{2}, -\frac{\sqrt{3}}{2}, \frac{1}{\sqrt{2}}\right), \tag{6.5}$$

so that the projection of $\mathbf{Q}_\eta$ onto the $q^x q^y$ plane matches $\mathbf{q}_\eta$, and $\mathbf{Q}_1$, $\mathbf{Q}_2$, and $\mathbf{Q}_3$ are orthogonal to each other as shown in Fig. 6.1a, without loss of generality. Then, we obtain the corresponding magnetic translation vectors $\mathbf{A}_\eta$ in the 3D hyperspace as

$$\mathbf{A}_1 = \frac{4\pi}{3q}\left(1, 0, \frac{1}{\sqrt{2}}\right), \tag{6.6}$$

$$\mathbf{A}_2 = \frac{4\pi}{3q}\left(-\frac{1}{2}, \frac{\sqrt{3}}{2}, \frac{1}{\sqrt{2}}\right), \tag{6.7}$$

$$\mathbf{A}_3 = \frac{4\pi}{3q}\left(-\frac{1}{2}, -\frac{\sqrt{3}}{2}, \frac{1}{\sqrt{2}}\right). \tag{6.8}$$

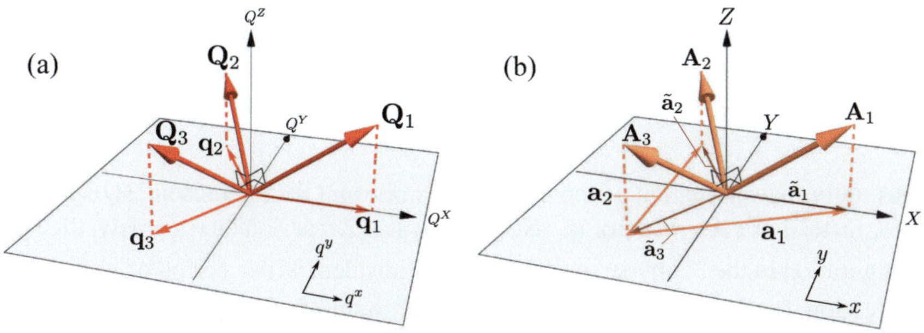

Fig. 6.1 **a** Schematic pictures of the relations between the wave vectors in the 3D reciprocal hyperspace, $\mathbf{Q}_\eta$, and those in the 2D reciprocal space, $\mathbf{q}_\eta$. **b** Corresponding relations between the magnetic translation vectors in the 3D hyperspace, $\mathbf{A}_\eta$, and those in the 2D real space, $\mathbf{a}_\eta$. The projections of $\mathbf{A}_\eta$ onto the xy plane denoted by $\tilde{\mathbf{a}}_\eta$ are also shown. Reproduced from Ref. [4]. Copyright (2022) by the American Physical Society

Figure 6.1b illustrates the relation between $\mathbf{A}_\eta$ and $\mathbf{a}_\eta$. Note that $\mathbf{a}_\eta$ is given by $\mathbf{a}_\eta = \tilde{\mathbf{a}}_\eta - \tilde{\mathbf{a}}_3$, where $\tilde{\mathbf{a}}_\eta$ is the projection of $\mathbf{A}_\eta$ onto the xy plane.

By using Eq. (4.20), the hyperspace positions are related with the real-space positions as

$$\begin{pmatrix} X \\ Y \\ Z \end{pmatrix} = V_3 \begin{pmatrix} x \\ y \\ 1 \end{pmatrix}, \tag{6.9}$$

where V_3 includes the phases as

$$V_3 = \begin{pmatrix} \mathbf{v}_1 \; \mathbf{v}_2 \; \mathbf{v}_3 \end{pmatrix} = \begin{pmatrix} 1 \; 0 \; \frac{2}{3q}\left(\varphi_1 - \frac{1}{2}(\varphi_2 + \varphi_3)\right) \\ 0 \; 1 \; \frac{1}{\sqrt{3}q}\left(\varphi_2 - \varphi_3\right) \\ 0 \; 0 \; \frac{\sqrt{2}}{3q}\left(\varphi_1 + \varphi_2 + \varphi_3\right) \end{pmatrix}. \tag{6.10}$$

It is worth noting that Eq. (6.9) can be rewritten as

$$\mathbf{R} = x\mathbf{v}_1 + y\mathbf{v}_2 + \mathbf{v}_3. \tag{6.11}$$

This means that the spin configuration on the original 2D xy plane is the same as the one on a slice of the hyperspace spin configuration spanned by $\mathbf{v}_1$ and $\mathbf{v}_2$ including the point at $\mathbf{v}_3$, namely, the horizontal plane with

$$Z = \frac{\sqrt{2}}{3q}\tilde{\varphi}, \tag{6.12}$$

where

$$\tilde{\varphi} = \sum_{\eta} \varphi_{\eta} = \varphi_1 + \varphi_2 + \varphi_3. \tag{6.13}$$

Thus, only the summation of the phases φ_η is relevant for the present $3Q$ spin textures, instead of each value of φ_η. Note that $\tilde{\varphi}$ has 2π periodicity; namely, the spin configuration in the 3D hyperspace becomes equivalent with a period of $\frac{2\sqrt{2}\pi}{3q}$ in the Z direction. In the original 2D space, the 2π phase shift in $\tilde{\varphi}$ ($\Delta\tilde{\varphi} = \sum_\eta \Delta\varphi_\eta = 2\pi$) corresponds to a spatial translation by

$$\Delta\mathbf{r}_\eta = \tilde{\mathbf{a}}_\eta - \sum_{\eta'=1}^{3} \frac{\Delta\varphi_{\eta'}}{2\pi}\tilde{\mathbf{a}}_{\eta'}, \tag{6.14}$$

where η may take any of 1, 2, and 3.

6.1.2 Screw 3*Q* State

In this subsection, we analyze the effect of phase shifts on the $3Q$ state composed of three proper screws, in which spins rotate on the plane perpendicular to $\mathbf{q}_\eta$. It includes a $3Q$-SkL found in a wide range of materials, as introduced in Sect. 1.4.2, and we call this type the screw $3Q$ state. The spin texture is given by Eq. (1.11) with $N_Q = 3$ and

$$\psi_\eta^{\mathrm{c}} = \psi_\eta^{\mathrm{s}} = \frac{1}{\sqrt{3}}, \tag{6.15}$$

$$\mathbf{e}_\eta^1 = \hat{\mathbf{z}},\ \mathbf{e}_\eta^2 = \mathbf{e}_\eta^0 \times \mathbf{e}_\eta^1,\ \mathbf{e}_\eta^0 = \left(\frac{q_\eta^x}{q}, \frac{q_\eta^y}{q}, 0\right)^{\mathsf{T}}, \tag{6.16}$$

$$\mathbf{m} = m\hat{\mathbf{z}}. \tag{6.17}$$

The explicit form is given as

$$\mathbf{S}(\mathbf{r}) \propto \begin{pmatrix} \frac{\sqrt{3}}{2}(\sin\mathcal{Q}_2 - \sin\mathcal{Q}_3) \\ -\sin\mathcal{Q}_1 + \frac{1}{2}(\sin\mathcal{Q}_2 + \sin\mathcal{Q}_3) \\ \cos\mathcal{Q}_1 + \cos\mathcal{Q}_2 + \cos\mathcal{Q}_3 + \sqrt{3}m \end{pmatrix}. \tag{6.18}$$

Before proceeding with the analysis of the topological properties, let us first examine the symmetry of the screw $3Q$ state in Eq. (6.18). Table 6.1 summarizes the symmetry operations of the spin texture, along with the decomposition of each operation into the change in the sum of phases $\tilde{\varphi}$, spatial translation, and magnetization change. We consider here the point group operations and time-reversal operation that do not

Table 6.1 Symmetry operations for the screw 3Q state in Eq. (6.18) and their decompositions into the change in the sum of phases, spatial translation in the original 2D space, and the magnetization change: $C_{n\alpha}$ represents an n-fold rotation about the axis in the α direction and $\mathcal{T}$ represents the time-reversal operation

Operation	Sum of phases	Translation	Magnetization
C_{6z}	$\tilde{\varphi} \to 2\pi - \tilde{\varphi}$	$\tilde{\mathbf{a}}_\eta + \tilde{\mathbf{a}}_{\eta'}$	$m \to m$
C_{2x}	$\tilde{\varphi} \to \pi + \tilde{\varphi}$	$\tilde{\mathbf{a}}_\eta$	$m \to -m$
$\mathcal{T}$	$\tilde{\varphi} \to \pi + \tilde{\varphi}$	$\tilde{\mathbf{a}}_\eta$	$m \to -m$

alter the wave vectors of the proper screws. Other symmetry operations are expressed as combinations of those listed in Table 6.1, for example, $C_{2y} = C_{2z}C_{2x} = C_{6z}^3 C_{2x}$. From Table 6.1, we can determine the symmetry operations that leave both $\tilde{\varphi}$ and m, namely the spin texture, unchanged. Specifically, we find that the screw 3Q state is invariant under C_{6z}, $\mathcal{T}C_{2x}$, and their combinations for $\tilde{\varphi} = 0$ and π, while for other cases, it remains symmetric under C_{3z}, $\mathcal{T}C_{2x}$, and their combinations. It is important to note that a spatial translation combined with the C_{6z} rotation can be represented by shifting the rotation axis.

Hedgehogs in Hyperspace

To examine the effect of phase shifts on the screw 3Q state in Eq. (6.18), we analyze the corresponding spin texture in the 3D hyperspace, whose reciprocal space is spanned by the wave vectors $\mathbf{Q}_\eta$ in Eqs. (6.3), (6.4), and (6.5). The schematic representation is shown in Fig. 6.2a. As discussed in Sect. 6.1.1, a real-space spin configuration for a given phase summation $\tilde{\varphi}$ on the original 2D plane corresponds to a hyperspace spin configuration on the horizontal plane with $Z = \frac{\sqrt{2}}{3q}\tilde{\varphi}$ [see Eq. (6.12)]. This correspondence is exemplified in Fig. 6.2b for $\tilde{\varphi} = (2n+1)\pi$. In this case, the intersections of the 3D cubic MUC with the horizontal planes at $Z = \frac{\sqrt{2}}{3q}(2n+1)\pi$ ($n = 0, 1, 2$) yield the hexagon and the two triangles, which together form the rhombus MUC in the original 2D plane.

Within the 3D hyperspace, the spin structure formed by the three proper screws may consist of hedgehogs and antihedgehogs, as introduced in Sect. 1.2.3. Indeed, by solving the equation $\mathbf{S}(\mathbf{r}) = 0$ for Eq. (6.18), we obtain the following eight solutions:

$$
\begin{aligned}
(\mathcal{Q}_1^*, \mathcal{Q}_2^*, \mathcal{Q}_3^*) = &\left(\pi + p_1^{\mathrm{scr}}(m), \pi + p_1^{\mathrm{scr}}(m), \pi + p_1^{\mathrm{scr}}(m)\right), \\
&\left(\pi - p_1^{\mathrm{scr}}(m), \pi - p_1^{\mathrm{scr}}(m), \pi - p_1^{\mathrm{scr}}(m)\right), \\
&\left(\pi - p_2^{\mathrm{scr}}(m), \pi - p_2^{\mathrm{scr}}(m), p_2^{\mathrm{scr}}(m)\right) \text{ and cyclic permutations,} \\
&\left(\pi + p_2^{\mathrm{scr}}(m), \pi + p_2^{\mathrm{scr}}(m), 2\pi - p_2^{\mathrm{scr}}(m)\right) \text{ and cyclic permutations,}
\end{aligned}
\tag{6.19}
$$

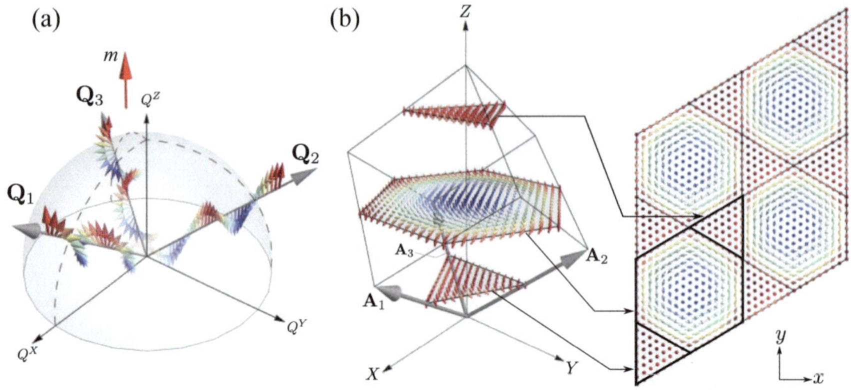

Fig. 6.2 **a** Schematic of the three proper screws in the 3D reciprocal hyperspace, whose wave vectors $\mathbf{Q}_\eta$ are defined by Eqs. (6.3), (6.4), and (6.5) for the screw $3Q$ state in Eq. (6.18). **b** Corresponding cubic MUC in the hyperspace and spin structures on the three horizontal planes with $Z = \frac{\sqrt{2}}{3q}(2n+1)\pi$ $(n = 0, 1, 2)$. The right panel in (**b**) shows the 2D spin texture on the three slices, which corresponds to the spin structure in Eq. (6.18) with $\tilde{\varphi} = \varphi_1 + \varphi_2 + \varphi_3 = (2n+1)\pi$, where n is an integer. The rhombus consisting of the two triangles and one hexagon denotes the 2D MUC in the original 2D plane. Reproduced from Ref. [4]. Copyright (2022) by the American Physical Society

where

$$p_1^{\rm scr}(m) = \arccos\left(\frac{m}{\sqrt{3}}\right), \quad p_2^{\rm scr}(m) = \arccos\left(\sqrt{3}m\right). \tag{6.20}$$

By using the relation

$$\mathbf{R}^* = \sum_\eta \frac{Q_\eta^*}{2\pi}\mathbf{A}_\eta, \tag{6.21}$$

we obtain the positions of the eight singular points in the hyperspace as

$$\begin{aligned}
&(X^*, Y^*, Z^*) = \\
&\frac{\sqrt{2}}{3q}\left(0, 0, 3(\pi + p_1^{\rm scr}(m))\right), \ \frac{\sqrt{2}}{3q}\left(0, 0, 3(\pi - p_1^{\rm scr}(m))\right), \\
&\left(\frac{\pi - 2p_2^{\rm scr}(m)}{3q}, \frac{\pi - 2p_2^{\rm scr}(m)}{\sqrt{3}q}, \frac{\sqrt{2}(2\pi - p_2^{\rm scr}(m))}{3q}\right) \text{ and } C_3^Z \text{ symmetric points,} \\
&\left(-\frac{\pi - 2p_2^{\rm scr}(m)}{3q}, -\frac{\pi - 2p_2^{\rm scr}(m)}{\sqrt{3}q}, \frac{\sqrt{2}(4\pi + p_2^{\rm scr}(m))}{3q}\right) \text{ and } C_3^Z \text{ symmetric points,}
\end{aligned} \tag{6.22}$$

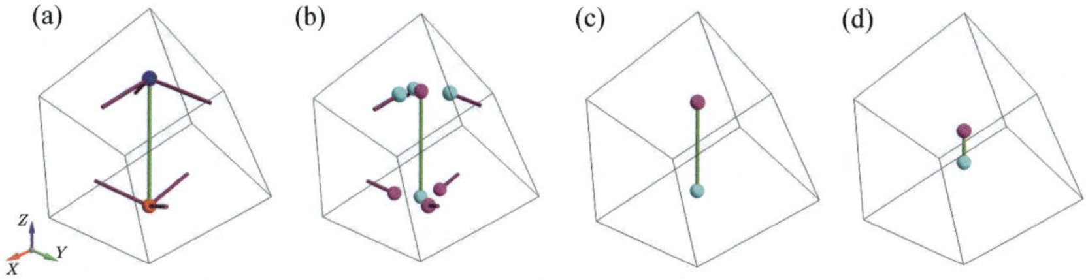

Fig. 6.3 Change of the positions of the hedgehogs and antihedgehogs, and the Dirac strings connecting them in the spin structure in the 3D hyperspace corresponding to the screw 3Q state in Eq. (6.18) while changing m: **a** $m = 0$, **b** $m = 0.4$, **c** $m = 0.8$, and **d** $m = 1.6$. The red, blue, magenta, and cyan spheres are the topological defects with $Q_{\mathrm{m}} = 2, -2, 1$, and -1, respectively. The green and purple lines denote the Dirac strings with the vorticity $\zeta = +1$ and -1, respectively. Reproduced from Ref. [4]. Copyright (2022) by the American Physical Society

where the C_3^Z symmetric points are obtained by $2\pi/3$ and $4\pi/3$ rotations about the Z axis.

Topological Transition in 3D Hyperspace

Figure 6.3 depicts the systematic evolution of topological defects as the magnetization along the Z direction, denoted by m, is varied. At $m = 0$, four out of the eight solutions in Eq. (6.22) become identical to $\frac{\sqrt{2}}{3q}\left(0, 0, \frac{9\pi}{2}\right)$, while the remaining four reduce to $\frac{\sqrt{2}}{3q}\left(0, 0, \frac{3\pi}{2}\right)$ due to $p_1^{\mathrm{scr}}(0) = \frac{\pi}{2}$ and $p_2^{\mathrm{scr}}(0) = \frac{\pi}{2}$. Hence, only two topological defects remain, both aligned along the Z axis, as illustrated in Fig. 6.3a. Following the methodology outlined in Refs. [5–9], we compute the monopole charge Q_{m} for these defects by substituting $\mathbf{r}$ in Eqs. (1.7) and (1.8) with $\mathbf{R}$. The defect at $\frac{\sqrt{2}}{3q}\left(0, 0, \frac{3\pi}{2}\right)$ in Fig. 6.3a is identified as a hedgehog with $Q_{\mathrm{m}} = +2$ (red sphere), whereas the defect at $\frac{\sqrt{2}}{3q}\left(0, 0, \frac{9\pi}{2}\right)$ is an antihedgehog with $Q_{\mathrm{m}} = -2$ (blue sphere).

Applying the framework in Sect. 7.2.2, we further identify four Dirac strings–topological structures connecting the hedgehog and antihedgehog via spins aligned antiparallel to the magnetization direction (see Sect. 1.3.1). The Dirac strings are characterized by their vorticity, defined as

$$\zeta = \frac{1}{2\pi} \oint d\mathbf{l} \cdot \nabla\phi(\mathbf{R}), \tag{6.23}$$

where $\phi(\mathbf{R})$ represents the azimuthal angle of $\mathbf{S}(\mathbf{R})$, and the contour integral is evaluated along a closed path encircling the Dirac string at $\mathbf{R}$ on a plane perpendicular to the Z axis [9, 10]. As shown in Fig. 6.3a, the Dirac strings consist of one segment along the Z axis (green line) and three additional segments traversing the MUC boundaries, linking topological defects in adjacent MUCs (purple lines). The former possesses a vorticity of $\zeta = +1$, whereas the latter three exhibit $\zeta = -1$.

By introducing a finite m, each topological defect splits into four topological defects with smaller Q_{m}: the hedgehog with $Q_{\mathrm{m}} = +2$ decomposes into three hedgehogs with $Q_{\mathrm{m}} = +1$ (magenta spheres) and one antihedgehog with $Q_{\mathrm{m}} = -1$ (cyan sphere), while the antihedgehog with $Q_{\mathrm{m}} = -2$ splits into three antihedgehogs with $Q_{\mathrm{m}} = -1$ and one hedgehog with $Q_{\mathrm{m}} = +1$, as depicted in Fig. 6.3b. Notably, the total monopole charge remains conserved throughout this process. Consequently, the system hosts a total of four hedgehogs with $Q_{\mathrm{m}} = +1$ and four antihedgehogs with $Q_{\mathrm{m}} = -1$. With a further increase of m, three pairs of defects connected by Dirac strings with $\zeta = -1$ undergo pair annihilation at $m = 1/\sqrt{3}$, leaving a single remaining pair linked by the Dirac string with $\zeta = +1$, as shown in Fig. 6.3c. Upon further increasing m, the surviving hedgehog and antihedgehog approach one another, as illustrated in Fig. 6.3d, and they finally annihilate at $m = \sqrt{3}$ by pair annihilation. Thus, the variation of m induces two distinct topological phase transitions, each governed by the pair annihilation of hedgehog-antihedgehog pairs at $m = 1/\sqrt{3}$ and $m = \sqrt{3}$.

Topological Transition on 2D Plane

Utilizing the results on topological defects and Dirac strings in the 3D hyperspace, we can systematically analyze the topological properties of the 2D spin texture described by Eq. (6.18) as a function of $\tilde{\varphi}$ and m. Figure 6.4 illustrates the correspondence between horizontal cross-sections in the 3D hyperspace at $Z = \frac{\sqrt{2}}{3q}\tilde{\varphi}$ and the real-space 2D spin textures with $\tilde{\varphi}$. In Fig. 6.4a, we depict the configurations of hedgehogs, antihedgehogs, and Dirac strings in the hyperspace at $m = 0$, along with the horizontal plane at $Z = \frac{\sqrt{2}}{3q}\frac{9}{4}\pi$. The corresponding spin configuration on this plane, shown in Fig. 6.4b, represents the 2D spin texture given by Eq. (6.18) for $\tilde{\varphi} = \frac{9}{4}\pi$. The topological characteristics of the spin texture in the 2D plane within the 3D hyperspace are quantified by the skyrmion number [11–13], expressed as

$$N_{\mathrm{sk}} = \frac{1}{4\pi}\int_{\text{2D MUC}} dXdY\, b_Z(\mathbf{R}), \tag{6.24}$$

where the integral extends over the 2D rhombic MUC.

The skyrmion number can alternatively be determined by summing the vorticities of the Dirac strings [9]:

$$N_{\mathrm{sk}} = -\sum_k \zeta_k(Z), \tag{6.25}$$

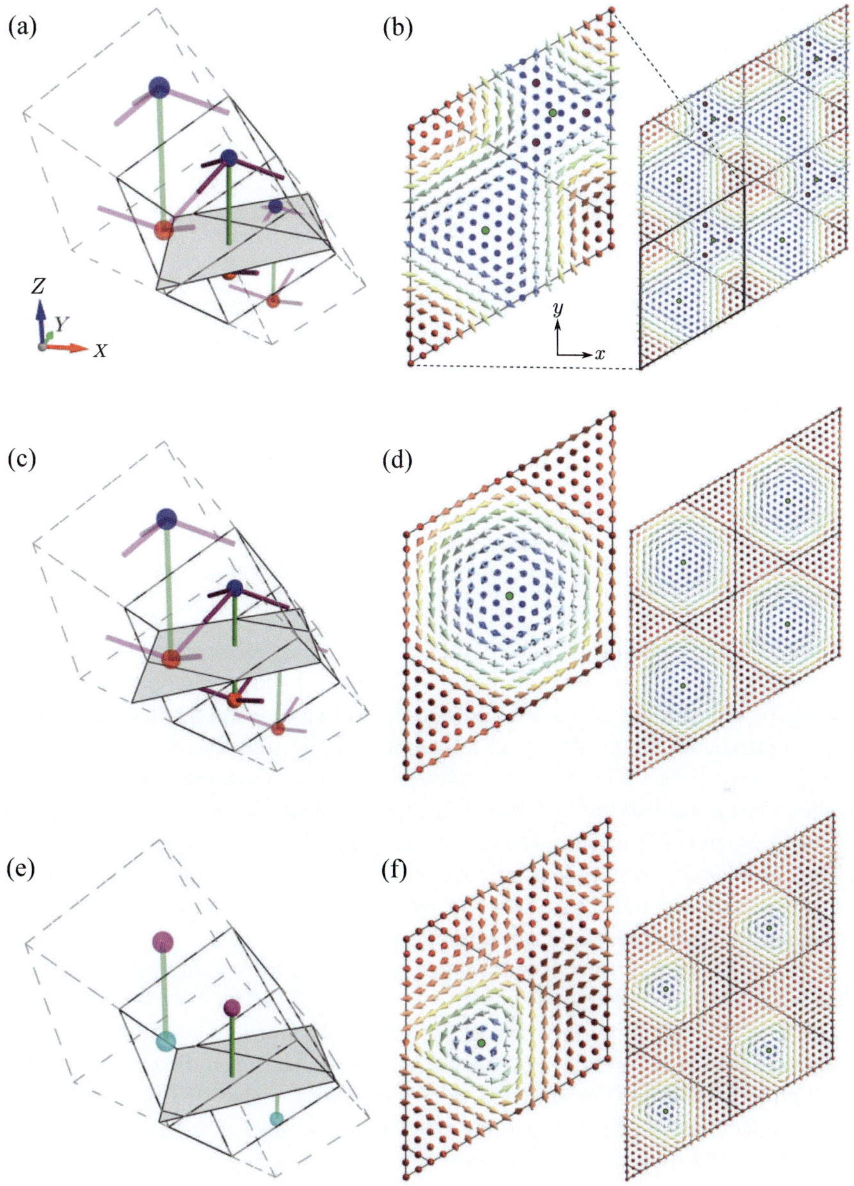

Fig. 6.4 **a** Hedgehogs and antihedgehogs, and Dirac strings for $m = 0$ in the 3D hyperspace. The pale colors are used for the objects in the neighboring MUCs denoted by the dashed cubes. The gray plane represents the 2D MUC on the intersection of the plane with $Z = \frac{\sqrt{2}}{3q}\frac{9}{4}\pi$. **b** The spin configuration on the gray plane in (**a**), which corresponds to the spin structure in Eq. (6.18) with $\tilde{\varphi} = \frac{9}{4}\pi$. The green and purple circles denote the intersections of the Dirac strings with $\zeta = +1$ and -1, respectively. The right panel displays the 2D real-space spin configuration by repeating the MUC. Similar figures for (**c**), (**d**) $m = 0$ and $\tilde{\varphi} = 3\pi$, and (**e**), (**f**) $m = 0.9$ and $\tilde{\varphi} = \frac{9}{4}\pi$. Reproduced from Ref. [4]. Copyright (2022) by the American Physical Society

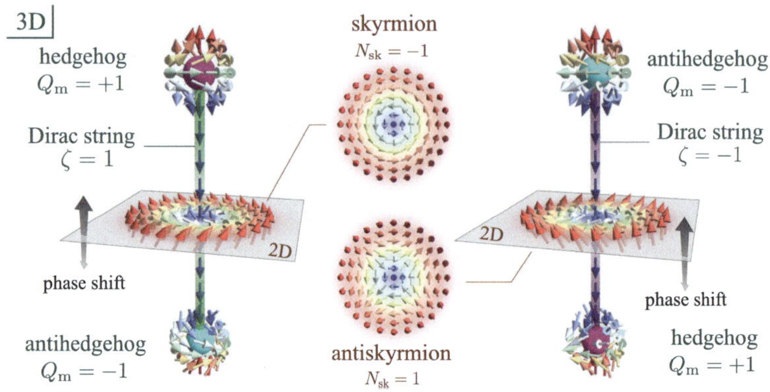

Fig. 6.5 Schematic figure for the correspondence between the original space and the hyperspace for the 2D $3Q$ case. The objects in the original space and the hyperspace are denoted by brown and green characters, respectively. The 2D $3Q$-SkL is obtained as a 2D intersection (gray plane) of the 3D HL in the 3D hyperspace. A phase shift in the original space corresponds to a shift of the intersection in the hyperspace. Presented here are the correspondence for the Dirac strings with $\zeta = 1$ (left) and $\zeta = -1$ (right), exhibiting the skyrmion and antiskyrmion on the intersection, respectively. Reproduced from Ref. [4]. Copyright (2022) by the American Physical Society

where $\zeta_k(Z)$ represents the vorticity of the kth Dirac string intersecting the 2D MUC.[1] For instance, in Fig. 6.4b, two Dirac strings with $\zeta = +1$ and three with $\zeta = -1$ intersect the plane, yielding $N_{sk} = +1$. This straightforward enumeration in the hyperspace representation thus allows us to identify the 2D spin texture at $m = 0$ with $\tilde{\varphi} = \frac{9}{4}\pi$ as a $3Q$-SkL with $N_{sk} = +1$, a conclusion that is independently confirmed by direct integration using Eq. (6.24). The correspondence between topological entities in the original 2D space and the 3D hyperspace is summarized in Fig. 6.5: when a hedgehog (antihedgehog) is positioned above an antihedgehog (hedgehog) along the Z axis, they are linked by a Dirac string with $\zeta = +1$ ($\zeta = -1$), manifesting as a skyrmion (antiskyrmion) in the intersection.

Figure 6.4c and d illustrate the results for $\tilde{\varphi} = 3\pi$ at $m = 0$. In this case, the 2D slice intersects a single Dirac string with $\zeta = +1$, resulting in a $3Q$-SkL with $N_{sk} = -1$ for $\tilde{\varphi} = 3\pi$. This highlights a topological transformation induced by a phase shift. Within the hyperspace framework, such transitions occur when the gray horizontal plane crosses hedgehogs or antihedgehogs located at the ends of Dirac strings. As topological defects shift their positions with varying m (see Fig. 6.3), the topological nature of the 2D spin structures likewise evolves. Such a change is demonstrated in Fig. 6.4e and f for $m = 0.9$. At this value, three hedgehog-antihedgehog pairs have already undergone pair annihilation, leaving only a single pair along the Z axis, as depicted in Fig. 6.4e. When the 2D slice intersects the Dirac string linking the remaining hedgehog-antihedgehog pair, the 2D spin structure forms a $3Q$-SkL with $N_{sk} = -1$, as exemplified in Fig. 6.4f for $\tilde{\varphi} = \frac{9}{4}\pi$.

[1] The skyrmion number is ill-defined when hedgehogs or antihedgehogs within the hyperspace participate in the 2D intersection.

This systematic approach enables a comprehensive investigation of the evolution of magnetic textures and the topological properties of 2D $3Q$ spin structures as functions of $\tilde{\varphi}$ and m. The procedure can be summarized as follows: (i) Construct the 3D spin texture in hyperspace using Eqs. (4.18) and (4.19), specifically Eqs. (6.3)–(6.8) in the present case. (ii) Identify the hedgehogs, antihedgehogs, and the Dirac strings connecting them in the 3D hyperspace spin structure. (iii) Extract a 2D slice of the 3D spin structure at a horizontal plane defined by $Z = \frac{\sqrt{2}}{3q}\tilde{\varphi}$, yielding the 2D spin texture corresponding to the phase summation $\tilde{\varphi}$. (iv) Compute the skyrmion number $N_{\rm sk}$ of the 2D spin texture using Eq. (6.25) by summing the vorticities of Dirac strings intersecting the plane.

6.1.2.1 Topological Phase Diagram

Figure 6.6 presents the topological phase diagram in the $(m, \tilde{\varphi})$ plane for Eq. (6.18), derived through the aforementioned procedure. The resulting phase diagram exhibits periodicity in the $\tilde{\varphi}$ direction with a period of 2π and is symmetric about $\tilde{\varphi} = \pi$. Notably, the phase diagram for $m < 0$ can be obtained by reflecting the diagram for $m > 0$ while simultaneously shifting $\tilde{\varphi}$ by π and inverting the sign of $N_{\rm sk}$. This follows from the fact that the spin texture with parameters (φ_η, m) is related to that with $(\varphi_\eta + \pi, -m)$ via time-reversal symmetry. The phase diagram reveals four distinct topologically nontrivial phases characterized by $N_{\rm sk} = -2$, -1, 1, and 2. The predominant regions are occupied by SkLs with $N_{\rm sk} = \pm 1$, while SkLs with $N_{\rm sk} = \pm 2$ emerge only in narrow regions at finite m, never appearing at $m = 0$, where the system always exhibits $N_{\rm sk} = \pm 1$. The black dots at $m = \pm 1/\sqrt{3}$ and $m = \pm\sqrt{3}$ mark topological transitions associated with the annihilation of hedgehog-anthedgehog pairs in hyperspace.

Figure 6.7 illustrates representative spin configurations of screw-type $3Q$-SkLs for different values of $N_{\rm sk}$. Figure 6.7a shows the spin texture of the SkL with $N_{\rm sk} = -2$ at $m = 0.7$ and $\tilde{\varphi} = 0$, where two small skyrmions per MUC form a honeycomb lattice. In this configuration, two Dirac strings with $\zeta = 1$ intersect the 2D plane, yielding $N_{\rm sk} = -2$. Figure 6.7b corresponds to the SkL with $N_{\rm sk} = -1$ at $m = 0.7$ and $\tilde{\varphi} = \pi$, featuring a single skyrmion at the center of the MUC. Here, a single Dirac string with $\zeta = 1$ passing through the skyrmion core results in $N_{\rm sk} = -1$.

Figure 6.7c and d depict spin configurations at $m = -0.7$ for $\tilde{\varphi} = 0$ and $\tilde{\varphi} = \pi$, respectively. These configurations are obtained by inverting all spins in Fig. 6.7b and a, leading to $N_{\rm sk} = 1$ and 2, respectively. Notably, spin textures at $\tilde{\varphi} = n\pi$ (where n is an integer) exhibit sixfold rotational symmetry, whereas those at $\tilde{\varphi} \neq n\pi$ possess threefold symmetry, in agreement with the symmetry considerations summarized in Table 6.1.

To conclude this subsection, we discuss the implications of our topological phase diagram for phase control. Previous experimental and theoretical studies on chiral magnets [14–19] have reported SkLs with $N_{\rm sk} = -1$ $(+1)$ stabilized under an external magnetic field along the $(-)\hat{\mathbf{z}}$ direction. In our phase diagram (Fig. 6.6),

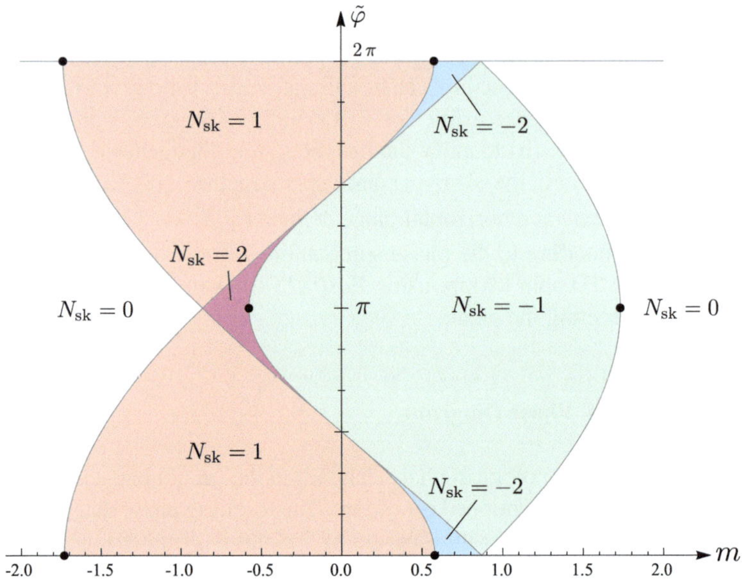

Fig. 6.6 Topological phase diagram for the screw $3Q$ state in Eq. (6.18) determined by the skyrmion number N_{sk} while changing m and $\tilde{\varphi}$. The colored regions are topologically nontrivial phases with nonzero N_{sk}, and the white areas in the left and right hand sides denote topologically trivial phases with $N_{sk} = 0$. The black dots represent the pair annihilation of the hedgehogs and antihedgehogs in the 3D hyperspace, which are located at $(m, \tilde{\varphi}) = (-\sqrt{3}, 0)$, $(-1/\sqrt{3}, \pi)$, $(1/\sqrt{3}, 0)$, and $(\sqrt{3}, \pi)$. The phase diagram is periodic in the $\tilde{\varphi}$ direction with a period of 2π. Reproduced from Ref. [4]. Copyright (2022) by the American Physical Society

this corresponds to $\tilde{\varphi} \approx \pi$ for $m > 0$ and $\tilde{\varphi} \approx 0$ for $m < 0$. However, SkLs with $N_{sk} = \pm 2$ have not yet been observed. Our phase diagram suggests that accessing the $N_{sk} = \pm 2$ states would require a phase shift of approximately π under an applied magnetic field. This is particularly intriguing, as the emergent magnetic field in the $N_{sk} = \pm 2$ state is twice as strong as that in the $N_{sk} = \pm 1$ state.

Furthermore, previous studies have indicated that SkLs with $N_{sk} = \pm 1$ typically exhibit a transition into a $1Q$ conical state or a uniformly polarized state with increasing field strength, rather than into the topologically trivial $3Q$ state with $N_{sk} = 0$ predicted in Fig. 6.6. This suggests that reaching the critical points where hedgehog-anthedgehog pairs annihilate in hyperspace (marked by black dots in Fig. 6.6) is challenging. The difficulty likely arises from the significant modulation of the spin texture in real space, which is energetically unfavorable under conventional two-spin interactions such as Heisenberg exchange. If one could circumvent the transition to the $1Q$ conical state, it might be possible to observe novel topological phenomena arising from singularities in the emergent electromagnetic fields due to pair annihilation. Recent studies have proposed potential methods for controlling the phase degree of freedom [2], which could provide viable pathways for exploring these novel topological states. We will revisit this issue in Sect. 6.1.4.

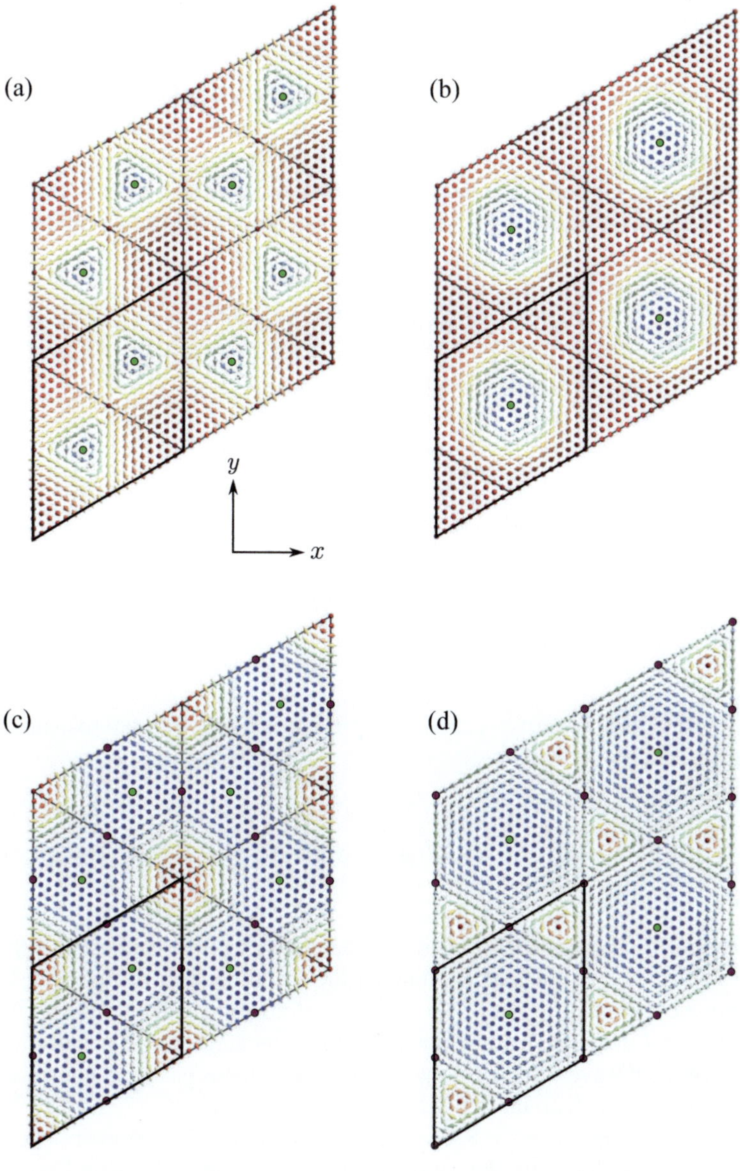

Fig. 6.7 Real-space spin configurations of the screw $3Q$ states: **a** $m = 0.7$ and $\tilde{\varphi} = 0$, **b** $m = 0.7$ and $\tilde{\varphi} = \pi$, **c** $m = -0.7$ and $\tilde{\varphi} = 0$, and **d** $m = -0.7$ and $\tilde{\varphi} = \pi$. Each spin configuration is topologically nontrivial with **a** $N_{\mathrm{sk}} = -2$, **b** $N_{\mathrm{sk}} = -1$, **c** $N_{\mathrm{sk}} = 1$, and **d** $N_{\mathrm{sk}} = 2$. The notations are common to those in Fig. 6.4

6.1.3 Sinusoidal 3 Q State

We next discuss the phase degree of freedom for another $3Q$-SkL composed of three sinusoidal waves, in which collinear spins are arranged on the plane perpendicular to $\mathbf{q}_\eta$. It includes a $3Q$-SkL with the skyrmion number of two found in the Kondo lattice system on the triangular lattice [3] and its effective spin model [20]. We call this type the sinusoidal $3Q$ state. Real-space spin configuration of the sinusoidal $3Q$ state is given by

$$\mathbf{S}(\mathbf{r}) \propto \begin{pmatrix} \frac{\sqrt{3}}{2} \sin\theta(-\cos\mathcal{Q}_2 + \cos\mathcal{Q}_3) \\ (-1)^\Gamma \frac{1}{2} \sin\theta\,(2\cos\mathcal{Q}_1 - \cos\mathcal{Q}_2 - \cos\mathcal{Q}_3) \\ \cos\theta(\cos\mathcal{Q}_1 + \cos\mathcal{Q}_2 + \cos\mathcal{Q}_3 + 3\tilde{m}) \end{pmatrix}, \tag{6.26}$$

which is obtained from Eq. (1.11) by taking $N_Q = 3$ and

$$\psi_\eta^{\mathrm{c}} = \frac{1}{\sqrt{3}}, \quad \psi_\eta^{\mathrm{s}} = 0, \tag{6.27}$$

$$\mathbf{e}_1^{\mathrm{I}} = \left(0, (-1)^\Gamma \sin\theta, \cos\theta\right)^{\mathsf{T}}, \quad \mathbf{e}_2^{\mathrm{I}} = \mathsf{R}_\Gamma \mathbf{e}_1^{\mathrm{I}}, \quad \mathbf{e}_3^{\mathrm{I}} = \mathsf{R}_\Gamma^2 \mathbf{e}_1^{\mathrm{I}}, \tag{6.28}$$

$$\mathbf{m} = m\hat{\mathbf{z}}, \quad \tilde{m} = \frac{m}{\sqrt{3}\cos\theta}, \tag{6.29}$$

where Γ takes 0 or $1, 0 < \theta < \frac{\pi}{2}$, and R_Γ represents a $(-1)^\Gamma \frac{2\pi}{3}$ rotation matrix about the z axis given by

$$\mathsf{R}_\Gamma = \begin{pmatrix} \cos\left((-1)^\Gamma \frac{2\pi}{3}\right) & -\sin\left((-1)^\Gamma \frac{2\pi}{3}\right) & 0 \\ \sin\left((-1)^\Gamma \frac{2\pi}{3}\right) & \cos\left((-1)^\Gamma \frac{2\pi}{3}\right) & 0 \\ 0 & 0 & 1 \end{pmatrix}. \tag{6.30}$$

In Eq. (6.26), Γ is a parameter to describe the chirality of the spin texture; the spin texture with $\Gamma = 1$ is obtained by flipping the S_y component of that with $\Gamma = 0$. Note that the spin texture with $\Gamma = 0$ has threefold rotational symmetry, whereas that with $\Gamma = 1$ does not. Meanwhile, θ describes the angle of the sinusoidal plane in the constituent waves. In the following, we mainly focus on the spin textures with $\Gamma = 0$, while we touch on those with $\Gamma = 1$ in Sect. 6.1.4.

Following the arguments in Sect. 6.1.2, we summarize the symmetry operations and their decompositions for the sinusoidal $3Q$ state in Eq. (6.26) with $\Gamma = 0$ and 1 in Tables 6.2 and 6.3, respectively. Other symmetry operations are expressed by the combinations of those in the tables. From Table 6.2, we find that the sinusoidal $3Q$ state with $\Gamma = 0$ is symmetric under $C_{3z}, \mathcal{I}, \mathcal{T}C_{2x}$, and their combinations for $\tilde{\varphi} = 0$ and π, otherwise $C_{3z}, \mathcal{T}C_{2x}$, and their combinations. Meanwhile, from Table 6.3, we find that the state with $\Gamma = 1$ is symmetric under $\mathcal{I}, \mathcal{T}C_{2x}$, and their combinations for $\tilde{\varphi} = 0, \pi$, otherwise $\mathcal{T}C_{2x}$.

Table 6.2 Similar table to Table 6.1 for the sinusoidal $3Q$ state with $\Gamma = 0$ in Eq. (6.26): $\mathcal{I}$ represents the spatial-inversion operation and the other notations are common to those in Table 6.1

Operation	Sum of phases	Translation	Magnetization
C_{3z}	$\tilde{\varphi} \to \tilde{\varphi}$	0	$m \to m$
C_{2x}	$\tilde{\varphi} \to \pi + \tilde{\varphi}$	$\tilde{\mathbf{a}}_\eta$	$m \to -m$
$\mathcal{I}$	$\tilde{\varphi} \to 2\pi - \tilde{\varphi}$	$\tilde{\mathbf{a}}_\eta + \tilde{\mathbf{a}}_{\eta'}$	$m \to m$
$\mathcal{T}$	$\tilde{\varphi} \to \pi + \tilde{\varphi}$	$\tilde{\mathbf{a}}_\eta$	$m \to -m$

Table 6.3 Similar table for the sinusoidal $3Q$ state with $\Gamma = 1$ in Eq. (6.26)

Operation	Sum of phases	Translation	Magnetization
C_{2x}	$\tilde{\varphi} \to \pi + \tilde{\varphi}$	$\tilde{\mathbf{a}}_\eta$	$m \to -m$
$\mathcal{I}$	$\tilde{\varphi} \to 2\pi - \tilde{\varphi}$	$\tilde{\mathbf{a}}_\eta + \tilde{\mathbf{a}}_{\eta'}$	$m \to m$
$\mathcal{T}$	$\tilde{\varphi} \to \pi + \tilde{\varphi}$	$\tilde{\mathbf{a}}_\eta$	$m \to -m$

Hedgehogs in Hyperspace

Following the procedure in the first subsubsection of Sect. 6.1.2, we can identify the hyperspace positions of the topological defects in the 3D spin texture corresponding to Eq. (6.26). Solving $\mathbf{S}(\mathbf{r}) = 0$, we obtain the following eight solutions:

$$
\begin{aligned}
(\mathcal{Q}_1^*, \mathcal{Q}_2^*, \mathcal{Q}_3^*) = &\left(\pi + p^{\sin}(\tilde{m}), \pi + p^{\sin}(\tilde{m}), \pi + p^{\sin}(\tilde{m})\right), \\
&\left(\pi - p^{\sin}(\tilde{m}), \pi - p^{\sin}(\tilde{m}), \pi - p^{\sin}(\tilde{m})\right), \\
&\left(\pi + p^{\sin}(\tilde{m}), \pi - p^{\sin}(\tilde{m}), \pi - p^{\sin}(\tilde{m})\right) \text{ and cyclic permutations}, \\
&\left(\pi - p^{\sin}(\tilde{m}), \pi + p^{\sin}(\tilde{m}), \pi + p^{\sin}(\tilde{m})\right) \text{ and cyclic permutations},
\end{aligned}
\tag{6.31}
$$

where

$$
p^{\sin}(\tilde{m}) = \arccos(\tilde{m}). \tag{6.32}
$$

By using the relation in Eq. (6.21), we obtain the positions of the eight singular points as

$$
\begin{aligned}
(X^*, Y^*, Z^*) = &\frac{\sqrt{2}}{3q}\left(0, 0, 3(\pi + p^{\sin}(\tilde{m}))\right), \ \frac{\sqrt{2}}{3q}\left(0, 0, 3(\pi - p^{\sin}(\tilde{m}))\right), \\
&\left(\frac{4p^{\sin}(\tilde{m})}{3q}, 0, \frac{\sqrt{2}}{3q}(3\pi - p^{\sin}(\tilde{m}))\right) \text{ and } C_3^Z \text{ symmetric points},
\end{aligned}
$$

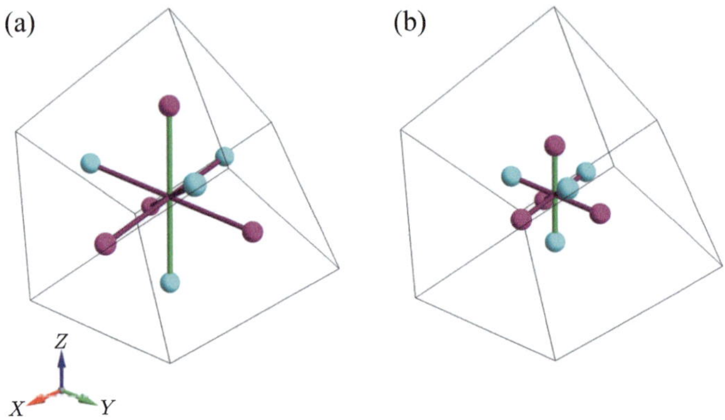

Fig. 6.8 Hedgehogs and antihedgehogs, and Dirac strings in the spin structure in the 3D hyperspace corresponding to the sinusoidal $3Q$ state in Eq. (6.26) for (**a**) $\tilde{m} = 0$ and (**b**) $\tilde{m} = 0.8$. The notations are common to those in Fig. 6.3. The results are obtained for $\Gamma = 0$ in Eq. (6.26); in the case of $\Gamma = 1$, all the monopole charges of the hedgehogs and antihedgehogs, and the vorticities of the Dirac strings reverse their signs. Reproduced from Ref. [4]. Copyright (2022) by the American Physical Society

$$\left(-\frac{4p^{\sin}(\tilde{m})}{3q}, 0, \frac{\sqrt{2}}{3q}(3\pi + p^{\sin}(\tilde{m}))\right) \text{ and } C_3^Z \text{ symmetric points.} \tag{6.33}$$

Topological Transition in 3D Hyperspace

Figure 6.8 depicts the evolution of topological defects and Dirac strings in hyperspace as $\tilde{m}$ in Eq. (6.26) is varied with $\Gamma = 0$. The monopole charges, Dirac strings, and their associated vorticities are determined using the same methodology outlined in the second subsubsection of Sect. 6.1.2; see Eqs. (1.7) and (6.23) for the definitions of Q_{m} and ζ, respectively. We note that the parameter θ in Eq. (6.26) does not influence the spatial positions of the topological entities in hyperspace.

In the absence of magnetization ($\tilde{m} = 0$), four out of the eight defects described in Eq. (6.33) are identified as hedgehogs with $Q_{\mathrm{m}} = 1$, while the remaining four correspond to antihedgehogs with $Q_{\mathrm{m}} = -1$. These eight topological defects form an NaCl-like lattice structure, as illustrated in Fig. 6.8a. We observe that each hedgehog-antihedgehog pair is interconnected by a Dirac string with vorticity $\zeta = \pm 1$, with all four Dirac strings intersecting at the center of the MUC.

Upon introducing $\tilde{m}$, the hedgehog-antihedgehog pairs migrate towards each other along the Dirac strings, leading to a contraction of the cubic arrangement formed by the eight defects, as demonstrated in Fig. 6.8b. Finally, at $\tilde{m} = 1$, all eight defects converge at the center of the MUC and undergo simultaneous pair annihilation.

In the case where $\Gamma = 1$, the spatial distribution of the topological defects remains unchanged; however, the signs of all monopole charges Q_{m} and vorticities ζ are reversed.

Topological Transition on 2D Plane

The two-dimensional spin structure with phase $\tilde{\varphi}$ in Eq. (6.26) corresponds to a cross section of the corresponding 3D HL at $Z = \frac{\sqrt{2}}{3q}\tilde{\varphi}$ (see Sect. 6.1.1). The skyrmion number N_{sk} is determined as the summation of the vorticities of the Dirac strings, as given in Eq. (6.25).

Figure 6.9a illustrates the configurations of hedgehogs, antihedgehogs, and Dirac strings in hyperspace for the spin structure described by Eq. (6.26) with $\Gamma = 0$ at $\tilde{m} = 0$, alongside the cross section at $Z = \frac{\sqrt{2}}{3q}3\pi$. The spin configuration extracted from this cross section is displayed in Fig. 6.9b, corresponding to the two-dimensional spin texture of Eq. (6.26) with $\tilde{\varphi} = (2n + 1)\pi$ (n being an integer) and $\Gamma = 0$. While the topological structures in hyperspace remain independent of the parameter θ, as noted in the previous subsubsection, the actual spin configuration in the original two-dimensional plane does depend on θ; in Fig. 6.9b, we adopt $\theta = \arccos\frac{1}{\sqrt{3}}$. In hyperspace, the four Dirac strings intersect at the center of the MUC, and since the slice contains this intersection point, the total vorticity at this point is given by $\zeta = 3 \times (-1) + 1 = -2$, as indicated by the black circles in the figure. Consequently, the two-dimensional spin texture at $\tilde{m} = 0$ and $\tilde{\varphi} = 3\pi$ is identified as the $3Q$-SkL with $N_{\mathrm{sk}} = 2$.

Figure 6.9c and d depict the result for $\tilde{\varphi} = 2\pi$ at $\tilde{m} = 0$. In this case, the two-dimensional slice contains two intersections of Dirac strings, both with vorticity $\zeta = +1$. As a result, the spin texture corresponding to $\tilde{\varphi} = 2\pi$ forms a $3Q$-SkL with $N_{\mathrm{sk}} = -2$, which serves as the time-reversal counterpart of the spin texture shown in Fig. 6.9b. Thus, analogous to the screw-type $3Q$ spin structures, a phase shift in the sinusoidal spin structures can induce topological transitions.

Topological Phase Diagram

Figure 6.10 presents the topological phase diagram for the two-dimensional spin texture described by Eq. (6.26). The obtained results hold for both $\Gamma = 0$ and $\Gamma = 1$, with the only distinction being the inversion of the skyrmion number N_{sk} in each phase: the upper (lower) signs correspond to $\Gamma = 0\,(1)$. Similar to Fig. 6.6, the phase diagram exhibits 2π periodicity and reflection symmetry about $\tilde{\varphi} = \pi$, while the result for $\tilde{m} < 0$ is obtained by mirroring that for $\tilde{m} > 0$ with a π shift in $\tilde{\varphi}$ and a sign reversal of N_{sk}.

Four distinct topologically nontrivial phases emerge, characterized by $N_{\mathrm{sk}} = -2$, -1, 1, and 2, akin to the proper screw case illustrated in Fig. 6.6, but with a different distributions of each phase. In the present sinusoidal case, the majority of the phase

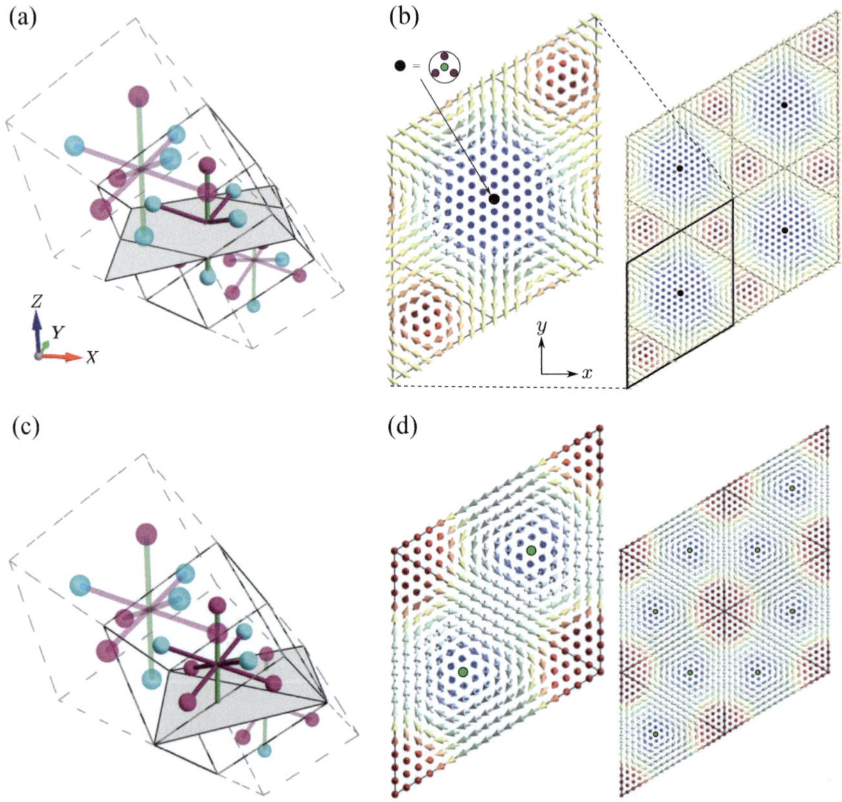

Fig. 6.9 **a** Hedgehogs and antihedgehogs, and Dirac strings for $\tilde{m} = 0$ in the 3D hyperspace for the sinusoidal $3Q$ state with Γ=0. The gray plane represents the intersection of the plane with $Z = \frac{\sqrt{2}}{3q}3\pi$ ($\tilde{\varphi} = 3\pi$). **b** The spin configuration on the gray plane in (**a**), which corresponds with the spin structure in Eq. (6.26) with $\tilde{\varphi} = 3\pi$ and $\Gamma = 0$. We take $\theta = \arccos \frac{1}{\sqrt{3}}$ in Eq. (6.28). The vorticity at the black circle takes $\zeta = -2$ (see the text for details). **c** and **d** Similar figures for $\tilde{m} = 0$ and $\tilde{\varphi} = 2\pi$. The notations are common to those in Fig. 6.4. Reproduced from Ref. [4]. Copyright (2022) by the American Physical Society

diagram is dominated by the SkL with $N_{\rm sk} = \pm 2$, whereas the SkL with $N_{\rm sk} = \pm 1$ appears only in intermediate regions for $\tilde{m} \neq 0$. Notably, the state at $\tilde{m} = 0$ is always characterized by $N_{\rm sk} = \pm 2$, in contrast to the case in Fig. 6.6. The $N_{\rm sk} = \pm 2\,(\mp 2)$ and $\mp 1\,(\pm 1)$ regions terminate at $\tilde{m} = 1\,(-1)$ and $\tilde{\varphi} = \pi\ (0)$, where the simultaneous pair annihilation of all hedgehogs and antihedgehogs occurs in hyperspace, as indicated by the black dots in Fig. 6.10.

Figure 6.11 illustrates representative spin configurations of the sinusoidal $3Q$ state for $\theta = \arccos \frac{1}{\sqrt{3}}$. Here, we take $\Gamma = 0$ in Fig. 6.11a–d and $\Gamma = 1$ in Fig. 6.11e–h. Figure 6.11a presents the spin configuration of the $N_{\rm sk} = -1$ state at $\tilde{m} = 0.5$ and $\tilde{\varphi} = \frac{\pi}{2}$. In this state, a single Bloch-type skyrmion with $N_{\rm sk} = -1$ per MUC is

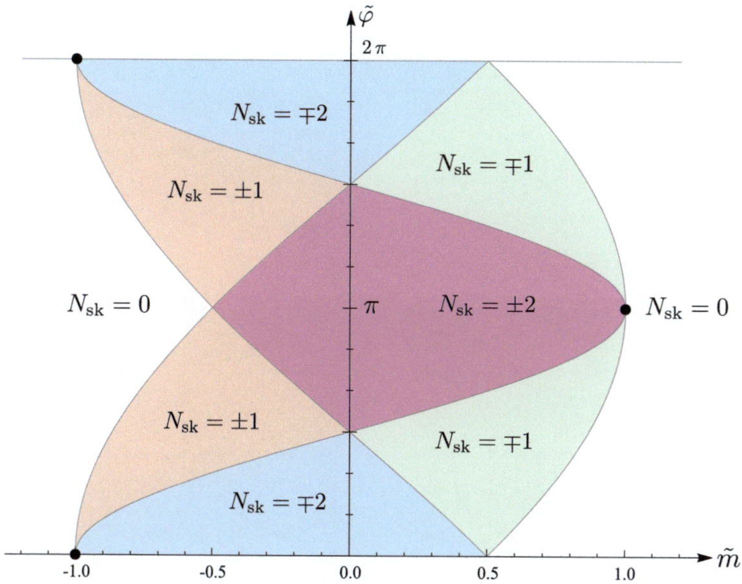

Fig. 6.10 Topological phase diagram for the sinusoidal $3Q$ state in Eq. (6.26) determined by the skyrmion number $N_{\rm sk}$ on the plane of $\tilde{m}$ and $\tilde{\varphi}$. The upper (lower) signs of $N_{\rm sk}$ are for $\Gamma = 0$ (1). The notations are common to those in Fig. 6.6. The black points represent the simultaneous annihilation of the four hedgehogs and the four antihedgehogs in the hyperspace, which are located at $(\tilde{m}, \tilde{\varphi}) = (-1, 0)$ and $(1, \pi)$. Reproduced from Ref. [4]. Copyright (2022) by the American Physical Society

observed, with a Dirac string of vorticity $\zeta = 1$ passing through the skyrmion core, thereby contributing to $N_{\rm sk} = -1$. Figure 6.11b depicts the $N_{\rm sk} = 2$ state at $\tilde{m} = 0.5$ and $\tilde{\varphi} = \pi$, which features a skyrmion with $N_{\rm sk} = 2$ at the center of the MUC. Here, the two-dimensional cross-section in hyperspace intersects the crossing point of four Dirac strings, yielding a total vorticity of -2 [see Fig. 6.9a]. Figure 6.11c and d illustrate the spin configurations for $\tilde{m} = -0.5$ at $\tilde{\varphi} = 0$ and $\tilde{\varphi} = \frac{\pi}{2}$, respectively. The former is obtained via time-reversal operation applied to Fig. 6.11b, whereas the latter results from a combined time-reversal and spatial-inversion operations to Fig. 6.11a. The corresponding configurations for $\Gamma = 1$ are displayed in Fig. 6.11e–h.

In contrast to the screw-type $3Q$ structures discussed in Sect. 6.1.2, all sinusoidal $3Q$ states with $\Gamma = 0$ retain threefold rotational symmetry regardless of $\tilde{\varphi}$ and $\tilde{m}$, whereas those with $\Gamma = 1$ break this symmetry; see Tables 6.2 and 6.3. Furthermore, spin textures with $\tilde{\varphi} = n\pi$ (n being an integer) exhibit inversion symmetry irrespective of Γ, in agreement with symmetry considerations.

The phase diagram in Fig. 6.10 suggests that the system undergoes a sequence of topological phase transitions as $\tilde{m}$ increases, progressing from $N_{\rm sk} = \pm 2$ to ± 1, and ultimately to $N_{\rm sk} = 0$. Such transitions were previously identified in numerical studies of the Kondo lattice model on a triangular lattice under an increasing magnetic field, as introduced in Sect. 3.3 [3]. Since the $N_{\rm sk} = \pm 2$ and ± 1 phases

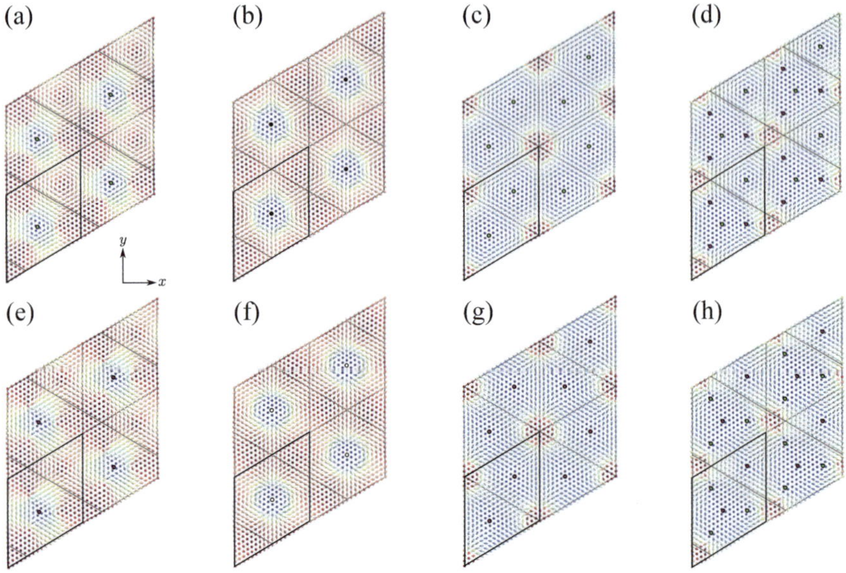

Fig. 6.11 Real-space spin configurations of the sinusoidal $3Q$ state in Eq. (6.26) with (**a**)–(**d**) $\Gamma = 0$ and (**e**)–(**h**) $\Gamma = 1$, and $\theta = \arccos \frac{1}{\sqrt{3}}$: (**a**)(**e**) $\tilde{m} = 0.5$ and $\tilde{\varphi} = \frac{\pi}{2}$, (**b**)(**f**) $\tilde{m} = 0.5$ and $\tilde{\varphi} = \pi$, (**c**)(**g**) $\tilde{m} = -0.5$ and $\tilde{\varphi} = 0$, (**d**)(**h**) $\tilde{m} = -0.5$ and $\tilde{\varphi} = \frac{\pi}{2}$. Each spin configuration is topologically nontrivial and has **a** $N_{\rm sk} = -1$, **b** $N_{\rm sk} = 2$, **c** $N_{\rm sk} = -2$, **d** $N_{\rm sk} = 1$, **e** $N_{\rm sk} = 1$, **f** $N_{\rm sk} = -2$, **g** $N_{\rm sk} = 2$, and **h** $N_{\rm sk} = -1$. The white circles in (**f**) denote the vorticity $\zeta = 2$. Other notations are common to those in Fig. 6.9. In (**b**) and (**f**), the four Dirac strings cross at the center of the MUC; see Fig. 6.8 and the text for details. Reproduced from Ref. [4]. Copyright (2022) by the American Physical Society

predominantly occupy different regions in the $\tilde{\varphi}$ parameter space, their transition may be accompanied by a phase shift. However, the phase degree of freedom was not explicitly investigated in prior studies. We will address this issue by analyzing the spin structures obtained in previous studies in Sect. 6.1.4.

6.1.4 *Numerical Analysis of Phase Shift in the Sinusoidal 3Q State*

Thus far, we elucidated the topological properties of 2D $3Q$-SkLs by systematically changing the phases of their constituent waves as well as the magnetization. In this section, we study how the actual phases of the superposed waves evolve while increasing an external magnetic field based on specific model Hamiltonians. To this end, we analyze the numerical data of the real-space spin configurations obtained for the Kondo lattice model in Ref. [3], extract the phases for two types of SkLs with $|N_{\rm sk}| = 1$ and 2, and discuss their magnetic field dependences in comparison with our result in Sect. 6.1.3. In the previous study, the ground state of the Kondo lattice model is given by Eq. (3.5) with the nearest-neighbor hopping $t_1 = 1$, the third-neighbor hopping $t_3 = -0.85$, and $J = 0.5$ was studied by using the numerical method based on the kernel polynomial method (KPM) and the Langevin dynamics (LD), which is called the modified KPM-LD method [3, 21, 22]. The ground state was obtained by minimizing the grand potential Ω given by $\Omega = \langle \mathcal{H} \rangle / N - \mu n_{\mathrm{e}}$, where N is the number of sites ($N = 96^2$), $\mu = -3.5$ is the chemical potential, and n_{e} is the electron density defined by $n_{\mathrm{e}} = \sum_{i,\sigma} \left\langle \hat{c}^{\dagger}_{i\sigma} \hat{c}_{i\sigma} \right\rangle / N$. The model was found to stabilize $3Q$ states whose wave vectors are dictated by the Fermi surfaces and given by Eq. (6.1) with $q = \frac{\pi}{3}$, exhibiting topological transitions with the chnages of the N_{sk} as $|N_{\mathrm{sk}}| = 2 \to 1 \to 0$, as shown in Fig. 3.2a. When $0 \leq h \lesssim 0.00325$, the ground state obtained by the numerical simulation is a $3Q$-SkL with $|N_{\mathrm{sk}}| = 2$ illustrated in Fig. 3.2b, which is well described by a superposition of three sinusoidal waves like in Eq. (6.26). On the other hand, it turns into a different $3Q$-SkL with $|N_{\mathrm{sk}}| = 1$ illustrated in Fig. 3.2c, similar to the typical SkL represented by the superposition of spin helices, for $h \gtrsim 0.00325$, and finally, to yet another $3Q$ state with $N_{\mathrm{sk}} = 0$ for $h \gtrsim 0.0065$.

In the following, we investigate the evolution of the moiré parameters of the $3Q$-SkL including the phases by following the procedure in Sect. 4.3. To this end, we focus on the region for $0 \leq h \leq 0.006$ since the $3Q$ state with $N_{\mathrm{sk}} = 0$ for $h \gtrsim 0.0065$ cannot be well represented by Eq. (6.26); the strong magnetic field largely modulates the spin structures, and the real-space spin configuration suggested in Ref. [3] for this state cannot be described by Eq. (6.26). First, we assume that the $3Q$-SkL for $0 \leq h \leq 0.006$ is represented by the superposition of the three sinusoidal waves given by Eq. (6.26), proposed as the ground state at zero field in Ref. [3]. By extracting the phases of the constituent sinusoidal waves, we compare the evolution of the phases and the topological properties with the phase diagram in Fig. 6.10. Then, we extend this analysis to encompass the superposition of three elliptical spirals, examining the evolution of moiré parameters accompanied by the phase shift.

Sinusoidal Case

From the spin configurations obtained in the previous study, we extract the phases of the constituent three sinusoidal waves in the two SkLs by assuming Eq. (6.26). We estimate the cost function (see Sect. 4.3)

$$U(\{\varphi_\eta\}, m, \theta, \Gamma; \hat{\mathbf{n}}, \xi) = \frac{1}{N} \sum_i \left(1 - \mathbf{S}_i^{\mathrm{KPM-LD}} \cdot \tilde{\mathbf{S}}_i^{\sin 3Q}(\{\varphi_\eta\}, m, \theta, \Gamma; \hat{\mathbf{n}}, \xi)\right), \tag{6.34}$$

where $\mathbf{S}_i^{\mathrm{KPM-LD}}$ is the spin configuration obtained by the modified KPM-LD simulation and $\tilde{\mathbf{S}}_i^{\sin 3Q}$ is that generated from Eq. (6.26) as

$$\tilde{\mathbf{S}}_i^{\mathrm{sin3Q}}(\{\varphi_\eta\}, m, \theta, \Gamma; \hat{\mathbf{n}}, \xi) = \mathsf{R}\left(\hat{\mathbf{n}}, \xi\right) \mathbf{S}_i^{\mathrm{sin3}Q}(\{\varphi_\eta\}, m, \theta, \Gamma). \tag{6.35}$$

Here, $\mathbf{S}_i^{\mathrm{sin3}Q}(\{\varphi_\eta\}, m, \theta, \Gamma)$ is given by Eq. (6.26) with $q = \frac{\pi}{3}$ and $\mathbf{r} = \mathbf{r}_i$, and $\mathsf{R}\left(\hat{\mathbf{n}}, \xi\right)$ denotes the 3D rotation matrix about the unit vector $\hat{\mathbf{n}}$ with the angle ξ. We note that the model in Eq. (3.5) does not change the energy by flipping all the y components of the localized spins, which corresponds to the change of Γ between 0 and 1 in Eq. (6.26) accompanied by the sign flip of N_{sk}; hence, the ground states with $\Gamma = 0$ and 1 are energetically degenerate. By minimizing U in Eq. (6.34) for the numerical data $\{\mathbf{S}_i^{\mathrm{KPM-LD}}\}$ at each value of h, we obtain the optimal values of the phases, φ_η^*, as well as the other parameters. For comparison with $\tilde{\varphi}$ in Sect. 6.1.3, we define the sum of φ_η^* in the form of

$$\tilde{\varphi}^* = \pi - \left| \pi - \mathrm{Mod}\left[\sum_{\eta=1}^{3} \varphi_\eta^*, 2\pi\right]\right|, \tag{6.36}$$

paying attention to the sixfold rotational symmetry of the model and the symmetry by the transformation from φ_η to $-\varphi_\eta$, in addition to the 2π periodicity.

We plot the h dependence of the optimal values of U in the upper panel of Fig. 6.12. We obtain sufficiently small values less than 4×10^{-4} for $h < 0.00325$. On the other hand, U suddenly increases to $\simeq 4 \times 10^{-2}$ at $h = 0.0035$ and shows a gradual increase while increasing h. The sudden increase of U is related with the topological phase transition at $h \simeq 0.00325$ as discussed below. The values for $h > 0.00325$ are, however, still very small, a few percent. The results indicate that the optimal spin state $\{\tilde{\mathbf{S}}_i^{\mathrm{sin3}Q}\}^*$ reproduces well $\{\mathbf{S}_i^{\mathrm{KPM-LD}}\}$ for all h.

In the lower panel of Fig. 6.12, we show the h dependences of the grand potential calculated from the numerically obtained ground state $\{\mathbf{S}_i^{\mathrm{KPM-LD}}\}$ and the optimal state $\{\tilde{\mathbf{S}}_i^{\mathrm{sin3}Q}\}^*$. We also plot the difference of the grand potential $\Delta\Omega = \Omega^{\mathrm{sin3}Q} - \Omega^{\mathrm{KPM-LD}}$. The results indicate that the optimal states $\{\tilde{\mathbf{S}}_i^{\mathrm{sin3}Q}\}^*$ almost perfectly reproduce $\{\mathbf{S}_i^{\mathrm{KPM-LD}}\}$ for $h < 0.00325$ ($\Delta\Omega < 9.0 \times 10^{-6}$, whose relative error is less than 3.2×10^{-3} %), while slight modifications could improve the results for $h > 0.00325$ ($2.0 \times 10^{-4} < \Delta\Omega < 4.0 \times 10^{-4}$, whose relative error is less than 0.14 %). We return to this point in the end of this subsection.

Figure 6.13a shows the h dependences of $\tilde{\varphi}^*$ and $|N_{\mathrm{sk}}|$ calculated from $\{\tilde{\mathbf{S}}_i^{\mathrm{sin3}Q}\}^*$. For comparison, we plot $|N_{\mathrm{sk}}|$ obtained by the modified KPM-LD simulations. For the $|N_{\mathrm{sk}}| = 2$ state at $h = 0$, we obtain $\tilde{\varphi}^* \simeq \pi$, but $\tilde{\varphi}^*$ is equivalent to $\tilde{\varphi}^* + \pi$ at $h = 0$ because of time-reversal symmetry; to show this explicitly, we plot the data points at both $\tilde{\varphi}^* \simeq 0$ and π. By increasing h, we find that $\tilde{\varphi}^* \simeq 0$ tends to be favored in the region with $|N_{\mathrm{sk}}| = 2$, whereas $\tilde{\varphi}^*$ becomes $\simeq \pi$ only at $h = 0.001$. This is presumably due to the insufficient optimization in the modified KPM-LD simulation in the small h region where the states with $\tilde{\varphi}^* = 0$ and π have a small energy difference. When h is increased above 0.003, $\tilde{\varphi}^*$ is shifted to $\sim \frac{\pi}{4}$, and becomes almost constant in the region with $|N_{\mathrm{sk}}| = 1$. This clearly shows that the topological transition with the change of $|N_{\mathrm{sk}}|$ caused by the magnetic field is

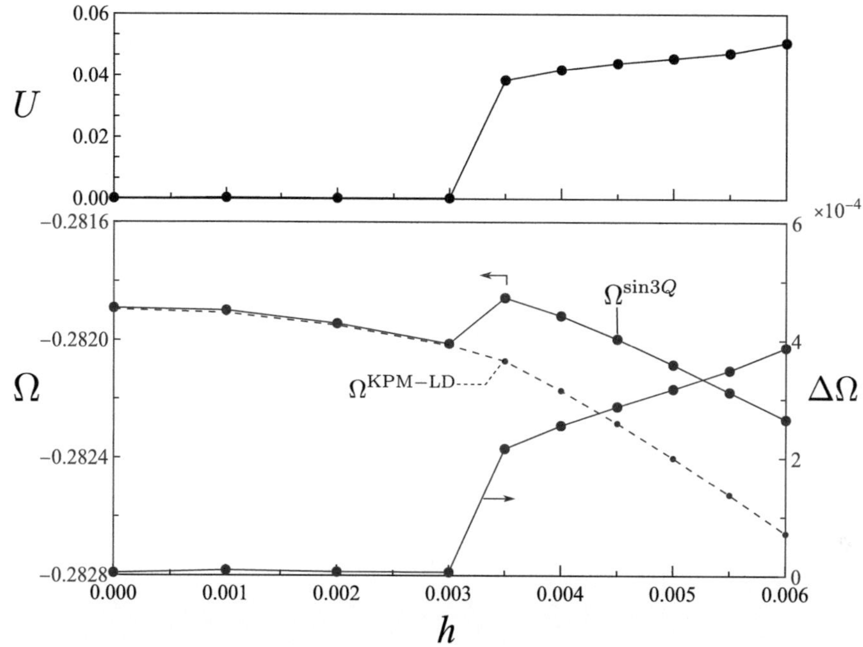

Fig. 6.12 Magnetic field dependences of (top) the cost function U in Eq. (6.34) and (bottom) the grand potential calculated from $\{\mathbf{S}_i^{\mathrm{KPM-LD}}\}$ and $\{\tilde{\mathbf{S}}_i^{\sin 3Q}\}^*$, and the difference of the grand potential, $\Delta\Omega$. Reproduced from Ref. [4]. Copyright (2022) by the American Physical Society

accompanied by the phase shift. We note that $|N_{\mathrm{sk}}|$ for the optimal spin configuration remains 2 at $h = 0.0035$ where the phase is shifted already, while $|N_{\mathrm{sk}}|$ is reduced to 1 for the modified KPM-LD result; see below.

Figure 6.13b shows the evolution of $\tilde{m}^*$ and $\tilde{\varphi}^*$ obtained for $\{\tilde{\mathbf{S}}_i^{\sin 3Q}\}^*$, plotted on the phase diagram in Fig. 6.10. Note that $\tilde{m}^*$ obtained by the fitting does not correspond to the magnetization in the simulation data; see also Eq. (6.29). The orange and green circles represent $|N_{\mathrm{sk}}| = 2$ and 1, respectively. The result indicates that $|N_{\mathrm{sk}}|$ calculated from $\{\tilde{\mathbf{S}}_i^{\sin 3Q}\}^*$ is consistent with the phase diagram shown in Fig. 6.10 and the topological transition from $|N_{\mathrm{sk}}| = 2$ to 1 is associated with the phase shift, except for $h = 0.0035$. When $h = 0.0035$, $\{\tilde{\mathbf{S}}_i^{\sin 3Q}\}^*$ gives $|N_{\mathrm{sk}}| = 2$, although the values of $\tilde{m}^*$ and $\tilde{\varphi}^*$ are in the $|N_{\mathrm{sk}}| = 1$ region, as shown in Fig. 6.13b. In Sect. 6.1.4, we will see that the discrepancy at $h = 0.0035$ is due to the insufficient approximation by Eq. (6.35). We here, however, restrict ourselves to Eq. (6.35) to discuss the phase shift within the same form of the constituent waves as in Sect. 6.1.3.

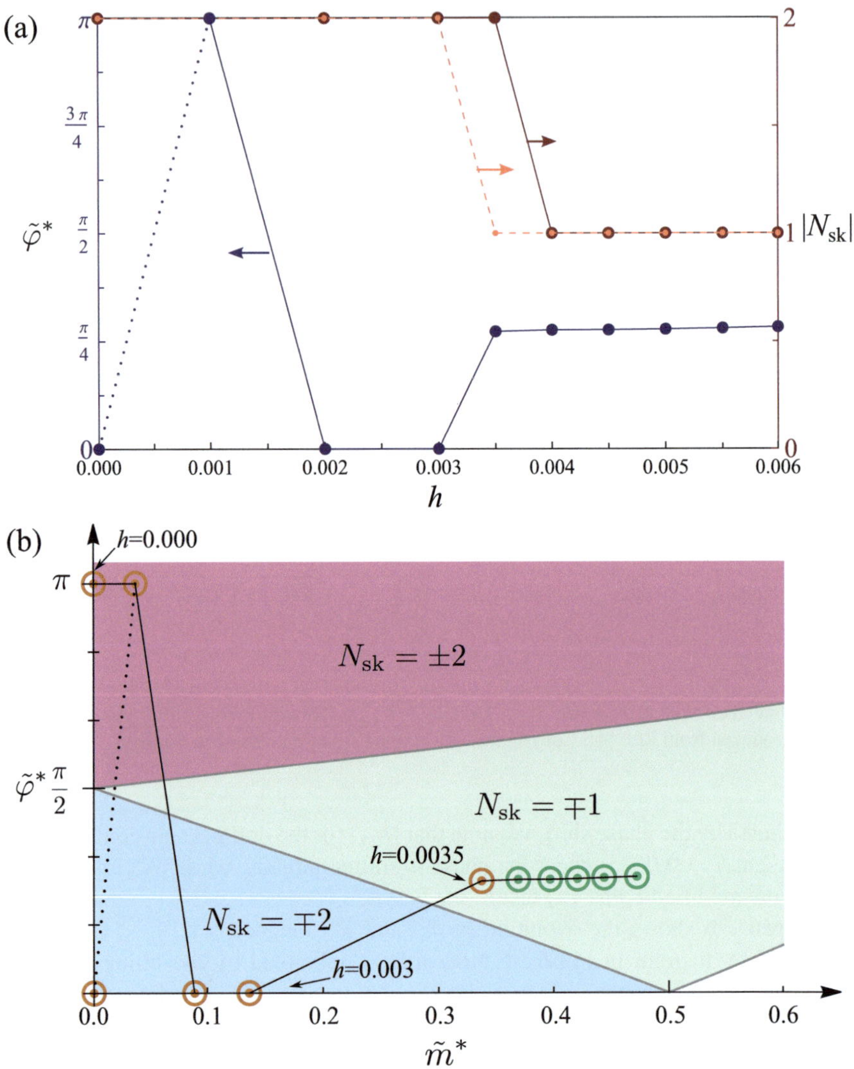

Fig. 6.13 **a** Magnetic field dependences of $\tilde{\varphi}^*$ and $|N_{\rm sk}|$ calculated from the optimal spin textures $\{\tilde{\mathbf{S}}_i^{\sin 3Q}\}^*$. For comparison, $|N_{\rm sk}|$ obtained by the modified KPM-LD simulations is plotted by the orange points connected by the dashed line. At $h = 0$, we plot the results of $\tilde{\varphi}^*$ at both $\tilde{\varphi}^* \simeq 0$ and π because of the time-reversal symmetry; the solid and dotted lines are the guides for the eye (see the text). **b** Evolution of $\tilde{m}^*$ and $\tilde{\varphi}^*$ obtained by the optimization. The background is a part of the phase diagram in Fig. 6.10. The orange and green circles represent that the data are for $|N_{\rm sk}| = 2$ and $|N_{\rm sk}| = 1$, respectively. Reproduced from Ref. [4]. Copyright (2022) by the American Physical Society

Elliptical Spiral Case

Here, we extend our analysis to the superposition of three elliptical spirals, investigating how moiré parameters other than the phase change accompanied by the phase shift. In the previous section, the cost function and the difference of the grand potential become relatively large in the SkL with $|N_{\rm sk}| = 1$ phase (see Fig. 6.12), indicating that the superposition of sinusoidal waves in Eq. (6.35) is insufficient. As the typical 3Q-SkL with $|N_{\rm sk}| = 1$ in chiral magnets is represented by the superposition of proper screws as Eq. (6.18), it is expected and indeed pointed out that both sinusoidal and cosinusoidal wave components are necessary for better representation [24]. Thus, we take into account the cosinusoidal components by considering Eq. (1.11) with $\psi^{\rm c}_\eta$, $\mathbf{e}^1_\eta$, and $\mathbf{m}$ in Eqs. (6.27)–(6.29), and

$$\psi^{\rm s}_\eta = \frac{\epsilon}{\sqrt{3}}, \tag{6.37}$$

$$\mathbf{e}^2_1 = \left(0, -(-1)^\Gamma \cos\theta, \sin\theta\right)^{\sf T}, \quad \mathbf{e}^2_2 = \mathsf{R}_\Gamma \mathbf{e}^2_1, \quad \mathbf{e}^2_3 = \mathsf{R}^2_\Gamma \mathbf{e}^2_1. \tag{6.38}$$

Here, the new parameter ϵ denotes the ellipticity of the constituent elliptical spirals, connecting the sinusoidal waves ($\epsilon = 0$) and proper screws ($\epsilon = 1$). By denoting the spin configuration generated with these parameters as $\mathbf{S}^{3Q}_i(\{\varphi_\eta\}, m, \theta, \epsilon, \Gamma)$, we estimate the cost function

$$U(\{\varphi_\eta\}, m, \theta, \epsilon, \Gamma; \hat{\mathbf{n}}, \xi) = \frac{1}{N}\sum_i \left(1 - \mathbf{S}^{\rm KPM-LD}_i \cdot \tilde{\mathbf{S}}^{3Q}_i(\{\varphi_\eta\}, m, \theta, \epsilon, \Gamma; \hat{\mathbf{n}}, \xi)\right), \tag{6.39}$$

with

$$\tilde{\mathbf{S}}^{3Q}_i(\{\varphi_\eta\}, m, \theta, \epsilon, \Gamma; \hat{\mathbf{n}}, \xi) = \mathsf{R}\left(\hat{\mathbf{n}}, \xi\right) \mathbf{S}^{3Q}_i(\{\varphi_\eta\}, m, \theta, \epsilon, \Gamma). \tag{6.40}$$

By minimizing Eq. (6.39), we obtain the optimal values of the phases, φ^*_η, as well as the other parameters.

Figure 6.14 shows h dependences of U and the difference of the grand potentials between the optimal state $\{\mathbf{S}^{3Q}_i\}^*$ and the numerically obtained ground state $\{\mathbf{S}^{\rm KPM-LD}_i\}$, $\Delta\Omega = \Omega^{3Q} - \Omega^{\rm KPM-LD}$. Surprisingly, we obtain significantly small values of $U < 0.006$ in the entire region of h, while U in the SkL with $|N_{\rm sk}| = 1$ is ~ 10 times larger in Fig. 6.12. In addition, $\Delta\Omega$ is also small enough, whose relative error is less than $\simeq 5 \times 10^{-5}$ for all h. Therefore, the ellipticity ϵ is an important parameter to precisely describe the evolution of the SkLs under the magnetic field and the optimal $\{\mathbf{S}^{3Q}_i\}^*$ gives a good approximation to $\{\mathbf{S}^{\rm KPM-LD}_i\}$. The optimal spin configuration at $h = 0.000$ and 0.005 are respectively shown in the left and right panels of Fig. 6.14, and indeed show a good agreement with the numerically obtained ones in Fig. 3.2b and c.

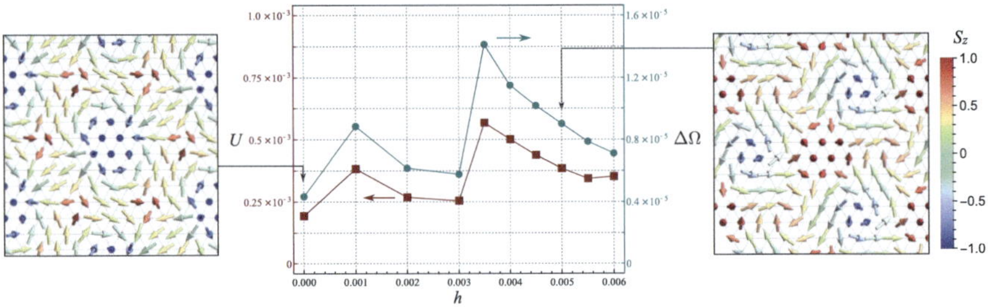

Fig. 6.14 Magnetic field dependences of U in (6.39) and the difference of the grand potentials between the optimal state $\{\mathbf{S}_i^{3Q}\}^*$ and the numerically-obtained ground state $\{\mathbf{S}_i^{\text{KPM-LD}}\}$, $\Delta\Omega$. The real-space spin configurations obtained by the optimization, $\{\mathbf{S}_i^{3Q}\}^*$, at $h = 0.000$ and 0.005 are plotted in the left and right panels, respectively. Reproduced from Ref. [23]. Copyright (2022) by the American Physical Society

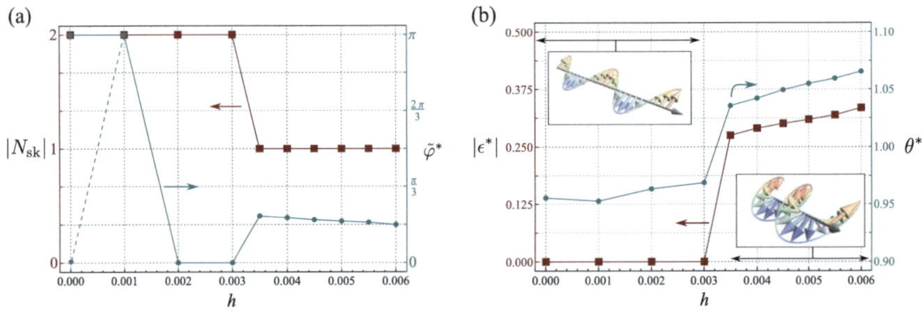

Fig. 6.15 **a** Magnetic field dependences of $|N_{\text{sk}}|$ and $\tilde{\varphi}^*$ calculated from the optimal spin textures $\{\tilde{\mathbf{S}}_i^{\sin 3Q}\}^*$. **b** Magnetic field dependences of the ellipticity ϵ^* and the angle θ^*. The insets in (**b**) represent the schematic illustrations of one of three constituent waves. Reproduced from Ref. [23]. Copyright (2022) by the American Physical Society

Figure 6.15a shows the h dependences of $|N_{\text{sk}}|$ and $\tilde{\varphi}^*$ calculated for the optimal spin configuration $\{\mathbf{S}_i^{3Q}\}^*$. The results well reproduce the topological transition at $h \simeq 0.00325$ for $\{\mathbf{S}_i^{\text{KPM-LD}}\}$ [3]. We also find that $\tilde{\varphi}^*$ exhibits a phase shift accompanied by the topological transition, similar to Sect. 6.1.4. For the SkL with $|N_{\text{sk}}| = 2$ at $h = 0$, we obtain $\tilde{\varphi}^* \simeq \pi$; $\tilde{\varphi}^*$ is equivalent to $\tilde{\varphi}^* - \pi$. While increasing h, we find that $\tilde{\varphi}^*$ remains $\simeq 0$ in the $|N_{\text{sk}}| = 2$ region for $h \lesssim 0.00325$, except for $\tilde{\varphi}^* \simeq \pi$ at $h = 0.001$ (see Sect. 6.1.4). When entering into the $|N_{\text{sk}}| = 1$ region for $h \gtrsim 0.00325$, $\tilde{\varphi}^*$ is suddenly shifted from 0 to $\sim \frac{\pi}{6}$ and gradually decreases while increasing h. We note that $\tilde{\varphi}^*$ in the $|N_{\text{sk}}| = 1$ phase is different from that in Fig. 6.15a of $\sim \frac{\pi}{4}$ since the analytical expression of the spin configuration $\tilde{\mathbf{S}}_i^{3Q}$ is refined from $\tilde{\mathbf{S}}_i^{\sin 3Q}$ through the introduction of ϵ.

Figure 6.15b shows the ellipticity ϵ^* and the angle θ^* as functions of h. In the $|N_{\text{sk}}| = 2$ state for $0 \leq h \lesssim 0.00325$, we obtain $\epsilon^* \simeq 0$ and $\theta^* \sim \arccos(1/\sqrt{3}) \simeq 0.955$ which indicates that the constituent waves are sinusoidal waves as schematically depicted in the upper-left inset and the vectors $\mathbf{e}_\eta^1$ are almost orthogonal to each other. Meanwhile, both ϵ^* and θ^* change suddenly at the topological transition at $h \simeq 0.00325$, and gradually increase while increasing h. The results indicate that the constituent waves in this region are elliptically modulated as exemplified in the lower-right inset and the angle between the major axes of the ellipses, $\mathbf{e}_\eta^1$, is larger than $\pi/2$. Thus, the topological transition from $|N_{\mathrm{sk}}| = 2$ to 1 is accompanied by not only the phase shift but also the changes in the ellipticity and the angle of the propagating planes of the constituent waves.

6.2 Concluding Remarks of this Chapter

To conclude this chapter, we discuss some implications of our results. In Sect. 6.1, we clarified the effects of phase shifts on the SkLs composed of three helices and sinusoidal waves. We unveiled the parameter regions of the phase and the uniform magnetization for the SkLs with high skyrmion number $|N_{\mathrm{sk}}| = 2$, in addition to those for the conventional ones with $|N_{\mathrm{sk}}| = 1$. Different values of N_{sk} bring about different emergent electromagnetic phenomena (see Sect. 1.3). Moreover, the dynamics of the skyrmions also shows different aspects depending on N_{sk}, as discussed in Refs. [25–29]. While it is usually difficult to directly measure the phase degree of freedom in experiments, our results indicate that the information of the phases can be obtained by such transport and optical responses. A crucial question is how to control the phase degree of freedom. In addition to the magnetic field studied in Sect. 6.1.4, it has been proposed that the phase shift can be caused by the temperature, multiple-spin interactions, and the competition between frustrated exchange interactions and itinerant frustration [2]. Meanwhile, a dynamical way of controlling the phase is yet to be unexplored. It is also intriguing to investigate the way to control the phase by using, e.g., an oscillating magnetic field and an electric current. Furthermore, in Sect. 6.1.4, we also clarified that other spin moiré parameters, such as ellipticity, exhibit changes accompanied by the phase shift by increasing the magnetic field. Even if the direct control of the phase is challenging, indirect control may be achievable if the phase interacts with other moiré parameters that are more easily tunable as different spin moiré parameters are not independent.

References

1. Kurumaji, T., Nakajima, T., Hirschberger, M., Kikkawa, A., Yamasaki, Y., Sagayama, H., Nakao, H., Taguchi, Y., Arima, T.-H., & Tokura, Y. (2019). *Science, 365*, 914.
2. Hayami, S., Okubo, T., & Motome, Y. (2021). *Nature Communications, 12*, 6927.
3. Ozawa, R., Hayami, S., & Motome, Y. (2017). *Physical Review Letters, 118*, 147205.
4. Shimizu, K., Okumura, S., Kato, Y., & Motome, Y. (2022). *Physical Review B, 105*, 224405.
5. Park, J.-H., & Han, J. H. (2011). *Physical Review B, 83*, 184406.
6. Zhang, X.-X., Mishchenko, A. S., De Filippis, G., & Nagaosa, N. (2016). *Physical Review B, 94*, 174428.

7. Kanazawa, N., Nii, Y., Zhang, X.-X., Mishchenko, A. S., De Filippis, G., Kagawa, F., Iwasa, Y., Nagaosa, N., & Tokura, Y. (2016). *Nature Communications, 7*, 11622.
8. Okumura, S., Hayami, S., Kato, Y., & Motome, Y. (2020). *Physical Review B, 101*, 144416.
9. Shimizu, K., Okumura, S., Kato, Y., & Motome, Y. (2021). *Physical Review B, 103*, 184421.
10. Tatara, G. (2019). *Physica E, 106*, 208.
11. Rajaraman, R. (1987). *Solitons and instantons: an introduction to solitons and instantons in quantum field theory*. North Holland.
12. Braun, H.-B. (2012). *Advances in Physics, 61*, 1.
13. Nagaosa, N., & Tokura, Y. (2013). *Nature Nanotechnology, 8*, 899.
14. Binz, B., & Vishwanath, A. (2006). *Physical Review B, 74*, 214408.
15. Mühlbauer, S., Binz, B., Jonietz, F., Pfleiderer, C., Rosch, A., Neubauer, A., Georgii, R., & Böni, P. (2009). *Science, 323*, 915.
16. Yu, X. Z., Onose, Y., Kanazawa, N., Park, J. H., Han, J. H., Matsui, Y., Nagaosa, N., & Tokura, Y. (2010). *Nature, 465*, 901.
17. Yu, X., Kanazawa, N., Onose, Y., Kimoto, K., Zhang, W., Ishiwata, S., Matsui, Y., & Tokura, Y. (2011). *Nature Materials, 10*, 106.
18. Han, J. H., Zang, J., Yang, Z., Park, J.-H., & Nagaosa, N. (2010). *Physical Review B, 82*, 094429.
19. Buhrandt, S., & Fritz, L. (2013). *Physical Review B, 88*, 195137.
20. Hayami, S., Ozawa, R., & Motome, Y. (2017). *Physical Review B, 95*, 224424.
21. Barros, K., & Kato, Y. (2013). *Physical Review B, 88*, 235101.
22. Ozawa, R., Hayami, S., Barros, K., & Motome, Y. (2017). *Physical Review B, 96*, 094417.
23. Shimizu, K., Okumura, S., Kato, Y., & Motome, Y. (2022). *Journal of Physics: Conference Series, 2164*, 012069.
24. Hayami, S., & Yambe, R. (2021). *Physical Review Research, 3*, 043158.
25. Thiele, A. A. (1973). *Physical Review Letters, 30*, 230.
26. Everschor, K., Garst, M., Duine, R. A., & Rosch, A. (2011). *Physical Review B, 84*, 064401.
27. Schulz, T., Ritz, R., Bauer, A., Halder, M., Wagner, M., Franz, C., Pfleiderer, C., Everschor, K., Garst, M., & Rosch, A. (2012). *Nature Physics, 8*, 301.
28. Seki, S., & Mochizuki, M. (2016). *Skyrmions in Magnetic Materials SpringerBriefs in Physics*. Cham: Springer.
29. Zhang, X., Xia, J., Zhou, Y., Liu, X., Zhang, H., & Ezawa, M. (2017). *Nature Communications, 8*, 1717.

Chapter 7
Spin Moiré Engineering in Three Dimensions

In this chapter, we investigate the spin moiré engineering in three dimensions. In Sect. 7.1, we consider the amplitude of the constituent waves as a parameter of controlling spin moirés. By using variational calculations, we demonstrate that the isotropic $3Q$- and $4Q$-HLs transform into anisotropic ones with variations of amplitudes depending on the constituent waves by introducing anisotropy. Furthermore, we clarify that such variations can lead to the topological transition associated with the creation and annihilation of hedgehogs and antihedgehogs. In Sect. 7.2, we consider the twist angle between the propagation directions of superposed waves as a parameter to modulate the spin moiré. For the $3Q$-HL, we systematically clarify how the spin textures, topological properties, and the EEMF are modulated by changing the twist angle and the magnetization. We also demonstrate such a modulation by using the variational calculation introduced in Sect. 4.1 for the phenomenological model in Eq. (3.1). In Sect. 7.3, we clarify the effect of another parameter, phase shift, on the two types of $4Q$-HLs composed of the helices and sinusoidal waves by using the hyperspace representation given in Sect. 4.6, similar to the 2D case for $3Q$-SkLs in Sect. 6.1. Furthermore, we also analyze the previous numerical data for the model Hamiltonian in Eq. (3.26) by using the technique in Sect. 4.3 and study the evolution of the phases under the magnetic field. In Sect. 7.4, we study the current-induced dynamics for the $3Q$-HL through numerical simulations based on the LLG equation. We clarify the velocity of the spin texture by changing the amplitude and direction of the electric current. Section 7.5 is devoted to the concluding remarks.

Supplementary Information The online version contains supplementary material available at https://doi.org/10.1007/978-981-96-5992-0_7.

K. Shimizu, *Spin Moiré Engineering and Emergent Electromagnetism in Topological Spin Crystals*, Springer Theses, https://doi.org/10.1007/978-981-96-5992-0_7

7.1 Amplitude Control in the 3Q and 4Q-Hedgehog Lattices

In this chapter, we investigate the effect of spatial anisotropy on spin moirés in chiral magnetic metals. Figure 7.1 showcases several spin textures with different dimensionality: a 1D helical structure in Fig. 7.1a, 2D 2Q-VC and 3Q-SkL in Fig. 7.1b and c, respectively, and two different types of 3D 3Q-and 4Q-HLs in Fig. 7.1d and e, respectively. The wave vectors of the constituent helices are shown in each inset. Figure 7.1b–e are spin moirés composed of more than one helices. They have different topological properties as well: The 3Q-SkL in Fig. 7.1c is a periodic array of the magnetic skyrmions characterized by the skyrmion number of $N_{\rm sk} = -1$ [Eq. (1.5)], and the HLs in Fig. 7.1d and e are those of the magnetic monopoles and antimonopoles characterized by the monopole charges of $Q_{\rm m} = \pm 1$ [Eq. (7.6)], while the 2Q-VC in Fig. 7.1b is topologically trivial (see Sect. 1.2.2). The question we address in this chapter is how to control the spin moirés with different dimensionality and topology. Once one can switch them from one to another, it may be able to manipulate not only the magnetism but also the electronic state, transport, and optical responses. In this chapter, we systematically study such a control of the spin moirés, focusing on the effect of spatial anisotropy in the magnetic interactions. Based on an effective spin model with long-range interactions stemming from itinerant nature of electrons, we clarify the ground-state phase diagram by variational calculations while changing the anisotropy and discuss various phase transitions in not only the magnetism but also the topological properties.

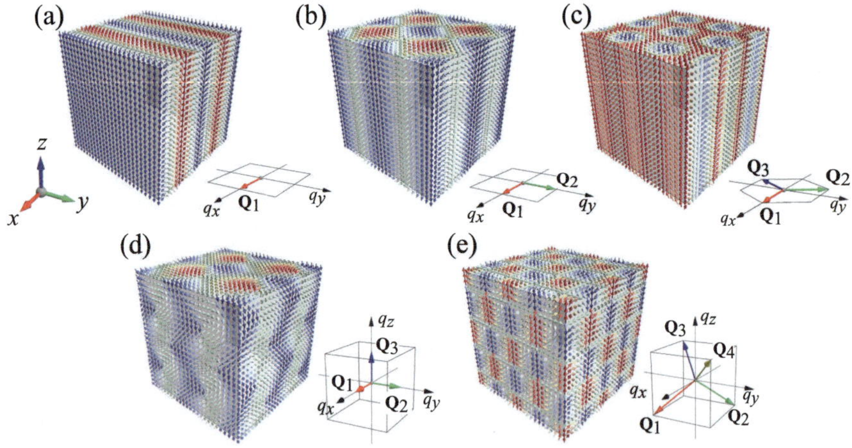

Fig. 7.1 Real-space spin configurations of (**a**) a helical state (1Q-H), (**b**) a vortex crystal composed of two helices (2Q-VC), (**c**) a skyrmion lattice composed of three (3Q-SkL), (**d**) a hedgehog lattice composed of three (3Q-HL) and (**e**) another hedgehog lattice composed of four (4Q-HL). The arrows represent the spins and their colors display the z component. Each inset shows the wave vectors of the constituent helices

The organization of this chapter is as follows. In Sect. 7.1.1, we describe the details of the variational calculation and introduce physical quantities characterizing magnetic and topological properties. In Sect. 7.1.2, we show the phase diagrams while changing the anisotropy. Next, in Sect. 7.1.3, we show how the magnetic and topological properties change through the phase transitions caused by changing the anisotropy. Finally, in Sect. 7.1.4, we show real-space distributions of hedgehogs and antihedgehogs, and the flow of the emergent magnetic field.

7.1.1 Identification of the Ground States

Variational States

We study the ground state of the model in Eq. (3.27) by a variational method described in Sect. 4.1. Specifically, we extend the variational states used in the previous study [13] to the anisotropic cases as follows. We consider the following three types of spin textures as the variational states. The first one is a superposition of sinusoidal waves given by

$$\mathbf{S}_{\mathbf{r}_l} \propto \sum_{\eta=1}^{n} a_\eta \mathbf{e}_\eta^0 \cos \mathcal{Q}_{\eta l}, \tag{7.1}$$

where $\mathbf{e}_\eta^0$ is the unit vector satisfying $\mathbf{e}_\eta^0 = \mathbf{Q}_\eta / |\mathbf{Q}_\eta|$, and $\mathcal{Q}_{\eta l} = \mathbf{Q}_\eta \cdot \mathbf{r}_l + \varphi_\eta$; $n = 3$ and 4 for the $3Q$ and $4Q$ cases, respectively. a_η and φ_η denote the amplitude and phase of the $\mathbf{Q}_\eta$ component, respectively. As this is the nonchiral state that has no energy gain from the DM-type interaction in Eq. (3.27), we call it the mQ nonchiral state (mQ-NC), where $m = 1, 2, \cdots, n$ is the number of the components with nonzero a_η in Eq. (7.1). The second variational state is given by

$$\mathbf{S}_{\mathbf{r}_l} \propto \sum_{\eta=1}^{n} a_\eta \left(\mathbf{e}_\eta^1 \cos \mathcal{Q}_{\eta l} + \mathbf{e}_\eta^2 \sin \mathcal{Q}_{\eta l} \right), \tag{7.2}$$

where $\mathbf{e}_\eta^1$ and $\mathbf{e}_\eta^2$ are the unit vectors which satisfy $\mathbf{e}_\eta^1 \times \mathbf{e}_\eta^2 = \mathbf{e}_\eta^0$; $n = 3$ and 4 for the $3Q$ and $4Q$ cases, respectively, as in Eq. (7.1). Here, we set $\mathbf{e}_1^1 = \hat{\mathbf{y}}$, $\mathbf{e}_2^1 = \hat{\mathbf{z}}$, and $\mathbf{e}_3^1 = \hat{\mathbf{x}}$ for the $3Q$ case, and $\mathbf{e}_\eta^1 = \hat{\mathbf{z}} \times \mathbf{e}_\eta^0 / |\hat{\mathbf{z}} \times \mathbf{e}_\eta^0|$ for the $4Q$ case, where $\hat{\mathbf{x}}$, $\hat{\mathbf{y}}$, and $\hat{\mathbf{z}}$ are the unit vectors in the xyz coordinates. This is a chiral multiple-Q state, which has an energy gain from the DM-type interaction in contrast to the nonchiral one in Eq. (7.1). We call it the $1Q$ helical state ($1Q$-H) when only one of a_η is nonzero, while the $2Q$ vortex crystal ($2Q$-VC) when two of them are nonzero. Meanwhile, for the states with more than two nonzero components, we have topologically different states depending on the existence of hedgehogs and antihedgehogs (see below); we call the states with hedgehogs and antihedgehogs the $3Q$ and $4Q$ hedgehog lattices ($3Q$-

and $4Q$-HLs), while otherwise simply the $3Q$ and $4Q$ states. The third variational state is called the double-Q chiral stripe ($2Q$-CS) found in the previous study [15], which is given by

$$\mathbf{S}_{\mathbf{r}_l} \propto \sqrt{1-u_{\eta' l}^2}\left(\mathbf{e}_\eta^1 \cos \mathcal{Q}_{\eta l}+\mathbf{e}_\eta^2 \sin \mathcal{Q}_{\eta l}\right)+u_{\eta' l} \mathbf{e}_{\eta'}^0, \tag{7.3}$$

where $u_{\eta' l} = v \sin \mathcal{Q}_{\eta' l}$; v denotes the amplitude of the sinusoidal wave, and η and η' take two of three and four components in the $3Q$ and $4Q$ cases, respectively.

In the variational calculations, a_η, φ_η, and v are treated as variational parameters that are optimized to find the lowest-energy state. a_η and v are varied from 0 to 1, while φ_η from 0 to Q for the $3Q$ case and 0 to 2π for the $4Q$ case.[1] We prepare each variational state with the normalization of the spin length at every site as $|\mathbf{S}_{\mathbf{r}_l}| = 1$, optimize the energy with respect to the variational parameters, and then compare the energies among different variational states. Note that the optimized energy is unchanged by exchanging η in the superposition, except for the $2Q$-CS state including the helix running along the anisotropic axis in η or η'. In the following analysis, we impose the following constraints on the variational parameters a_η from the symmetry point of view: $a_1 = a_2$ for all the multiple-Q states in the $3Q$ case, and $a_1 = a_2$ (or equivalently, $a_2 = a_3$ or $a_3 = a_1$) for the $2Q$-NC and VC in the $4Q$ case. In addition, we set $a_1 = a_2 = a_3$ for all the $3Q$ and $4Q$ states including the HLs for the $4Q$ case. In all cases, $\sum_\eta a_\eta^2 = 1$. We perform the variational calculations on the cubic lattice with $N = 24^3$ sites under the periodic boundary conditions in all three directions. Note that there is no finite-size effects as $Q = \pi/12$ induces magnetic orders with period of 24 sites.

Physical Quantities

In order to identify the magnetic and topological properties of the optimized states obtained by the variational calculations, we calculate the following quantities. First, we calculate the magnetic moment with wave vector $\mathbf{Q}_\eta$,[2]

$$m_{\mathbf{Q}_\eta} = \sqrt{\frac{S(\mathbf{Q}_\eta)+S(-\mathbf{Q}_\eta)}{N}}, \tag{7.4}$$

---

[1] We note that the range of φ_η was taken from 0 to Q in Ref. [13], which is not sufficient to cover all possible textures in the $4Q$ case. We find, however, that the present calculations for 0 to 2π do not alter the phase diagrams from the previous ones; see Fig. 7.3b and e. The effect on the topological properties from the phase shift for the $4Q$ case will be discussed in Sect. 7.3.

[2] We adopt a slightly different definition for $m_{\mathbf{Q}_\eta}$ from that in Ref. [13]; we take a symmetric summation with respect to $\pm\mathbf{Q}_\eta$, which results in just a difference of factor $\sqrt{2}$.

where $S(\mathbf{q})$ is the spin structure factor defined by

$$S(\mathbf{q}) = \frac{1}{N}\sum_{l,l'} \mathbf{S}_{\mathbf{r}_l} \cdot \mathbf{S}_{\mathbf{r}_{l'}} e^{-i\mathbf{q}\cdot(\mathbf{r}_l - \mathbf{r}_{l'})}. \tag{7.5}$$

The topological properties of the 3D spin moirés are characterized by the hedgehogs and antihedgehogs introduced in Sect. 1.1.3. To identify the topological aspect of the spin structure, we also calculate the monopole charge Q_{m} in Eq. (1.7), which is defined for each unit cube composed of the neighboring eight sites on the cubic lattice as [13, 14]

$$Q_{\mathrm{m}}(\mathbf{R}_j) = \frac{1}{4\pi} \sum_{p \in j\text{th unit cube}} \Omega_p, \tag{7.6}$$

where $\mathbf{R}_j$ is the center position of jth unit cube and Ω_p is the solid angle spanned by four spins on the pth square plaquette surrounding the unit cube ($p = 1, 2, \ldots, 6$). Ω_p is calculated by dividing the plaquette into two triangles $\triangle_i$ ($i = 1, 2$) and taking the sum of the solid angles of three spins in the triangles as $\Omega_p = \Omega_{\triangle_1} + \Omega_{\triangle_2}$, where

$$\Omega_{\triangle_i} = 2\tan^{-1}\left[\frac{\mathbf{S}_1 \cdot (\mathbf{S}_2 \times \mathbf{S}_3)}{1 + \mathbf{S}_1 \cdot \mathbf{S}_2 + \mathbf{S}_2 \cdot \mathbf{S}_3 + \mathbf{S}_3 \cdot \mathbf{S}_1}\right]. \tag{7.7}$$

Here, $\mathbf{S}_1$, $\mathbf{S}_2$, and $\mathbf{S}_3$ are three spins on the triangle $\triangle_i$ in the clockwise order viewed from the center of the cube. The argument in $\tan^{-1}$ is taken in the range of $[-\pi, \pi)$. The schematic picture for calculating Ω_p is shown in Fig. 7.2a. The monopole charge in Eq. (7.6) signals the position of hedgehogs and antihedgehogs by $Q_{\mathrm{m}}(\mathbf{R}_j) = +1$ and -1, respectively. Typical spin configurations with nonzero $Q_{\mathrm{m}}(\mathbf{R}_j)$ are

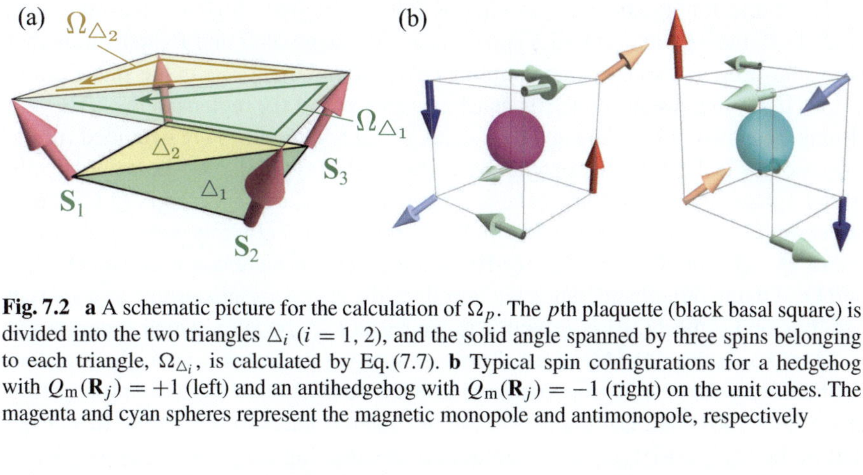

Fig. 7.2 **a** A schematic picture for the calculation of Ω_p. The pth plaquette (black basal square) is divided into the two triangles $\triangle_i$ ($i = 1, 2$), and the solid angle spanned by three spins belonging to each triangle, $\Omega_{\triangle_i}$, is calculated by Eq. (7.7). **b** Typical spin configurations for a hedgehog with $Q_{\mathrm{m}}(\mathbf{R}_j) = +1$ (left) and an antihedgehog with $Q_{\mathrm{m}}(\mathbf{R}_j) = -1$ (right) on the unit cubes. The magenta and cyan spheres represent the magnetic monopole and antimonopole, respectively

exemplified in Fig. 7.2b. We also compute the total number of hedgehogs and antihedgehogs in the MUC by

$$N_{\mathrm{m}} = \sum_{j \in \mathrm{MUC}} |Q_{\mathrm{m}}(\mathbf{R}_j)|. \tag{7.8}$$

We also calculate the emergent magnetic field $\mathbf{b}_{\mathbf{r}_l} = (b^x_{\mathbf{r}_l}, b^y_{\mathbf{r}_l}, b^z_{\mathbf{r}_l})$ at each site, whose component is defined by the summation of the spin triple products on the adjacent four triangles as [13, 14]

$$b^{\alpha}_{\mathbf{r}_l} = \frac{1}{4} \sum_{\beta\gamma\nu_\beta\nu_\gamma} \varepsilon^{\alpha\beta\gamma} \nu_\beta \nu_\gamma \mathbf{S}_{\mathbf{r}_l} \cdot (\mathbf{S}_{\mathbf{r}_l+\nu_\beta \mathbf{e}_\beta} \times \mathbf{S}_{\mathbf{r}_l+\nu_\gamma \mathbf{e}_\gamma}). \tag{7.9}$$

Here, β and γ are the directions perpendicular to α, and $\nu_{\beta(\gamma)} = \pm 1$. Real-space mapping of the positions of the monopoles and antimonopoles with $\mathbf{b}_{\mathbf{r}_l}$ manifests the flow of the emergent magnetic field; see Sect. 7.1.4.

7.1.2 Phase Diagram

We show the phase diagrams obtained by the variational calculations for the model in Eq. (3.27) for the $3Q$ case in Fig. 7.3a–c and for the $4Q$ case in Fig. 7.3d–f: (a) and (d) are for the hard-axis anisotropy with $J'/J = 0.8$, (b) and (e) are for the isotropic cases with $J'/J = 1.0$, and (c) and (f) are for the easy-axis anisotropy with $J/J' = 0.8$. We plot the results for D and K when $J'/J \leq 1.0$, while for D' and K' when $J'/J > 1.0$, with the relations in Eq. (3.30). Note that the $2Q$-VC, $2Q$-CS, $3Q$-HL, and $3Q$ are different states between the $3Q$ and $4Q$ cases due to the different sets of $\mathbf{Q}_\eta$.

The results for the isotropic cases in Fig. 7.3b and e agree with those in the previous study [13]; the $3Q$- and $4Q$-HLs appear in a wide range of D and K, indicating that both spin-orbit and spin-charge couplings play an important role in the stabilization of the HLs. When we introduce the hard-axis anisotropy, the regions of these HLs are shrunk as shown in Fig. 7.3a and d. Instead, the $2Q$-VC and CS are extended for the $3Q$ case, while the $2Q$-CS is extended and the $2Q$-VC, $3Q$-HL, and the topologically trivial $3Q$ are induced for the $4Q$ case; in the $3Q$ case, the $2Q$-VC and $2Q$-CS states are composed of $\mathbf{Q}_1$ and $\mathbf{Q}_2$, while in the $4Q$ case, the $2Q$-CS is composed of two out of $\mathbf{Q}_1$, $\mathbf{Q}_2$, and $\mathbf{Q}_3$, and the $3Q$-HL and the $3Q$ states are composed of $\mathbf{Q}_1$, $\mathbf{Q}_2$, and $\mathbf{Q}_3$. On the other hand, when we introduce the easy-axis anisotropy, as shown in Fig. 7.3c and f, the $1Q$-H is extended to the regions of the HLs, which is composed of $\mathbf{Q}_3$ ($\mathbf{Q}_4$) in the $3Q$ ($4Q$) case. These results are understood from the fact that the hard(easy)-axis anisotropy tends to destabilize (stabilize) the helix propagating along $\mathbf{Q}_3 \parallel [001]$ and $\mathbf{Q}_4 \parallel [111]$ for the $3Q$ and $4Q$ cases, respectively. Besides, a part of the $3Q$($4Q$)-HL region is replaced by the topologically trivial $3Q$($4Q$) state; see Sect. 7.1.3.

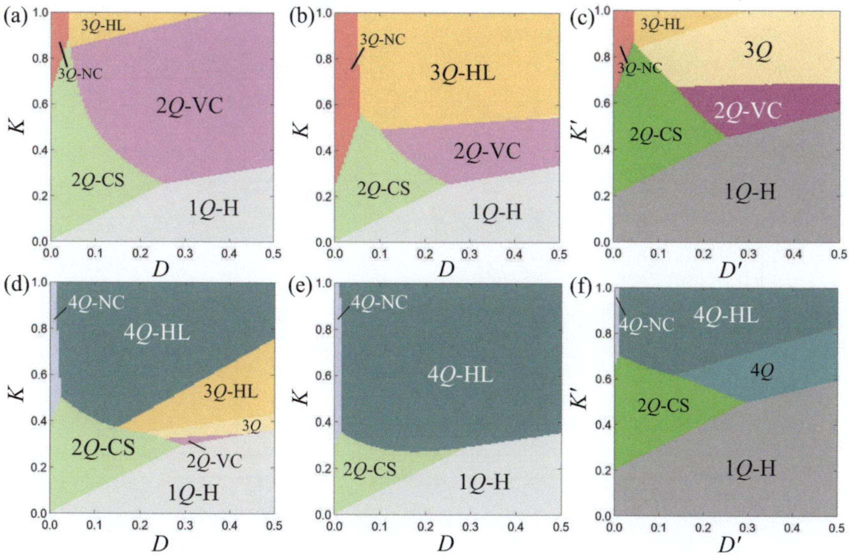

Fig. 7.3 Phase diagrams of the model in Eq. (3.27) obtained by the variational calculations for (**a**)–(**c**) the 3Q case and (**d**)–(**f**) the 4Q case: (**a**), (**d**) $J'/J = 0.8$ (hard-axis), (**b**), (**e**) $J'/J = 1.0$ (isotropic), and (**c**), (**f**) $J/J' = 0.8$ (easy-axis). 1Q-H, 2Q-VC, 2Q-CS, 3Q(4Q)-HL, 3Q(4Q), and 3Q(4Q)-NC represent the 1Q helical, the 2Q vortex crystal, the 2Q chiral stripe, the 3Q(4Q) hedgehog lattice, the topologically trivial 3Q(4Q), and the 3Q(4Q) nonchiral states, respectively. The light and dark colors for 2Q-VC, 2Q-CS, and 1Q-H denote the differences in the constituent wave vectors. Reproduced from Ref. [17]. Copyright (2021) by the American Physical Society

7.1.3 *Phase Transitions Driven by Anisotropy*

The results in Fig. 7.3 indicate that the system exhibits phase transitions from the isotropic HLs to other magnetic states by introducing the anisotropy. We discuss such phase transitions in detail, for $D = 0.3$ and $K = 0.8$ where the 3Q- and 4Q-HLs are stable in the isotropic 3Q and 4Q cases, respectively. Figures 7.4 and 7.5 show the results for the 3Q and 4Q cases, respectively: anisotropy dependences of the optimized energy per site, E, the variational parameters a_3 and a_4, the magnetic moments $m_{\mathbf{Q}_\eta}$ in Eq. (7.4), and the total number of hedgehogs and antihedgehogs in the MUC, N_{m}, in Eq. (7.8).

Let us first discuss the 3Q case in Fig. 7.4. In this case, we identify five different phases from the energy comparison and the topological nature. The energy is plotted in the upper panel of Fig. 7.4 for the 1Q-H state with $a_3 = 1$ (gray), 2Q-VC with $a_3 = 0$ (light pink), 2Q-VC with $a_1 = 0$ (or $a_2 = 0$) (dark pink), and 3Q (orange). The values of a_3 and $m_{\mathbf{Q}_\eta}$ for the lowest-energy state are shown in the middle and lower panels, respectively. In the region where the 3Q variational state has the lowest energy, we distinguish two topologically different phases by N_{m} plotted in the lower panel. We describe the details below.

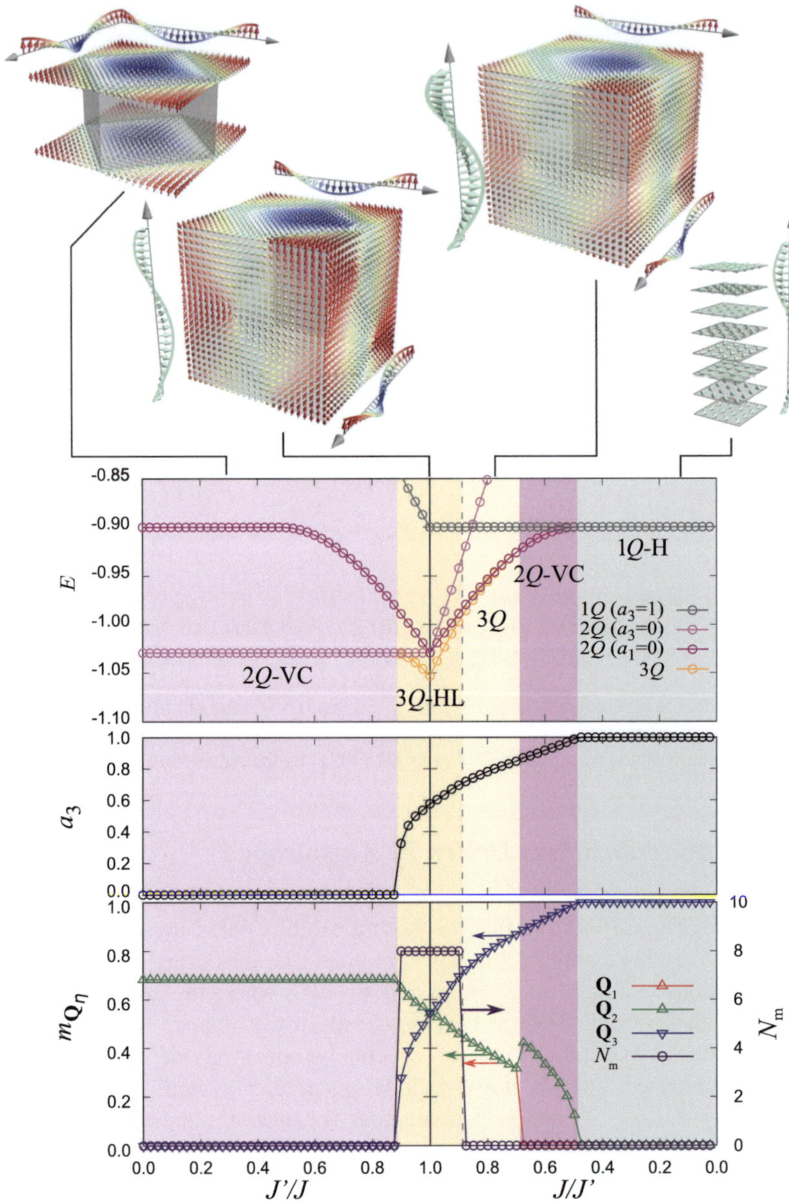

Fig. 7.4 Anisotropy dependences of (top) the real-space spin configuration of the lowest energy state with its constituent helices and the variational energy per site for each state, E, (middle) the amplitude of the helix in the J' direction, a_3, and (bottom) the magnetic moment with the wave vector $\mathbf{Q}_\eta$, $m_{\mathbf{Q}_\eta}$ [Eq. (7.4)], and the number of hedgehog and antihedgehog, N_m [Eq. (7.8)], for the $3Q$ case. The dashed lines represent the topological transitions with the changes of N_m. The results are calculated for $D = 0.3$ and $K = 0.8$ in Eq. (3.27). We take $J(J')$ as the energy unit for $J'/J < 1(> 1)$. Reproduced from Ref. [17]. Copyright (2021) by the American Physical Society

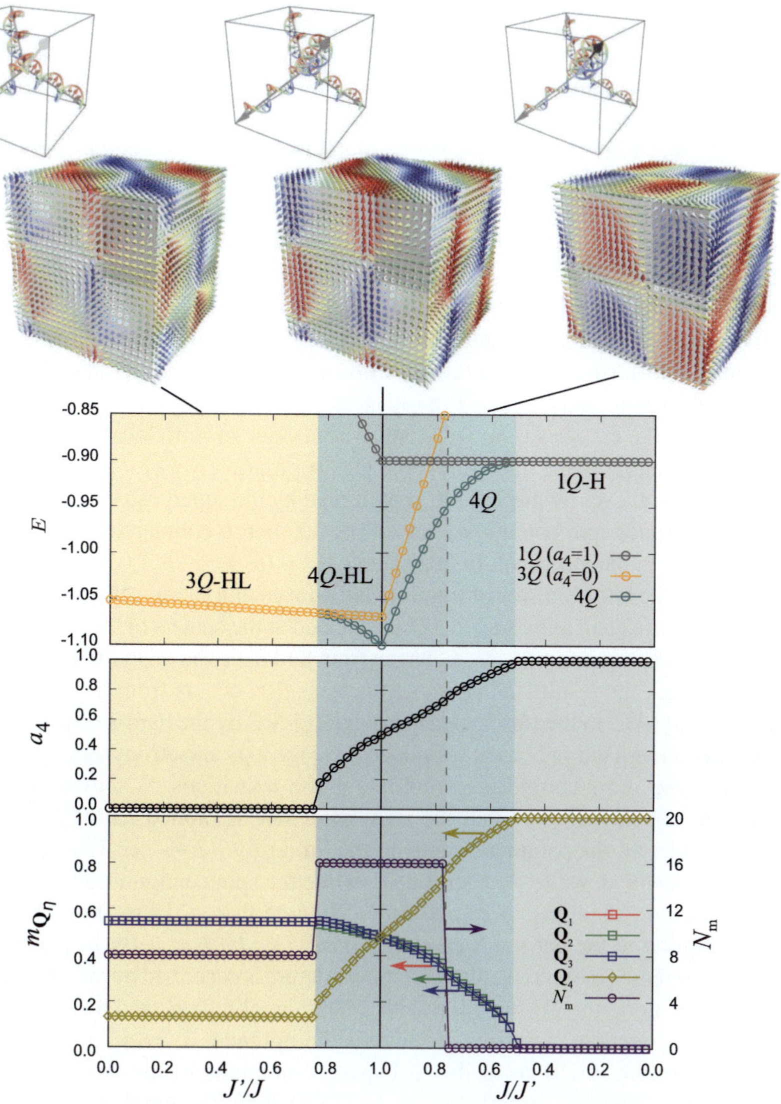

Fig. 7.5 Similar plots for the 4Q case. The notations and plotted quantities are common to those in Fig. 7.4: the variational energy per site for each state, E, the amplitude of the helix in the J' direction, a_4, the magnetic moment with the wave vector $\mathbf{Q}_\eta$, $m_{\mathbf{Q}_\eta}$ [Eq. (7.4)], and the number of hedgehog and antihedgehog, N_m [Eq. (7.8)]. Reproduced from Ref. [17]. Copyright (2021) by the American Physical Society

In the isotropic case, the system exhibits the $3Q$-HL with the same amplitude as indicated by $a_3 = 1/\sqrt{3}$ ($= a_1 = a_2$) and $m_{\mathbf{Q}_1} = m_{\mathbf{Q}_2} = m_{\mathbf{Q}_3}$ as plotted in the middle and lower panels of Fig. 7.4, respectively. The corresponding real-space spin configuration and the schematic illustration of the superposition of three helices with the same amplitudes are displayed in the top panel. This state is topologically nontrivial as it includes four pairs of hedgehogs and antihedgehogs in the MUC, i.e., $N_{\mathrm{m}} = 8$, as plotted in the lower panel of Fig. 7.4. We find that the phases of the three ordering vectors are $\varphi_1 = \varphi_2 = \varphi_3 = Q/2$, which indicates that the hedgehogs and antihedgehogs are located at the centers of the cubes so as to avoid the lattice sites [13].

When we introduce the hard-axis anisotropy by taking $J'/J < 1$, the $3Q$-HL remains stable down to $J'/J \simeq 0.8875$ while the spin texture is modulated by reflecting the anisotropy as $a_3 < 1/\sqrt{3}$ ($a_1 = a_2 > 1/\sqrt{3}$) and $m_{\mathbf{Q}_1} = m_{\mathbf{Q}_2} > m_{\mathbf{Q}_3}$. In the anisotropic HL, φ_η takes 0 or $Q/2$, which is chosen so as to locate the hedgehogs and antihedgehogs at the interstitial positions appropriately. For $J'/J \lesssim 0.8875$, the $3Q$-HL is taken over by the $2Q$-VC represented by the superposition of two helices, whose real-space spin configuration exhibits the meron-antimeron lattice structure, as illustrated in the top panel. In this state, $a_3 = 0$ ($a_1 = a_2 = 1/\sqrt{2}$) and $m_{\mathbf{Q}_3} = 0$, as shown in the left hand side of Fig. 7.4, and both φ_1 and φ_2 are optimized to be $Q/2$ so that the vortex and antivortex cores are located at the centers of the square plaquettes. At this phase transition, N_{m} changes from 8 to 0, as shown in the lower panel of Fig. 7.4. Thus, the results indicate that the transition occurs from the topologically nontrivial $3Q$-HL to the topologically trivial $2Q$-VC by the hard-axis anisotropy.

On the other hand, when we introduce the easy-axis anisotropy as $J'/J > 1$, the system shows more complicated multiple phase transitions, as shown in the right hand side of Fig. 7.4. Note that the right hand side is plotted for J/J' instead of J'/J to cover all the parameter range to the limit of $J'/J \to \infty$. The $3Q$-HL with $N_{\mathrm{m}} = 8$ is stable down to $J/J' \simeq 0.8875$ while the spin configuration is modulated as $a_3 > a_1 = a_2$ and $m_{\mathbf{Q}_3} > m_{\mathbf{Q}_1} = m_{\mathbf{Q}_2}$. We note that φ_η takes 0 or $Q/2$ in this region, similar to the hard-axis case with $0.8875 \lesssim J'/J < 1$. By further reducing J/J', a_3 increases monotonically. The spin structure is described by the superposition of the helix along the $\mathbf{Q}_3$ direction with a large amplitude a_3 and helices along the $\mathbf{Q}_1$ and $\mathbf{Q}_2$ directions with smaller but equal amplitudes $a_1 = a_2$, as shown in the top panel. While variations of the real-space spin configuration from the isotropic case are less apparent, remarkably, at $J/J' \simeq 0.8875$, the topologically nontrivial $3Q$-HL changes into the topologically trivial $3Q$ state with $N_{\mathrm{m}} = 0$. In the $3Q$ state, while φ_η takes 0 or $Q/2$ for $0.7875 \lesssim J/J' \lesssim 0.8875$ similar to the $3Q$-HL, all φ_η become zero for $0.6875 \lesssim J/J' \lesssim 0.7875$; the origin of the phase shifts is unclear. With a further decrease of J/J' to $\simeq 0.6875$, the $3Q$ state is taken over by the $2Q$-VC with $a_1 = 0$ and $a_3 > a_2 > 0$, and $m_{\mathbf{Q}_1} = 0$ and $m_{\mathbf{Q}_3} > m_{\mathbf{Q}_2} > 0$, as shown in the middle and lower panels of Fig. 7.4. This state is different from the $2Q$-VC stable in the hard-axis case for $J'/J \lesssim 0.8875$ as it is composed of two helices with different amplitude and different wave vectors, and breaks fourfold rotational symmetry spontaneously. We note that φ_η varies with J'/J in this state. Finally, for

$J/J' \lesssim 0.4875$, the $2Q$-VC is replaced by the $1Q$-H with $a_3 = 1$ and $a_1 = a_2 = 0$, and $m_{\mathbf{Q}_3} = 1$ and $m_{\mathbf{Q}_1} = m_{\mathbf{Q}_2} = 0$, which is stable to the easy-axis limit, $J/J' \to 0$.

Next, we discuss the $4Q$ case in Fig. 7.5. In this case, we identify four phases. The energy is plotted in the top panel of Fig. 7.5 for $1Q$-H with $a_4 = 1$ (gray), $3Q$ with $a_4 = 0$ (orange), and $4Q$ (green). As in the $3Q$ case above, we obtain two topologically different phases in the $4Q$ region from the values of $N_{\rm m}$ plotted in the lower panel of Fig. 7.5, as described below.

In the isotropic case, the system exhibits the $4Q$-HL with equal amplitude of the four helices, i.e., $a_4 = 1/2$ $(= a_1 = a_2 = a_3)$ and $m_{\mathbf{Q}_1} = m_{\mathbf{Q}_2} = m_{\mathbf{Q}_3} = m_{\mathbf{Q}_4}$; the corresponding real-space spin configuration and the constituent four helices are depicted in the top panel. This state is topologically nontrivial as it includes eight pairs of hedgehogs and antihedgehogs, i.e., $N_{\rm m} = 16$, as shown in the lower panel of Fig. 7.5. We find that φ_η satisfy the relation $\tilde{\varphi} = \mathrm{Mod}\left[\sum_{\eta=1}^{4} \varphi_\eta, 2\pi\right] = \pi$ to place the hedgehogs and antihedgehogs at the centers of the cubes [13]. We note that the spin configurations with $\tilde{\varphi} = \frac{\pi}{3}$ and $\frac{5\pi}{3}$ should give the same energy as these three configurations are transformed each other by threefold rotation operation about the [111] axis (see Sect. 7.3.2). When we introduce the hard-axis anisotropy by taking $J'/J < 1$, the $4Q$-HL remains stable down to $J'/J \simeq 0.7625$, while the spin texture is modulated as $a_4 < 1/2$ $(a_1 = a_2 = a_3 > 1/2)$ and $0 < m_{\mathbf{Q}_4} < m_{\mathbf{Q}_1} \simeq m_{\mathbf{Q}_2} \simeq m_{\mathbf{Q}_3}$.[3] For $J'/J \lesssim 0.7625$, the $4Q$-HL is taken over by the $3Q$-HL with $a_1 = a_2 = a_3 = 1/\sqrt{3}$, $a_4 = 0$, and $m_{\mathbf{Q}_1} = m_{\mathbf{Q}_2} = m_{\mathbf{Q}_3} > m_{\mathbf{Q}_4}$, as shown in the left hand side of Fig. 7.5. This $3Q$-HL is different from one in the $3Q$ model since the constituent helices are propagating in the tetrahedral directions, as shown in the top panel. The nonzero small value of $m_{\mathbf{Q}_4}$ despite $a_4 = 0$ is a consequence of the normalization of the spin length. In this state, φ_η are optimized as $\varphi_1 = \varphi_2 = \varphi_3 = Q/2$. At this phase transition, $N_{\rm m}$ changes from 16 to 8. Note that the MUC of the $3Q$-HL becomes four times smaller than that of the $4Q$-HL; it is rhombohedral since the three ordering wave vectors, $\mathbf{Q}_1$, $\mathbf{Q}_2$, and $\mathbf{Q}_3$, are not orthogonal to each other. This means that this transition doubles the total number of hedgehogs and antihedgehogs (see Sect. 7.1.4).

On the other hand, when we introduce the easy-axis anisotropy by $J'/J > 1$, the system shows two phase transitions as shown in the right hand side of Fig. 7.5. The $4Q$-HL is stable down to $J/J' \simeq 0.7625$, while the spin configuration is modulated reflecting the easy-axis anisotropy as $a_1 = a_2 = a_3 < 1/2 < a_4$, and $m_{\mathbf{Q}_1} \simeq m_{\mathbf{Q}_2} \simeq m_{\mathbf{Q}_3} < m_{\mathbf{Q}_4}$. At $J/J' \simeq 0.7625$, the topologically nontrivial $4Q$-HL with $N_{\rm m} = 16$ is replaced by the topologically trivial $4Q$ state with $N_{\rm m} = 0$, as shown in the lower panel of Fig. 7.5. In these anisotropic $4Q$-HL and $4Q$ states, the values of φ_η depend on J'/J, but the sum always satisfies $\sum_{\eta=1}^{4} \varphi_\eta = (2k+1)\pi$, as for the isotropic $4Q$-HL. While further decreasing J/J', the $4Q$ state is taken over by the $1Q$-H with $a_1 = a_2 = a_3 = 0$ and $a_4 = 1$, and $m_{\mathbf{Q}_1} = m_{\mathbf{Q}_2} = m_{\mathbf{Q}_3} = 0$ and $m_{\mathbf{Q}_4} = 1$ for $J/J' \lesssim 0.5125$.

[3] The slight differences between $m_{\mathbf{Q}_1}$, $m_{\mathbf{Q}_2}$, and $m_{\mathbf{Q}_3}$ are probably due to insufficient numerical precision in the variational calculations, in particular, in the optimization of the phases φ_η which is relevant in this $4Q$-HL region.

7.1.4 Topological Properties

In the previous subsection, we found multiple phase transitions by changing the anisotropy J'/J, including the topological ones accompanied by the changes of N_{m}. In this subsection, we discuss the topological aspects in detail, by tracing the real-space positions of the monopoles and antimonopoles corresponding to the hedgehogs and antihedgehogs, respectively, as well as the emergent magnetic field from the noncoplanar spin configurations.

Figures 7.6 and 7.7 display the real-space pictures of the monopoles and antimonopoles (magenta and cyan balls) for the $3Q$ and $4Q$ cases, respectively. The positions are determined by calculating the monopole charge in Eq. (7.6). We also plot $\mathbf{b}_{\mathbf{r}_l}$ in Eq. (7.9) by arrows to show the flows of the emergent magnetic field. For better visibility, we display the arrows whose norm is larger than a certain threshold in each case.

First, we discuss the $3Q$ case in Fig. 7.6. In the $2Q$-VC under the hard-axis anisotropy, although there are no monopoles and antimonopoles, weak but nonzero $\mathbf{b}_{\mathbf{r}_l}$ appears locally around the antivortex cores, as exemplified in Fig. 7.6a for $J'/J =$

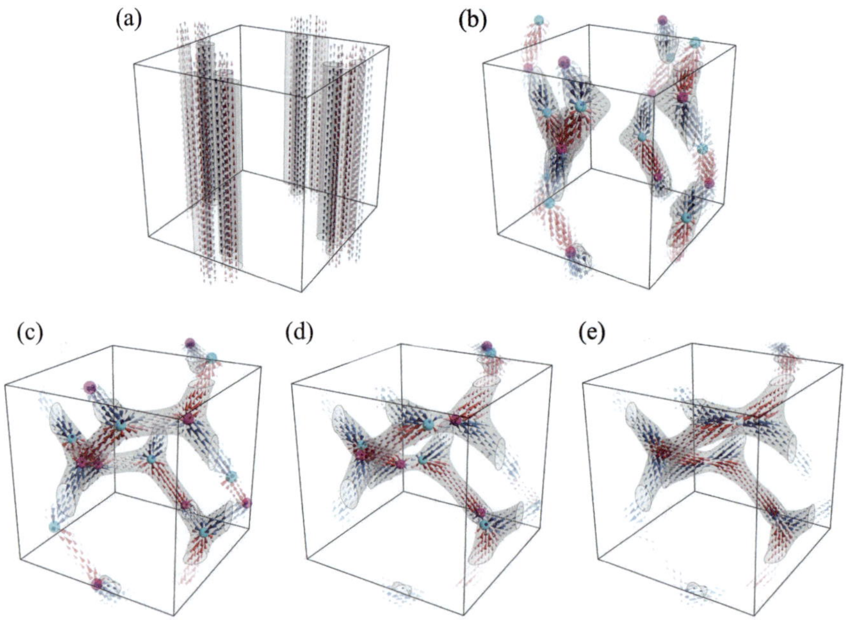

Fig. 7.6 Real-space configurations of the emergent magnetic field $\mathbf{b}_{\mathbf{r}_l}$ (arrows), the isosurfaces of $|\mathbf{b}_{\mathbf{r}_l}|$ (gray surfaces), and the monopoles and antimonopoles (magenta and cyan balls, respectively) in a MUC obtained from the variational calculations for the $3Q$ case: **a** $J'/J = 0.875$, **b** $J'/J = 0.9$, **c** $J'/J = 1.0$, **d** $J/J' = 0.9$, and **e** $J/J' = 0.875$. $\mathbf{b}_{\mathbf{r}_l}$ are plotted when their lengths are longer than a certain threshold for better visibility. The color of the arrows represents the z component of $\mathbf{b}_{\mathbf{r}_l}$; pale color is used for the arrows in the neighboring MUCs. Reproduced from Ref. [17]. Copyright (2021) by the American Physical Society

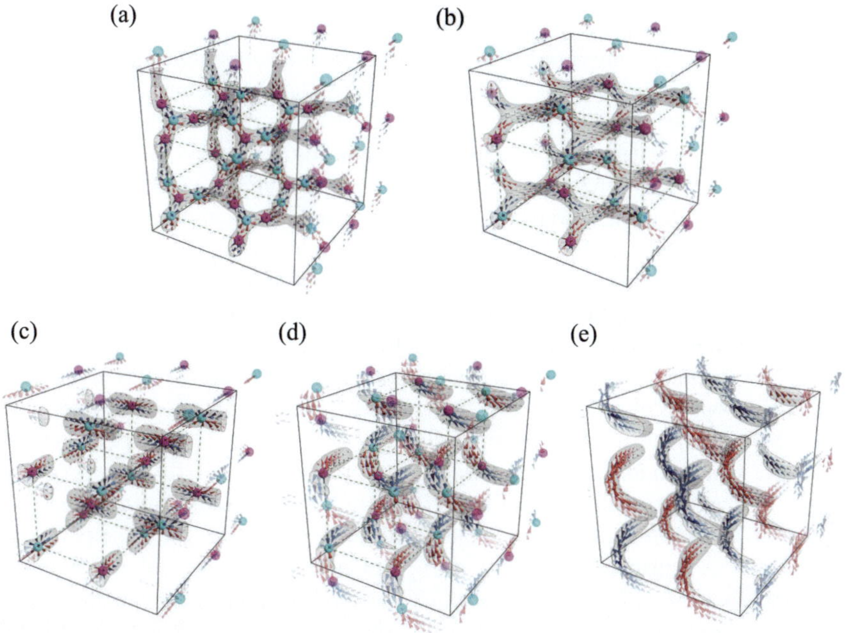

Fig. 7.7 Similar plots for the $4Q$ case: **a** $J'/J = 0.75$, **b** $J'/J = 0.775$, **c** $J'/J = 1.0$, **d** $J/J' = 0.775$, and **e** $J/J' = 0.75$. The notations are common to those in Fig. 7.6. Note that the MUC is one fourth of the whole cube for the $3Q$-HL in (**a**). The green dashed cubes are the guides for eyes; see the text for details. Reproduced from Ref. [17]. Copyright (2021) by the American Physical Society

0.875. $\mathbf{b}_{\mathbf{r}_l}$ are all parallel to the [001] direction, while they form staggered flows around each core. While increasing J'/J above $\simeq 0.8875$, the system undergoes a phase transition from the $2Q$-VC to the $3Q$-HL. At this phase transition, monopoles and antimonopoles are created in pair from the flows of $\mathbf{b}_{\mathbf{r}_l}$, and accordingly, the flows change qualitatively; $\mathbf{b}_{\mathbf{r}_l}$ starts to flow from the monopoles (sources) to the antimonopoles (sinks), as shown in Fig. 7.6b. While further increasing J'/J, the monopoles and antimonopoles change their relative positions with keeping their z coordinates, and for the isotropic case with $J'/J = 1.0$ shown in Fig. 7.6c, they form spiral structures in all the x, y, and z directions [9, 13]. When we introduce weak easy-axis anisotropy, the monopoles and antimonopoles move again with keeping the z coordinates and approach each other as shown in Fig. 7.6d, and eventually annihilate in pair at the transition to the $3Q$ state at $J/J' \simeq 0.8875$. Even in this topologically trivial state with no monopoles and antimonopoles, there remain nonzero flows of $\mathbf{b}_{\mathbf{r}_l}$, especially near the pair annihilation points, as shown in Fig. 7.6e for $J/J' = 0.875$. We note that the total emergent magnetic field is zero in all the phases as the magnetic field is not applied. The effect from the magnetic field in the [111] direction for the isotropic case will be discussed in Sect. 7.2.

Next, we discuss the $4Q$ case in Fig. 7.7. In the $3Q$-HL state under the hard-axis anisotropy, there are four pairs of monopoles and antimonopoles in the MUC, as exemplified in Fig. 7.7a for $J'/J = 0.75$. Note that the volume of the MUC is four times smaller than the whole cube in Fig. 7.7a, as mentioned above. $\mathbf{b}_{\mathbf{r}_l}$ are distributed so as to interconnect the monopoles and antimonopoles. While increasing J'/J above $\simeq 0.7625$, the system undergoes a phase transition from the $3Q$-HL to the $4Q$-HL. At this phase transition, monopoles and antimonopoles connected by the green dotted lines in Fig. 7.7a remain and the others disappear. The remaining ones change their relative positions while keeping the sizes and shapes of the two green cubes, as exemplified in Fig. 7.7b for $J'/J = 0.775$, and finally comprise the body-centered-cubic lattice for the isotropic case with $J'/J = 1.0$ shown in Fig. 7.7c. When we introduce weak easy axis anisotropy, the monopoles of one cube and antimonopoles of the other cube approach each other, and $\mathbf{b}_{\mathbf{r}_l}$ becomes relatively strong between them, as shown in Fig. 7.7d for $J/J' = 0.775$. By further decreasing J/J', the system undergoes the phase transition from the $4Q$-HL to the $4Q$ state caused by the pair annihilation of the remaining monopoles and antimonopoles at $J/J' \simeq 0.7625$. Similar to the $3Q$ state in Fig. 7.6e, even after the monopoles and antimonopoles vanish, nonzero flows of $\mathbf{b}_{\mathbf{r}_l}$ remain near the pair annihilation points, as exemplified in Fig. 7.7e for $J/J' = 0.75$. The total emergent magnetic field is zero in all the phases like the $3Q$ model; their evolution under the magnetic field in the [001] direction from the isotropic case will be discussed in Sect. 7.3.

7.2 Twist Angle Control in $3Q$-HL

In this section, we discuss the effect of the twist angle in a 3D spin moiré. The twist angle plays a pivotal role when discussing the control of moiré patterns in atomic layered materials (see Sect. 2.1). Here, we extend this parameter to the $3Q$-HL and analyze how the spin textures, topological properties, and the EEMF can be controlled. Specifically, we consider the twist angle Θ for three helices (see Sect. 7.2.1). Similar modulation has been achieved experimentally in MnGe, e.g., by changing the sample thickness and an external magnetic field [8, 10]. Furthermore, the enhanced electrical magnetochiral effect associated with the twist angle modulation was reported [11].

In the following, we systematically clarify how the spin textures, topological properties, and the EEMF are modulated by the twist angle and the magnetization. We also demonstrate such a modulation based on the phenomenological model for 3D chiral magnets. In Sect. 7.2.1, we introduce the setup of the 3D spin moiré and exemplify the spin patterns by varying the twist angle spanning the helical directions and the uniform magnetization. In Sect. 7.2.2, we introduce the hedgehogs and anti-hedgehogs, the Dirac strings, and the EEMFs. In Sect. 7.2.3, we discuss how these topological properties evolve through the spin moiré modulation by the twist angle.

Finally, in Sect. 7.2.4, we present the results of variational calculations based on the Ginzburg-Landau free energy in Eq. (3.1) to discuss how the twist angle evolves with the magnetization.

7.2.1 Spin Configurations

We consider 3D spin moiré patterns generated by a superposition of three spin helices in continuous space, focusing on the control by the angle spanning the helical directions and the magnetization induced by an external magnetic field. The setup is schematically shown in Fig. 7.8a. By taking $\psi_\eta^{\rm c} = \psi_\eta^{\rm s} = \frac{1}{\sqrt{3}}$ and $\mathbf{m} = m\hat{\mathbf{z}}$ in Eq. (1.11), the superposition is represented by

$$\mathbf{S}(\mathbf{r}) = \sum_{\eta=1}^{3} \frac{1}{\sqrt{3}} \left(\mathbf{e}_\eta^1 \cos \mathcal{Q}_\eta + \mathbf{e}_\eta^2 \sin \mathcal{Q}_\eta \right) + m\hat{\mathbf{z}}, \tag{7.10}$$

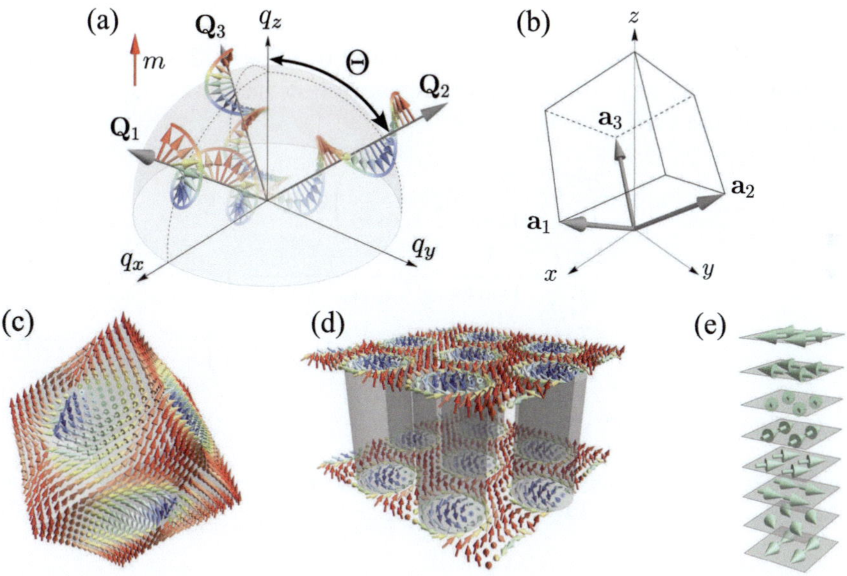

Fig. 7.8 **a** Schematic picture of the superposition of three helices in Eq. (7.10). Θ represents the twist angle to control the propagation direction of the helices. **b** Corresponding rhombohedral magnetic unit cell. **c–e** Real-space spin configurations of (**c**) the 3D HL, (**d**) the 2D SkL, (**e**) and the 1D helical state, which are obtained at $\Theta = \Theta_{\rm cubic}$ [Eq. (7.21)], $\pi/2$, and 0 in Eq. (7.10) with $m = 0$, respectively. The arrows represent the normalized spins $\mathbf{n}(\mathbf{r})$. Reproduced from Ref. [18]. Copyright (2021) by the American Physical Society

where $\mathcal{Q}_\eta = \mathbf{Q}_\eta \cdot \mathbf{r}$[4] and $\mathbf{Q}_\eta = Q\mathbf{e}^0_\eta$ with

$$\mathbf{e}^0_1 = (\sin\Theta, 0, \cos\Theta), \tag{7.11}$$

$$\mathbf{e}^0_2 = \left(-\frac{1}{2}\sin\Theta, \frac{\sqrt{3}}{2}\sin\Theta, \cos\Theta\right), \tag{7.12}$$

$$\mathbf{e}^0_3 = \left(-\frac{1}{2}\sin\Theta, -\frac{\sqrt{3}}{2}\sin\Theta, \cos\Theta\right). \tag{7.13}$$

In the following, we consider the superposition of helices, in each of which the spins rotate on the plane perpendicular to $\mathbf{Q}_\eta$, by taking

$$\mathbf{e}^1_1 = (-\cos\Theta, 0, \sin\Theta), \tag{7.14}$$

$$\mathbf{e}^1_2 = \left(\frac{1}{2}\cos\Theta, -\frac{\sqrt{3}}{2}\cos\Theta, \sin\Theta\right), \tag{7.15}$$

$$\mathbf{e}^1_3 = \left(\frac{1}{2}\cos\Theta, \frac{\sqrt{3}}{2}\cos\Theta, \sin\Theta\right), \tag{7.16}$$

and $\mathbf{e}^2_\eta = \mathbf{e}^0_\eta \times \mathbf{e}^1_\eta$. The MUC of this spin texture for $0 < \Theta < \pi/2$ is a rhombohedral one shown in Fig. 7.8b whose translation vectors $\mathbf{a}_\eta$ are defined by $\mathbf{a}_\eta \cdot \mathbf{Q}_{\eta'} = 2\pi\delta_{\eta,\eta'}$. The explicit forms are given as

$$\mathbf{a}_1 = \frac{2\pi}{Q}\left(\frac{2}{3\sin\Theta}, 0, \frac{1}{3\cos\Theta}\right), \tag{7.17}$$

$$\mathbf{a}_2 = \frac{2\pi}{Q}\left(-\frac{1}{3\sin\Theta}, \frac{1}{\sqrt{3}\sin\Theta}, \frac{1}{3\cos\Theta}\right), \tag{7.18}$$

$$\mathbf{a}_3 = \frac{2\pi}{Q}\left(-\frac{1}{3\sin\Theta}, -\frac{1}{\sqrt{3}\sin\Theta}, \frac{1}{3\cos\Theta}\right). \tag{7.19}$$

The volume of the MUC is given by

$$V = \mathbf{a}_1 \cdot (\mathbf{a}_2 \times \mathbf{a}_3) = \left(\frac{2\pi}{Q}\right)^3 \frac{2}{3\sqrt{3}\sin^2\Theta\cos\Theta}. \tag{7.20}$$

We choose Eq. (7.10) so as to satisfy threefold rotational symmetry about the z axis and to interpolate the three known spin textures introduced in Sect. 1.2 while changing Θ: (i) the 3D HL for $0 < \Theta < \frac{\pi}{2}$, (ii) the 2D SkL at $\Theta = \pi/2$ where the

[4] In this section, we ignore the phase degrees of freedom except for the 2D SkL with $\Theta = \pi/2$ in the variational calculations; see Sect. 7.2.4. To make the 1D helical and conical state with $\Theta = 0$, we need to choose proper phases, but this is irrelevant for the variational calculations.

three wave vectors $\mathbf{Q}_\eta$ lie on the $q_x q_y$ plane with 120° relations, and (iii) 1D helical state ($m = 0$) or conical state ($m > 0$) at $\Theta = 0$. For $\Theta = \Theta_{\rm cubic}$ with

$$\Theta_{\rm cubic} = \arccos\left(\frac{1}{\sqrt{3}}\right), \tag{7.21}$$

Equation (7.10) represents the $3Q$-HL with the superposition of three orthogonal helices introduced in Sect. 1.2.3. The real-space spin structures of the 3D HL with $\Theta = \Theta_{\rm cubic}$, 2D SkL, and 1D helical state are shown in Fig. 7.8c, d, and e, respectively.

The superposition in Eq. (7.10) allows a continuous modulation of the spin moiré patterns between these known spin textures by varying Θ and m. Figure 7.9a–d exemplify such modulations while changing Θ at $m = 0$. In this case, the MUC changes its shape and volume, together with the modulation of the spin moiré. Figure 7.9e–h display the cases while changing m at $\Theta = \Theta_{\rm cubic}$ where the MUC is cubic. We also present the spin patterns on the horizontal plane through the center of the MUC in the right panels. The plane comprises a SkL; the skyrmion number for the 2D MUC is -1 per the 2D MUC shown by the black rhombus in Fig. 7.9e [12]. While increasing m, the spin-up region is extended on the verge of skyrmions [Fig. 7.9e–g], and eventually, the system turns into a topologically trivial state when the spins at the cores are flipped [Fig. 7.9h]. Note that similar topological transitions to a trivial state are observed at the black points in Figs. 6.6 and 6.10, where the hedgehogs and antihedgehogs exhibit pair annihilation in the 3D hyperspace.

7.2.2 *Topological Properties*

As exemplified in Fig. 7.9, the angle Θ and the magnetization m modulate the spin moiré pattern in Eq. (7.10). Accordingly, they modulate the topological nature, which is characterized by the hedgehogs and antihedgehogs introduced in Sect. 1.1.3 and the resultant EEMF. Before going into the spin moiré engineering of these properties, we review their fundamental aspects for the superposition given by Eq. (7.10). We discuss the topological properties by examining the real-space positions of the hedgehogs and antihedgehogs, and introduce their relation to the Dirac string. Then, we give definitions of the net EMF and EEF.

Hedgehog and Antihedgehog

Following the procedure in the first subsubsection of Sect. 6.1.2, we examine the real-space positions of the hedgehogs and antihedgehogs. Solving the equation $\mathbf{S}(\mathbf{r}) = 0$ for Eq. (7.10), we obtain the following eight solutions:

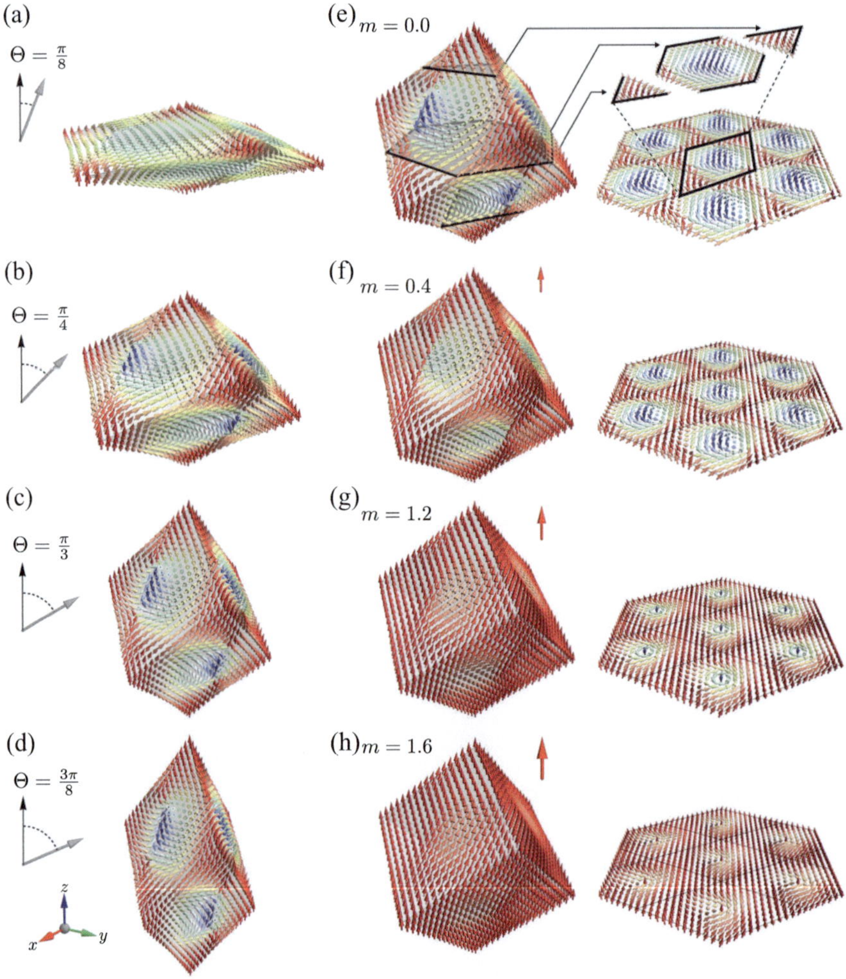

Fig. 7.9 Spin moirés in Eq. (7.10) for **a** $\Theta = \frac{\pi}{8}$, **b** $\Theta = \frac{\pi}{4}$, **c** $\Theta = \frac{\pi}{3}$, and **d** $\Theta = \frac{3\pi}{8}$ at $m = 0$, and **e** $m = 0$, **f** $m = 0.4$, **g** $m = 1.2$, and **h** $m = 1.6$ at $\Theta = \Theta_{\text{cubic}}$. The bold black rhombus in (**e**) represents the 2D MUC for the slices. Reproduced from Ref. [18]. Copyright (2021) by the American Physical Society

$$\begin{aligned}
\mathcal{Q}^*_\eta = &\left(\frac{\pi}{2}, \frac{\pi}{2}, \frac{\pi}{2}\right), \left(\frac{3\pi}{2}, \frac{3\pi}{2}, \frac{3\pi}{2}\right), \\
&\left(\frac{\pi}{2}, \frac{\pi}{2} + 2p_0(\Theta), \frac{\pi}{2} - 2p_0(\Theta)\right) \quad \text{and cyclic permutations,} \\
&\left(\frac{3\pi}{2}, \frac{3\pi}{2} + 2p_0(\Theta), \frac{3\pi}{2} - 2p_0(\Theta)\right) \quad \text{and cyclic permutations,} \qquad (7.22)
\end{aligned}$$

where $\tan p_0(\Theta) = \sqrt{3}\cos\Theta$. By using the relation

$$\mathbf{r}^* = (x^*, y^*, z^*) = \sum_\eta \frac{Q_\eta^*}{2\pi}\mathbf{a}_\eta, \tag{7.23}$$

we obtain the real-space positions of the eight defects as

$$\begin{aligned}\mathbf{r}^* = &\frac{2\pi}{Q}\left(0, 0, \frac{1}{4\cos\Theta}\right), \frac{2\pi}{Q}\left(0, 0, \frac{3}{4\cos\Theta}\right),\\ &\frac{2\pi}{Q}\left(0, \frac{2p_0(\Theta)}{\sqrt{3}\pi\sin\Theta}, \frac{1}{4\cos\Theta}\right) \text{ and } C_3^z \text{ symmetric points,}\\ &\frac{2\pi}{Q}\left(0, \frac{2p_0(\Theta)}{\sqrt{3}\pi\sin\Theta}, \frac{3}{4\cos\Theta}\right) \text{ and } C_3^z \text{ symmetric points,}\end{aligned} \tag{7.24}$$

where the C_3^z symmetric points are obtained by $2\pi/3$ and $4\pi/3$ rotations about the z axis. The first two out of the eight solutions are located on the z axis, while the rest six form a triangular prism about the z axis [the locations will be depicted in Figs. 7.11a and 7.12a]. As shown in Sect. 7.2.3, these defects are classified into four hedgehogs and four antihedgehogs according to the values of the monopole charge Q_m in Eq. (1.7).

For nonzero m, these defects change their positions. We find the analytical solutions for two out of the eight defects as

$$\begin{aligned}Q_\eta^* = &\left(\frac{\pi}{2} + p(\Theta, m), \frac{\pi}{2} + p(\Theta, m), \frac{\pi}{2} + p(\Theta, m)\right),\\ &\left(\frac{3\pi}{2} - p(\Theta, m), \frac{3\pi}{2} - p(\Theta, m), \frac{3\pi}{2} - p(\Theta, m)\right),\end{aligned} \tag{7.25}$$

where $p(\Theta, m) = \arcsin\left(\frac{m}{\sqrt{3}\sin\Theta}\right)$ ($|m| < \sqrt{3}\sin\Theta$). The real-space positions of these two defects are given by

$$\begin{aligned}\mathbf{r}^* = &\frac{2\pi}{Q}\left(0, 0, \frac{1}{4\cos\Theta}\left(1 + \frac{2}{\pi}p(\Theta, m)\right)\right),\\ &\frac{2\pi}{Q}\left(0, 0, \frac{3}{4\cos\Theta}\left(1 - \frac{2}{\pi}p(\Theta, m)\right)\right).\end{aligned} \tag{7.26}$$

Note that these two defects correspond to the first two in Eq. (7.24) on the z axis for $m = 0$. Equation (7.26) indicates that they move along the z axis while increasing m and pair annihilate at $|m| = \sqrt{3}\sin\Theta$. We obtain the other six solutions corresponding to the latter six in Eq. (7.24) numerically. As in the case with $m = 0$, the eight defects are classified into four hedgehogs and four antihedgehogs. We confirm that any other solutions do not appear in the following calculations.

Dirac String

These topological defects are connected by the Dirac string, on which spins point downward, as seen in Sect. 1.3.1. Furthermore, the Dirac strings give important information about the EMF, as will be discussed in the next subsubsection. Before going into the discussion for the EMF, we introduce the Dirac string in more detail. For this purpose, we define the vector potential $\mathbf{a}(\mathbf{r})$ given by $\mathbf{b}(\mathbf{r}) = \nabla \times \mathbf{a}(\mathbf{r})$ and show that the anomaly of $\mathbf{a}(\mathbf{r})$ corresponds to the Dirac string. For $\mathbf{a}(\mathbf{r})$, we choose the representation

$$\mathbf{a}(\mathbf{r}) = \begin{cases} \mathbf{a}^{\mathrm{S}}(\mathbf{r}) = \big(1 - \cos\theta(\mathbf{r})\big)\nabla\phi(\mathbf{r}) \\ \mathbf{a}^{\mathrm{N}}(\mathbf{r}) = -\big(1 + \cos\theta(\mathbf{r})\big)\nabla\phi(\mathbf{r}), \end{cases} \tag{7.27}$$

where $\theta(\mathbf{r})$ and $\phi(\mathbf{r})$ are defined by Eq. (1.2). Note that these two vector potentials are related to each other by the gauge transformation given by

$$\mathbf{a}^{\mathrm{N}}(\mathbf{r}) = \mathbf{a}^{\mathrm{S}}(\mathbf{r}) - 2\nabla\phi(\mathbf{r}). \tag{7.28}$$

In Eq. (7.27), $\mathbf{a}^{\mathrm{S}}(\mathbf{r})$ and $\mathbf{a}^{\mathrm{N}}(\mathbf{r})$ have singularities at $\theta(\mathbf{r}) = \pi$ and 0, respectively. To define $\mathbf{b}(\mathbf{r})$ properly, we avoid these singularities by analytic continuation using $\mathbf{a}^{\mathrm{S}}(\mathbf{r})$ for $0 \leq \theta(\mathbf{r}) \leq \pi - \delta$ and $\mathbf{a}^{\mathrm{N}}(\mathbf{r})$ otherwise, where δ is an infinitesimal.[5]

The singular points of the vector potentials form lines in real space and are called the Dirac string [21]. In the present system with the magnetization parallel to $\hat{\mathbf{z}}$, the singularity of $\mathbf{a}^{\mathrm{S}}(\mathbf{r})$ where the spins point downward defines the Dirac strings connecting the hedgehog and antihedgehog. The situation is exemplified for $\Theta = \Theta_{\mathrm{cubic}}$ and $m = 0$ in Fig. 7.10, by taking the pair of hedgehog and antihedgehog on the z axis given by the first two solutions in Eq. (7.24). Figure 7.10a shows the xz slice of the spin configuration through the center of the MUC; the Dirac string is denoted by the green vertical line connecting the hedgehog and antihedgehog denoted by the magenta and cyan spheres, respectively. Figure 7.10b shows the xy slice through the center of the MUC.

Importantly, the plane perpendicular to the Dirac string can exhibit a skyrmion, and the intersection of the Dirac string and the plane corresponds to the core of the skyrmion, as exemplified in Fig. 7.10b. This observation motivates us to characterize the Dirac string by the vorticity ζ given by Eq. (1.4), whose integral is taken on a closed path surrounding the string, as already introduced in Sect. 6.1.2.[6] For the example in Fig. 7.10, the Dirac string has $\zeta = +1$, and hence the skyrmion on the plane shows $N_{\mathrm{sk}} = -1$ (see the second subsubsection of Sect. 1.1.2). The value of ζ is, however, known without the direct calculation of Eq. (1.4), once we identify the hedgehog and antihedgehog connected by the Dirac string; it is given by their

---

[5] δ is introduced just to define the gauge connection. It does not appear in the numerical calculations of the monopole charge and the velocity of the Dirac string.

[6] In the numerical calculations of ζ, we define a small square surrounding the Dirac string and count how many times the projections of the four spins on the corners onto the xy plane wrap a unit circle.

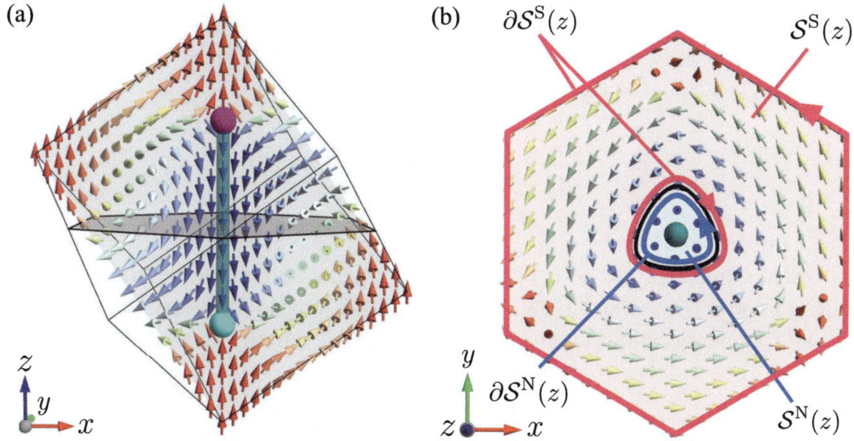

Fig. 7.10 **a** Spin configuration on the xz slice through the center of the MUC for Eq. (7.10) with $\Theta = \Theta_{\mathrm{cubic}}$ and $m = 0$. The Dirac string with the vorticity $\zeta = +1$ (pale green vertical line) and the associated hedgehog and antihedgehog (magenta and cyan spheres) are also shown. Note that only a single hedgehog-antihedgehog pair out of four is shown for simplicity. **b** Spin configuration on the xy slice through the center of the MUC. The green sphere at the center denotes the intersection of the Dirac string. The pink and blue areas denote $\mathcal{S}^{\mathrm{S}}(z)$ and $\mathcal{S}^{\mathrm{N}}(z)$ comprising $\mathcal{S}(z)$ in Eq. (7.30). We take $\delta = \arccos(0.9)$ for better visibility. The closed paths encircling the areas represent $\partial\mathcal{S}^{\mathrm{S}}(z)$ and $\partial\mathcal{S}^{\mathrm{N}}(z)$ in the first line of Eq. (7.31). Reproduced from Ref. [18]. Copyright (2021) by the American Physical Society

relative positions and monopole charges Q_{m}. For instance, in Fig. 7.10, the hedgehog is located above the antihedgehog in the z direction (see Sect. 7.2.3), from which we can immediately assign $\zeta = +1$ for the Dirac string connecting them. We confirm that the numerical estimates are consistent with such topological assignments for all the Dirac strings in our analysis.

Emergent Magnetic Field

The hedgehogs and antihedgehogs correspond to the emergent magnetic monopoles and antimonopoles, respectively, and their periodic arrangement determines the real-space distribution of the EMF, as we will discuss in Sect. 7.2.3. The net value of $\mathbf{b}(\mathbf{r})$ in the MUC is important for the bulk properties of the system, e.g., the topological Hall conductivity. In the present case with the magnetization along the z axis, as the x and y components of $\mathbf{b}(\mathbf{r})$ cancel out due to threefold rotational symmetry about the z axis, we focus on the spatially averaged value of the z component as

$$\bar{b}_z = \frac{1}{4\pi L_z} \int dV b_z(\mathbf{r}), \tag{7.29}$$

where $L_z = |\mathbf{a}_1 + \mathbf{a}_2 + \mathbf{a}_3|$ and the integral is taken within the MUC; see Fig. 7.8b. Note that $\bar{b}_z$ has the dimension of the magnetic flux. Using the vector potential in Eq. (7.27), we can write Eq. (7.29) as

$$\bar{b}_z = \frac{1}{4\pi L_z} \int dz \int_{\mathcal{S}(z)} d\mathcal{S}\hat{\mathbf{z}} \cdot (\nabla \times \mathbf{a}(\mathbf{r})), \tag{7.30}$$

whose integral is taken for a 2D horizontal slice of the MUC, $\mathcal{S}(z)$. Note that $\mathcal{S}(z)$ is divided into $\mathcal{S}^{\mathrm{S}}(z)$ and $\mathcal{S}^{\mathrm{N}}(z)$ corresponding to the regions where $\mathbf{a}(\mathbf{r})$ is defined as $\mathbf{a}^{\mathrm{S}}(\mathbf{r})$ and $\mathbf{a}^{\mathrm{N}}(\mathbf{r})$, respectively; see Fig. 7.10b. Using Stokes' theorem under the periodic boundary conditions and the definition of the vorticity in Eq. (1.4), we obtain

$$\begin{aligned}\bar{b}_z &= \frac{1}{4\pi L_z} \int dz \left[\oint_{\partial\mathcal{S}^{\mathrm{S}}(z)} d\mathbf{l} \cdot \mathbf{a}^{\mathrm{S}}(\mathbf{r}) + \oint_{\partial\mathcal{S}^{\mathrm{N}}(z)} d\mathbf{l} \cdot \mathbf{a}^{\mathrm{N}}(\mathbf{r}) \right] \\ &= -\frac{1}{L_z} \sum_k \int dz \zeta_k, \end{aligned} \tag{7.31}$$

where k labels the Dirac strings and ζ_k is the vorticity of the kth Dirac string; the last integral is taken in the z range where ζ_k is nonzero.

In the present system, the vorticity takes $\zeta_k = +1$ or -1 for all the Dirac strings, and ζ_k is unique for each string, as discussed later. Thus, we can rewrite Eq. (7.31) into

$$\bar{b}_z = -\frac{1}{L_z} \sum_k \mathbf{R}_k^{\pm} \cdot \hat{\mathbf{z}}, \tag{7.32}$$

where $\mathbf{R}_k^{\pm}$ denotes the vector directed from the antihedgehog to the hedgehog at the ends of the kth Dirac string. Equation (7.32) shows that $\bar{b}_z$ is given by the sum of the lengths of the Dirac strings projected onto the z axis. We will discuss the case where ζ_k in Eq. (7.31) takes 1 and -1 depending on z in Sect. 7.3.

Emergent Electric Field

In addition, we also introduce the net EEF by assuming time (t) dependence in m and Θ as $\mathbf{e}^m(\mathbf{r})\partial_t m + \mathbf{e}^{\Theta}(\mathbf{r})\partial_t \Theta$, whose $i = x, y, z$ component is defined by

$$e_i^{\nu}(\mathbf{r}) = \mathbf{n}(\mathbf{r}) \cdot (\partial_i \mathbf{n}(\mathbf{r}) \times \partial_{\nu} \mathbf{n}(\mathbf{r})), \tag{7.33}$$

where $\nu = m$ or Θ [12, 23]. We also calculate the average of the z component as

$$\bar{e}_z = \frac{L_z}{4\pi V} \int d\mathcal{V} \left[e_z^m(\mathbf{r}) + e_z^{\Theta}(\mathbf{r}) \partial_m \Theta \right], \tag{7.34}$$

where V is the volume of the MUC in Eq. (7.20) and the integral is taken within the MUC. Note that $\bar{e}_z$ has the dimension of the electric voltage.

7.2.3 Topological Phase Diagram

Using the definitions in the previous subsection, we can identify the topological properties and the EEMF of the spin moiré in Eq. (7.10), and investigate their evolution while changing Θ and m. Figure 7.11 illustrates the systematic evolution of the topological objects and the EMF while increasing m at $\Theta = \Theta_{\text{cubic}}$ in Eq. (7.10) [see the corresponding spin configurations in Fig. 7.9e–h]. For $m = 0$, the system has eight topological defects, as discussed in Eq. (7.22). By computing the monopole charge Q_{m}, we identify four out of them as hedgehogs with $Q_{\text{m}} = +1$ and the rest four as antihedgehogs with $Q_{\text{m}} = -1$: The upper (lower) defect on the z axis and

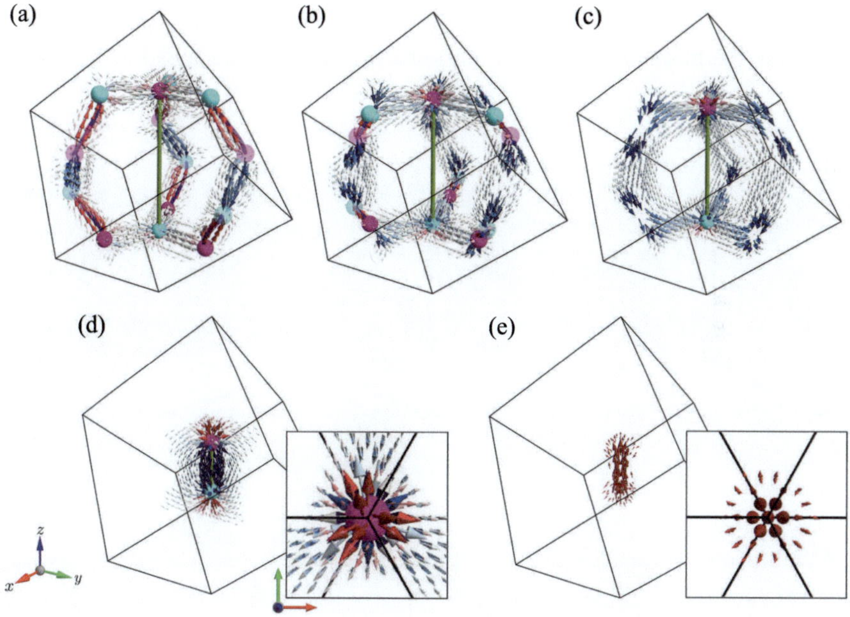

Fig. 7.11 Hedgehogs and antihedgehogs, Dirac strings, and the EMF, in the spin moiré in Eq. (7.10) while changing m at $\Theta = \Theta_{\text{cubic}}$: **a** $m = 0$, **b** $m = 0.24$, **c** $m = 0.4$, **d** $m = 1.2$, and **e** $m = 1.6$. The arrows in **a**–**e** display the EMF $\mathbf{b}(\mathbf{r})$. Only $\mathbf{b}(\mathbf{r})$ with large amplitude are shown and the divergent lengths close to the hedgehogs and antihedgehogs are normalized for clarity. The magenta and cyan spheres are the hedgehogs and antihedgehogs, and the green and purple lines denote the Dirac strings connecting them with the vorticity $\zeta = +1$ and -1, respectively. The pale colors represent the objects in the neighboring MUCs. The colors of the arrows are guides for the z components. The insets in (**d**) and (**e**) are the projections of the central part onto the xy plane. Reproduced from Ref. [18]. Copyright (2021) by the American Physical Society

the three on the lower (upper) plane of the triangular prism are (anti)hedgehogs. They are connected by the Dirac strings, as shown in Fig. 7.11a. Note that one of the Dirac strings runs straight along the z axis, while the rest three are not straight and run through the boundaries of the MUC to connect with the (anti)hedgehogs in the neighboring MUCs; the former one has the vorticity $\zeta = +1$, as already seen in Fig. 7.10a, and the latter three have $\zeta = -1$. The EMF $\mathbf{b}(\mathbf{r})$ flows from hedgehogs to antihedgehogs. While increasing m, the hedgehogs and antihedgehogs move toward their counterparts along the Dirac strings, as exemplified in Fig. 7.11b. With a further increase of m, three pairs connected by the Dirac strings with $\zeta = -1$ disappear with pair annihilation, leaving one pair connected by the Dirac string with $\zeta = +1$, as shown in Fig. 7.11c. Strong flows of $\mathbf{b}(\mathbf{r})$ around the positions where the hedgehogs and antihedgehogs pair annihilate are the remnants associated with the pair annihilation. Finally, the remaining pair annihilates, leaving remnants of $\mathbf{b}(\mathbf{r})$, as shown in Fig. 7.11d and e. These pair annihilations define topological transitions between different topological phases with different numbers of the hedgehog-antihedgehog pairs (see Fig. 7.13).

The corresponding results for the EEF $\mathbf{e}(\mathbf{r})$ in Eq. (7.33) are shown in Fig. 7.12. Note that $\mathbf{e}^{\Theta}(\mathbf{r}) = 0$ as Θ is fixed here and the EEF presented here arises from the changes in the magnetization m. As $\mathbf{e}(\mathbf{r})$ and $\mathbf{b}(\mathbf{r})$ follow Faraday's law [21], $\mathbf{e}(\mathbf{r})$

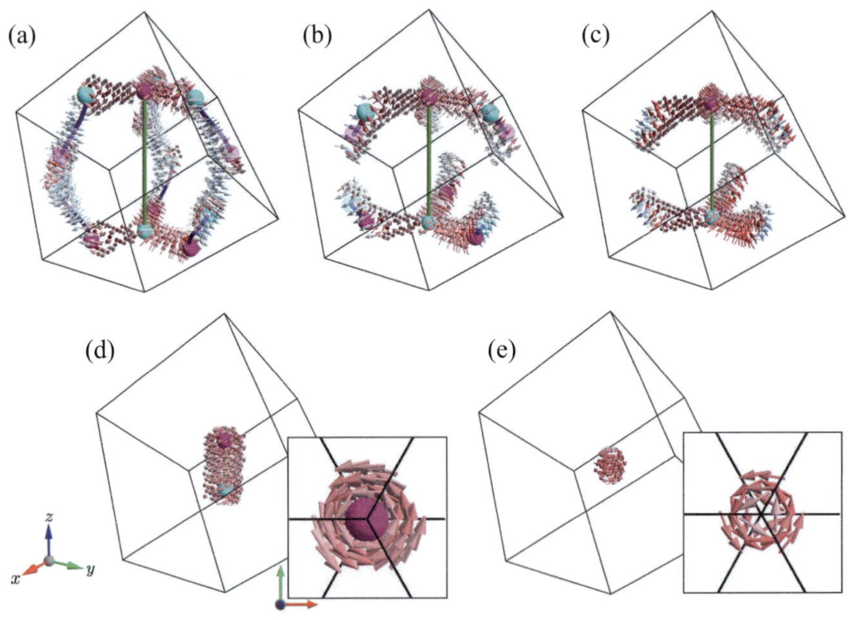

Fig. 7.12 Similar figure to Fig. 7.11 for the EEF: **a** $m = 0$, **b** $m = 0.24$, **c** $m = 0.4$, **d** $m = 1.2$, and **e** $m = 1.6$. The arrows represent the EEF and their colors are guides for the z components. Other notations are common to those in Fig. 7.11. Reproduced from Ref. [18]. Copyright (2021) by the American Physical Society

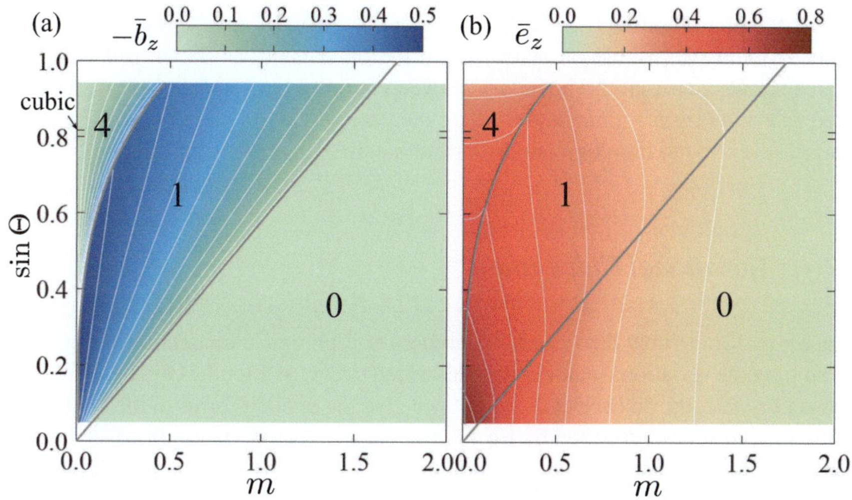

Fig. 7.13 Topological phase diagram determined by the number of hedgehog-antihedgehog pairs within the MUC, N_{pair}, with the contour plots of **a** $-\bar{b}_z$ and **b** $\bar{e}_z$ as functions of m and $\sin\Theta$. N_{pair} is indicated in each phase divided by the gray boundaries. The white lines denote the contours drawn every (**a**) 0.05 and (**b**) 0.1. The blanks near $\sin\Theta = 0$ and 1 indicate the regions where sufficient numerical precision is not guaranteed. Reproduced from Ref. [18]. Copyright (2021) by the American Physical Society

becomes large where $\mathbf{b}(\mathbf{r})$ changes rapidly while increasing m. As a consequence, $\mathbf{e}(\mathbf{r})$ appears along the flows of $\mathbf{b}(\mathbf{r})$ with a swirling texture around them. As discussed in Fig. 7.11, the hedgehogs and antihedgehogs change their positions along the Dirac string by increasing m. Hence, in the vicinity of hedgehogs and antihedgehogs, the EEF circulates on the plane perpendicular to the direction of the Dirac string. This is compatible with the discussion in Fig. 1.10.

Figure 7.13 summarizes the topological phase diagram determined by the number of hedgehog-antihedgehog pairs within the MUC, N_{pair}, on the plane of m and $\sin\Theta$. We find three different phases with $N_{\text{pair}} = 4$, 1, and 0. We also plot the net EMF $\bar{b}_z$ in Eq. (7.29) and the net EEF $\bar{e}_z$ in Eq. (7.34), in Fig. 7.13a and b, respectively. Note that we plot $-\bar{b}_z$ in Fig. 7.13a and the contribution from $e_z^m(\mathbf{r})$ in Fig. 7.13b for simplicity [we include $e_z^{\Theta}(\mathbf{r})$ in the variational calculations in Sect. 7.2.4]. As shown in Fig. 7.13a, $-\bar{b}_z$ rapidly increases with m in the $N_{\text{pair}} = 4$ region, while decreases in the $N_{\text{pair}} = 1$ region after showing a sharp cusp on the boundary, and eventually vanishes in the $N_{\text{pair}} = 0$ region. In the $N_{\text{pair}} = 1$ region, using the relation in Eq. (7.32) and the real-space positions of the remaining pair in Eq. (7.26), $\bar{b}_z$ is explicitly obtained as a function of m and Θ as

$$\bar{b}_z = -\frac{1}{2}\left[1 - \frac{2}{\pi}\arcsin\left(\frac{m}{\sqrt{3}\sin\Theta}\right)\right]. \tag{7.35}$$

Hence, the phase boundary from $N_{\text{pair}} = 1$ to 0 is given by $m = \sqrt{3}\sin\Theta$. See also Eq. (7.26) and the discussion below it. On the other hand, as shown in Fig. 7.13b, $\bar{e}_z$ also shows a cusp at the boundary between the $N_{\text{pair}} = 4$ and 1 regions, but it is large near the origin and decreases while increasing m and Θ. Note that $\bar{e}_z$ remains nonzero even in the topologically trivial state with $N_{\text{pair}} = 0$, in contrast to $\bar{b}_z$.

7.2.4 *Variational Calculation*

In order to discuss how the system undergoes the topological transitions in Fig. 7.13, we employ the variational calculations described in Sect. 4.1 for the Ginzburg-Landau free energy introduced in Sect. 3.1. First, we give the explicit forms of the energy for different spin states described by Eq. (7.10). Then, we show the results of the variational calculations which show the evolution of Θ, N_{pair}, $\bar{b}_z$, and $\bar{e}_z$ while changing m.

Energy for Each State

Using the energy functional in Eq. (3.1), we estimate the energy for different spin states expressed by Eq. (7.10): the 3D HL for $0 < \Theta < \pi/2$, the 2D SkL at $\Theta = \pi/2$, and the 1D helical or conical state at $\Theta = 0$. We note that Eq. (7.10) cannot represent a FFM state which is expected to compete with Eq. (7.10) in the large m region. For the energy comparison to the FFM state, we need specific values of J and D in Eq. (3.2), which are irrelevant to the other three spin states. Because of this complexity, we omit the FFM state in the following analysis.

For the four-spin interactions in Eq. (3.3), we employ the simplifications and assumptions following Refs. [1, 2]. Given that the helical wave number $Q_0 = D/J$ is set by F_2^{GL} in Eq. (3.2), hereafter, we consider any sets of the four wave vectors $\mathbf{Q}_\eta$ satisfying $\mathbf{Q}_1 + \mathbf{Q}_2 + \mathbf{Q}_3 + \mathbf{Q}_4 = 0$ with $|\mathbf{Q}_\eta| = Q_0$ or 0, and omit the energy contributions from other wave vectors. When $|\mathbf{Q}_\eta| = Q_0$, we assume that $U(\mathbf{Q}_1, \mathbf{Q}_2, \mathbf{Q}_3)$ is given by

$$U(\alpha, \beta) = U_0 + U_{11}\sin\alpha\cos\beta + U_{20}(3\cos^2\alpha - 1) + U_{22}\sin^2\alpha\cos 2\beta, \quad (7.36)$$

with two parameters

$$\alpha = \frac{1}{2}\arccos\left(\hat{\mathbf{Q}}_1 \cdot \hat{\mathbf{Q}}_2\right), \quad (7.37)$$

$$\beta = 2\arccos\left[\frac{(\hat{\mathbf{Q}}_2 - \hat{\mathbf{Q}}_1) \cdot \hat{\mathbf{Q}}_3}{1 - \hat{\mathbf{Q}}_1 \cdot \hat{\mathbf{Q}}_2}\right]. \quad (7.38)$$

Here, α and β parametrize the relative angles of the four wave vectors $\mathbf{Q}_\eta$. In addition, we take into account the components $U(0,0,0)$, $U(\mathbf{Q}_\eta, -\mathbf{Q}_\eta, 0)$, and $U(\mathbf{Q}_\eta, 0, 0)$, etc. as the cases where some of $\mathbf{Q}_\eta$ are 0, and adopt the approximation $U(0,0,0) \approx U(\mathbf{Q}_\eta, -\mathbf{Q}_\eta, 0)$. Noting that $U(\mathbf{Q}_\eta, -\mathbf{Q}_\eta, 0) \approx U(\alpha = \pi/2, \beta = 0)$, we introduce the parameter U_s defined as

$$U_s \equiv U\left(\alpha = \frac{\pi}{2}, \beta = 0\right) = U_0 + U_{11} + U_{22}. \tag{7.39}$$

Then, by taking U_0 as the energy unit, the energy functional has three independent parameters, U_{11}, U_{22}, and $U(\mathbf{Q}_\eta, 0, 0)$, which are denoted as

$$t = \frac{U_{11}}{U_0}, \quad u = \frac{U_{22}}{U_0}, \quad v = \frac{U(\mathbf{Q}_\eta, 0, 0)}{U_0}. \tag{7.40}$$

In the following, we calculate the energies for the 3D HL, 2D SkL, and 1D helical state.

A. Hedgehog lattice

First, we calculate the energy for the 3D HL given by Eq. (7.10) with $0 < \Theta < \frac{\pi}{2}$. The energy is obtained as

$$F_{\mathrm{HL}} = \sum_\eta [f_2(\mathbf{Q}_\eta) + f_s(\mathbf{Q}_\eta)] + \sum_{\eta<\eta'} f_p(\mathbf{Q}_\eta, \mathbf{Q}_{\eta'}) + f_m^{(1)}(m) + \sum_\eta f_m^{(2)}(\mathbf{Q}_\eta, m), \tag{7.41}$$

where

$$f_2(\mathbf{Q}_\eta) = \left(r_0 - J\tilde{q}^2\right) |\psi_{\mathbf{Q}_\eta}|^2, \tag{7.42}$$
$$f_s(\mathbf{Q}_\eta) = U_s |\psi_{\mathbf{Q}_\eta}|^4, \tag{7.43}$$
$$f_p(\mathbf{Q}_\eta, \mathbf{Q}_{\eta'}) = 2V_p\left(\alpha_{\eta\eta'}\right) |\psi_{\mathbf{Q}_\eta}|^2 |\psi_{\mathbf{Q}_{\eta'}}|^2, \tag{7.44}$$
$$f_m^{(1)}(m) = r_0 m^2 + U_s m^4, \tag{7.45}$$
$$f_m^{(2)}(\mathbf{Q}_\eta, m) = 2\left[U_s m^2 + U(\mathbf{Q}_\eta, 0, 0)\mathbf{m}_{\perp,\eta}^2\right] |\psi_{\mathbf{Q}_\eta}|^2, \tag{7.46}$$

with

$$\psi_{\mathbf{Q}_\eta} = \sqrt{2}|\mathbf{S}_{\mathbf{Q}_\eta}|, \tag{7.47}$$
$$V_p(\alpha) = U\left(\frac{\pi}{2}, 4\alpha\right) + U(\alpha, 0)\sin^4\alpha + U\left(\frac{\pi}{2} - \alpha, 0\right)\cos^4\alpha, \tag{7.48}$$
$$\alpha_{\eta\eta'} = \frac{1}{2}\arccos\left(\hat{\mathbf{Q}}_\eta \cdot \hat{\mathbf{Q}}_{\eta'}\right), \tag{7.49}$$
$$\mathbf{m}_{\perp,\eta} = m\left[\hat{\mathbf{z}} - \left(\hat{\mathbf{z}} \cdot \hat{\mathbf{Q}}_\eta\right)\hat{\mathbf{Q}}_\eta\right]. \tag{7.50}$$

In this case, the following relations hold:

$$|\psi_{\mathbf{Q}_\eta}| = \frac{1}{\sqrt{3}}, \quad \sin\alpha_{\eta\eta'} = \frac{\sqrt{3}}{2}\sin\Theta, \tag{7.51}$$

where $0 < \alpha_{\eta\eta'} < \pi/3$. In this energy functional, v in Eq. (7.40) couples m and Θ; Eq. (7.46) indicates that a positive (negative) v favors (disfavors) helices propagating in the z direction, which tends to decreases (increases) Θ as m increases.

We note that the HL turns into a topologically trivial state with $N_{\text{pair}} = 0$ for $m > \sqrt{3}\sin\Theta$, as shown in Sect. 7.2.3, while the energy can be estimated by the same equation in Eq. (7.41); see also the results in the next subsubsection. We call such a topologically trivial state with a 3D spin texture the 3D $3Q$ state in the following analysis to distinguish it from the 2D one derived from the SkL below.

B. Skyrmion lattice

Next, we calculate the energy for the 2D SkL realized in Eq. (7.10) for $\Theta = \frac{\pi}{2}$. In this state, the energy depends on the relative phases between the constituent helices, as $\mathbf{Q}_\eta$ are not linearly independent (see Sect. 6.1). Thus, we need to extend Eq. (7.10) by replacing $\mathbf{Q}_\eta \cdot \mathbf{r}$ with $\mathbf{Q}_\eta \cdot \mathbf{r} + \varphi_\eta$. For this state, there is an additional energy contribution from the quartic interaction given by

$$f_W = 9U_0 m|\psi_{\mathbf{Q}_1}||\psi_{\mathbf{Q}_2}||\psi_{\mathbf{Q}_3}|\cos\sum_\eta \varphi_\eta. \tag{7.52}$$

Thus, the energy of the SkL is given as

$$F_{\text{SkL}} = F_{\text{HL}} + f_W. \tag{7.53}$$

Note that the relations in Eq. (7.51) hold also in the present case. We find that the energy is always optimized with $\sum_\eta \varphi_\eta = (2n+1)\pi$, where n is an integer.

Similar to the HL case, the SkL turns into a topologically trivial state with $N_{\text{pair}} = 0$ for $m > \sqrt{3}$, while the energy can be estimated by Eq. (7.53); see also the results in the next subsubsection. We call such a topologically trivial state the 2D $3Q$ state in the following analysis.

C. Helical or conical state

Finally, we calculate the energy for the 1D helical or conical state in Eq. (7.10) with $\Theta = 0$, where all three $\mathbf{Q}_\eta$ merge into one. The spin texture is given by

$$\mathbf{S}(\mathbf{r}) = \mathbf{e}_1^1\cos\mathbf{Q}_1\cdot\mathbf{r} + \mathbf{e}_1^2\sin\mathbf{Q}_1\cdot\mathbf{r} + m\hat{\mathbf{z}}, \tag{7.54}$$

where $\mathbf{Q}_1 \parallel \hat{\mathbf{z}}$. For this state, the energy is given by

$$F_{1Q} = f_2(\mathbf{Q}_1) + f_s(\mathbf{Q}_1) + f_m^{(1)}(m) + f_m^{(2)}(\mathbf{Q}_1, m), \tag{7.55}$$

where $|\psi_{\mathbf{Q}_1}| = 1$.

Results

Comparing the energies in Eqs. (7.41), (7.53), and (7.55), we determine the variational ground state. We find a variety of modulations of the spin moiré depending on the parameters in the energy functional, t, u, and v in Eq. (7.40). Figure 7.14 shows two representative results starting from the cubic configuration with $\Theta = \Theta_{\text{cubic}}$ at $m = 0$. The parameters t, u, and v defined in Eq. (7.40) are taken as $(t, u, v) = (3.0, -2.0, -2.0)$, and $(t, u, v) = (3.0, 0.25, 2.0)$ for these two cases. The energy comparison corresponding to Fig. 7.14 is shown in Fig. 7.15.

For the case with $(t, u, v) = (3.0, -2.0, -2.0)$ denoted by the blue lines in Fig. 7.14, Θ gradually increases with m and changes discontinuously to $\Theta = \frac{\pi}{2}$ at $m \simeq 1.054$. As the three $\mathbf{Q}_\eta$ become coplanar for $\Theta = \frac{\pi}{2}$, this is a magnetic phase transition to the 2D SkL; see the inset of Fig. 7.14a. In this case, N_{pair} changes from 4 to 1 at $m \simeq 0.295$, and from 1 to 0 at the magnetic transition to the SkL, as shown in Fig. 7.14b. The former is a topological transition within the HL phase, where $\bar{b}_z$ shows a cusp, as shown in Fig. 7.14c. We note that the energy does not show any anomalies at this topological transition, as shown in Fig. 7.15a. In the magnetic transition at $m \simeq 1.054$, $\bar{b}_z$ jumps to a quantized value -1 in the SkL state, while it vanishes for $m > \sqrt{3}$, as shown in Fig. 7.14c; the system undergoes another transition to the 2D 3Q state at $m = \sqrt{3}$, where the z component of the core spins are turned from negative to positive and the spin texture becomes topologically trivial. Meanwhile, $\bar{e}_z$ also shows a cusp at the topological transition, but it vanishes at the magnetic transition to the SkL with showing a negative δ-function anomaly, as shown in Fig. 7.14d. Note that $\bar{e}_z$ includes both $e_z^m(\mathbf{r})$ and $e_z^\Theta(\mathbf{r})$ in Eq. (7.34), as Θ changes as m, in contrast to the calculations in Sect. 7.2.3. This anomaly comes from the discontinuous change of Θ.

Meanwhile, for the case with $(t, u, v) = (3.0, 0.25, 2.0)$ denoted by the red lines in Fig. 7.14, Θ decreases gradually until the sudden drop to 0 at $m \simeq 1.164$. As all $\mathbf{Q}_\eta$ overlap for $\Theta = 0$, this is a magnetic phase transition to a 1D conical state specified by the single $\mathbf{Q} \parallel \hat{\mathbf{z}}$; see the inset of Fig. 7.14a. In this case, N_{pair} changes from 4 to 1 at $m \simeq 0.280$ and from 1 to 0 at $m \simeq 1.139$; namely, the system undergoes two successive topological transitions before the magnetic transition to the conical state. Similar to the previous case, the energy smoothly changes at the topological transitions for the 3D HL, as shown in Fig. 7.15b. The narrow window for $1.139 \lesssim m \lesssim 1.164$ is a topologically-trivial 3D 3Q state with $N_{\text{pair}} = 0$. This is in contrast to the above case with $(t, u, v) = (3.0, -2.0, -2.0)$ where the topological transition with $N_{\text{pair}} = 1 \to 0$ occurs simultaneously with the magnetic transition to the SkL. In addition to the cusp at the topological transition at $m \simeq 0.280$, $\bar{e}_z$ shows a small

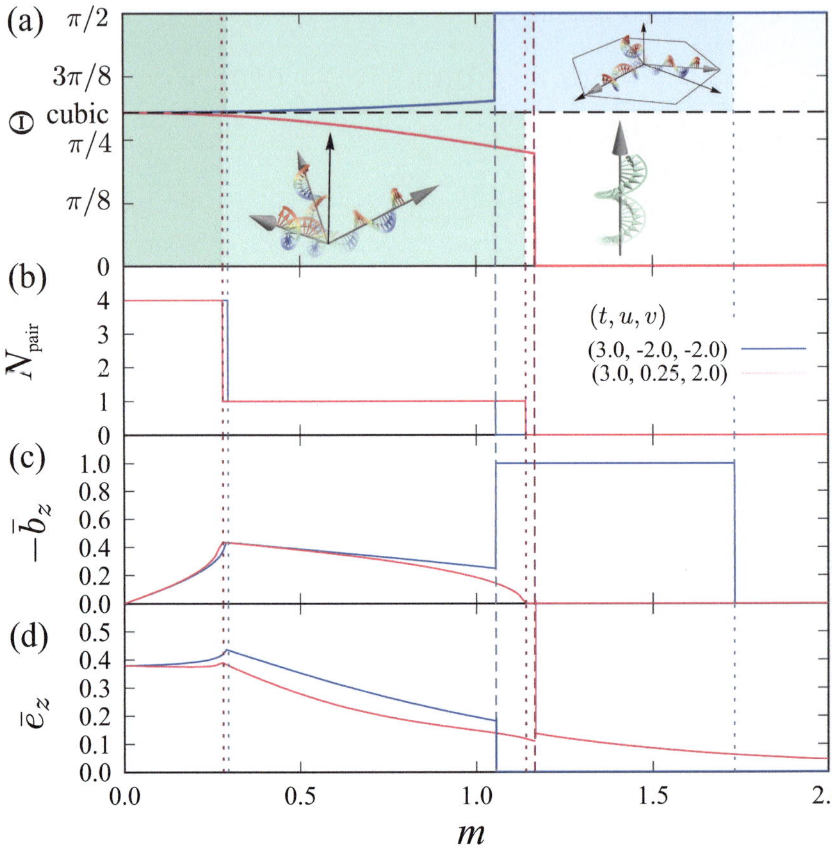

Fig. 7.14 **a** Optimized value of Θ obtained by the variational calculations for the model in Eq. (3.1) while changing m. We show two cases starting from the 3D HL at $\Theta = \Theta_{\rm cubic}$ for $m = 0$: One with $(t, u, v) = (3.0, -2.0, -2.0)$ exhibits a transition to $\Theta = \frac{\pi}{2}$ (2D SkL with coplanar $\mathbf{Q}_\eta \perp \hat{\mathbf{z}}$), and the other with $(t, u, v) = (3.0, 0.25, 2.0)$ to $\Theta = 0$ (1D conical state with $\mathbf{Q} \parallel \hat{\mathbf{z}}$). Schematic figures of each spin state are shown in the inset. Corresponding values of **b** $N_{\rm pair}$, **c** $-\bar{b}_z$, and **d** $\bar{e}_z$. The green regions in (**a**) represent the states with $0 < \Theta < \pi/2$; the dark, normal, and pale colors are for the HLs with $N_{\rm pair} = 4$ and 1, and the 3D 3Q state with $N_{\rm pair} = 0$, respectively [the 3D 3Q state is limited to the narrow region of $1.139 \lesssim m \lesssim 1.164$ for the case with $(t, u, v) = (3.0, 0.25, 2.0)$]. The dark and pale blue regions for the case with $(t, u, v) = (3.0, 0.25, 2.0)$ represent the SkL and the 2D 3Q states, respectively, at $\Theta = \pi/2$. The red region for the case with $(t, u, v) = (3.0, 0.25, 2.0)$ represents the 1Q conical state at $\Theta = 0$. The vertical dashed and dotted lines denote the magnetic and topological transitions, respectively. Reproduced from Ref. [18]. Copyright (2021) by the American Physical Society

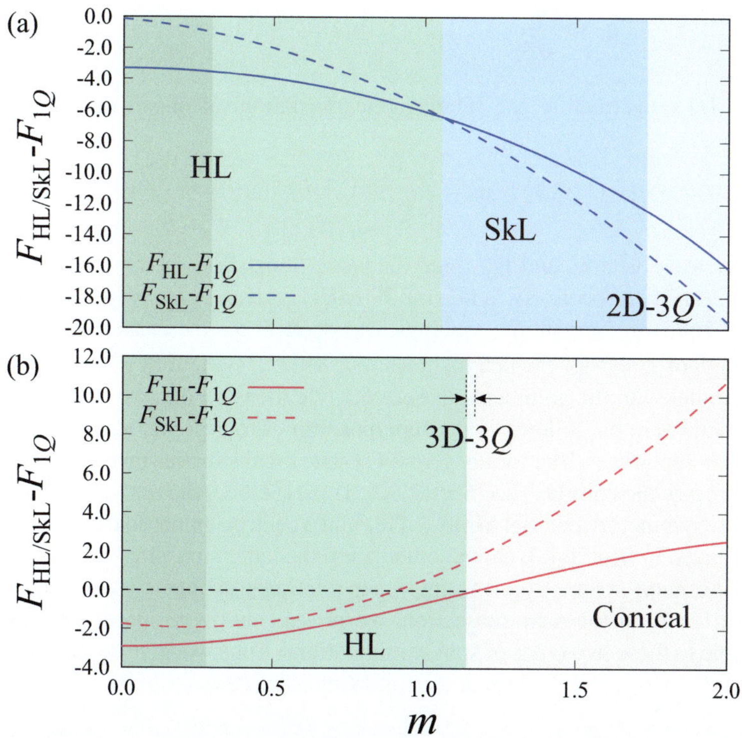

Fig. 7.15 Comparison of the energy between different states for **a** $(t, u, v) = (3.0, -2.0, -2.0)$ and **b** $(t, u, v) = (3.0, 0.25, 2.0)$. The solid (dashed) lines are the energy for the HL in Eq. (7.41) [SkL in Eq. (7.53)] measured from that for the 1Q conical state in Eq. (7.55). The background colors are common to those in Fig. 7.14a. Reproduced from Ref. [18]. Copyright (2021) by the American Physical Society

hump at $m \simeq 1.1$ before the second topological transition, which is ascribed to the contribution from $e_z^{\Theta}(\mathbf{r})$. $\bar{e}_z$ remains nonzero in the conical state after showing a positive δ-function anomaly at the magnetic phase transition.

7.3 Phase Shift in 4Q-HL

In this section, we discuss the effect of phase shifts on the 4Q-HLs composed of four wave vectors in three dimensions, i.e., $N_Q = 4$ and $d = 3$, by using the hyperspace representation introduced in Sect. 4.6. Specifically, we consider Eq. (1.11) with $\mathcal{Q}_\eta = \mathbf{q}_\eta \cdot \mathbf{r} + \varphi_\eta$ and four $\mathbf{q}_\eta$ given by

$$\mathbf{q}_1 = \frac{q}{\sqrt{3}}(1, 1, 1), \quad \mathbf{q}_2 = \frac{q}{\sqrt{3}}(-1, -1, 1),$$

$$\mathbf{q}_3 = \frac{q}{\sqrt{3}}(-1, 1, -1), \quad \mathbf{q}_4 = \frac{q}{\sqrt{3}}(1, -1, -1). \tag{7.56}$$

For this $4Q$ spin structure, the 3D magnetic translation vectors are defined as

$$\mathbf{a}_1 = \frac{\sqrt{3}\pi}{q}(1, 0, 1), \quad \mathbf{a}_2 = \frac{\sqrt{3}\pi}{q}(1, 1, 0), \quad \mathbf{a}_3 = \frac{\sqrt{3}\pi}{q}(0, 1, 1). \tag{7.57}$$

The four wave vectors and the three magnetic translation vectors are depicted in Fig. 7.16a and b, respectively. While the 3D MUC is given by the gray rhombohedron in Fig. 7.16b, we compute the topological properties for the dashed cube with the side length of $L = \frac{2\sqrt{3}\pi}{q}$, which includes four MUCs, in the following analyses.

In parallel with the arguments in Sect. 6.1, we focus on the two types of the $4Q$ spin structures in the following: the superposition of four proper screws and that of four sinusoidal waves. The former screw $4Q$ state breaks spatial-inversion symmetry and possesses the chirality; see Sect. 7.3.2. This type of spin structure is found in a noncentrosymmetric material $MnSi_{1-x}Ge_x$ and a centrosymmetric material $SrFeO_3$, as introduced in Sect. 1.4.3. On the other hand, the latter sinusoidal $4Q$ state retains spatial-inversion symmetry or S_4 improper rotational symmetry about the z axis; see Sect. 7.3.3. In the next subsection, we present the hyperspace representation applicable to these two types of spin configurations. Then, we give the results for the screw and sinusoidal $4Q$ states in the following subsections.

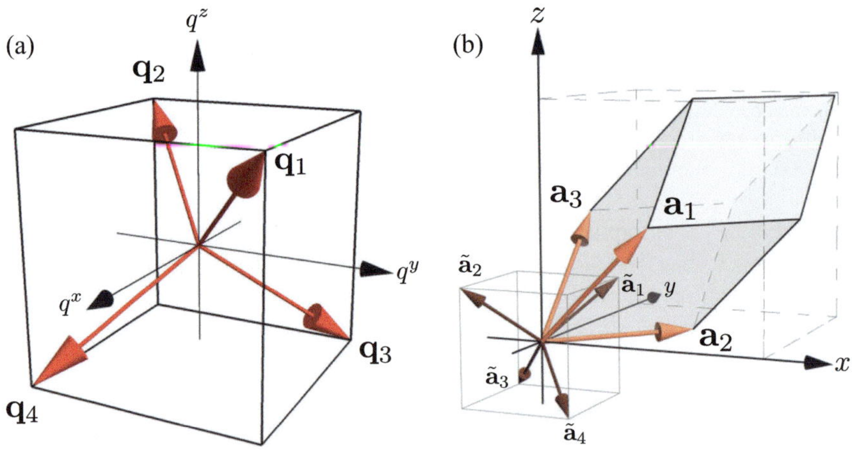

Fig. 7.16 **a** Schematic picture of the wave vectors in 3D reciprocal space, $\mathbf{q}_\eta$. **b** Corresponding magnetic translation vectors in the 3D real space, $\mathbf{a}_\eta$, represented by orange arrows. The brown arrows represent the projections of $\mathbf{A}_\eta$ onto the xyz space denoted by $\tilde{\mathbf{a}}_\eta$. The gray rhombohedron is the MUC, and the dashed cube with the side length of $L = \frac{2\sqrt{3}\pi}{q}$ includes four MUCs. The small gray cube has the side length of $\frac{L}{2}$. Reproduced from Ref. [19]. Copyright (2022) by the American Physical Society

7.3.1 Hyperspace Representation of the 4Q States

Following the same procedure as for the 3Q states in Sect. 6.1.1, we construct the hyperspace representation of the 4Q spin structures. Considering 4D reciprocal hyperspace, we set the wave vectors $\mathbf{Q}_\eta = (Q_\eta^X, Q_\eta^Y, Q_\eta^Z, Q_\eta^W)$ in Eq. (4.18) without loss of generality:

$$\mathbf{Q}_1 = q\left(\frac{1}{\sqrt{3}}, \frac{1}{\sqrt{3}}, \frac{1}{\sqrt{3}}, \frac{1}{q}\right), \qquad \mathbf{Q}_2 = q\left(-\frac{1}{\sqrt{3}}, -\frac{1}{\sqrt{3}}, \frac{1}{\sqrt{3}}, \frac{1}{q}\right),$$
$$\mathbf{Q}_3 = q\left(-\frac{1}{\sqrt{3}}, \frac{1}{\sqrt{3}}, -\frac{1}{\sqrt{3}}, \frac{1}{q}\right), \quad \mathbf{Q}_4 = q\left(\frac{1}{\sqrt{3}}, -\frac{1}{\sqrt{3}}, -\frac{1}{\sqrt{3}}, \frac{1}{q}\right), \quad (7.58)$$

where Q_η^X, Q_η^Y, and Q_η^Z are taken to be proportional to q_η^x, q_η^y, and q_η^z in Eq. (7.56), respectively. In this setting, $\mathbf{q}_\eta$ is a projection of $\mathbf{Q}_\eta$ onto the $q^x q^y q^z$ space. Then, we obtain the corresponding magnetic translation vectors $\mathbf{A}_\eta$ in the 4D hyperspace as

$$\mathbf{A}_1 = \frac{\pi}{2q}\left(\sqrt{3}, \sqrt{3}, \sqrt{3}, q\right), \quad \mathbf{A}_2 = \frac{\pi}{2q}\left(-\sqrt{3}, -\sqrt{3}, \sqrt{3}, q\right),$$
$$\mathbf{A}_3 = \frac{\pi}{2q}\left(-\sqrt{3}, \sqrt{3}, -\sqrt{3}, q\right), \quad \mathbf{A}_4 = \frac{\pi}{2q}\left(\sqrt{3}, -\sqrt{3}, -\sqrt{3}, q\right). \quad (7.59)$$

Note that $\mathbf{a}_1$, $\mathbf{a}_2$, and $\mathbf{a}_3$ are given by $\tilde{\mathbf{a}}_1 - \tilde{\mathbf{a}}_3$, $\tilde{\mathbf{a}}_1 - \tilde{\mathbf{a}}_2$, and $\tilde{\mathbf{a}}_1 - \tilde{\mathbf{a}}_4$, respectively, where $\tilde{\mathbf{a}}_\eta$ are the projections of $\mathbf{A}_\eta$ onto the xyz space; see Fig. 7.16b.

By using Eq. (4.20), the hyperspace positions $\mathbf{R} = (X, Y, Z, W)$ are related with the real-space positions $\mathbf{r}$ as

$$\begin{pmatrix} X \\ Y \\ Z \\ W \end{pmatrix} = V_4 \begin{pmatrix} x \\ y \\ z \\ 1 \end{pmatrix}, \quad (7.60)$$

where V_4 includes the phases as

$$V_4 = \begin{pmatrix} 1\,0\,0\; \frac{\sqrt{3}}{4q}(\varphi_1 - \varphi_2 - \varphi_3 + \varphi_4) \\ 0\,1\,0\; \frac{\sqrt{3}}{4q}(\varphi_1 - \varphi_2 + \varphi_3 - \varphi_4) \\ 0\,0\,1\; \frac{\sqrt{3}}{4q}(\varphi_1 + \varphi_2 - \varphi_3 - \varphi_4) \\ 0\,0\,0\; \frac{1}{4}(\varphi_1 + \varphi_2 + \varphi_3 + \varphi_4) \end{pmatrix}. \quad (7.61)$$

Equation (7.60) tells that the spin configuration in the original 3D xyz space is the same as the one on a hyperplane in the 4D hyperspace with

$$W = \frac{1}{4}\tilde{\varphi}, \quad (7.62)$$

where

$$\tilde{\varphi} = \sum_{\eta} \varphi_\eta = \varphi_1 + \varphi_2 + \varphi_3 + \varphi_4. \tag{7.63}$$

Similar to the $3Q$ case in Sect. 6.1.1, only the summation of the phases φ_η is relevant, instead of each value of φ_η, and $\tilde{\varphi}$ has 2π periodicity. Due to this periodicity, the 2π phase shift in $\tilde{\varphi}$ ($\Delta\tilde{\varphi} = \sum_\eta \Delta\varphi_\eta = 2\pi$) corresponds to a spatial translation by

$$\Delta\mathbf{r}_\eta = \tilde{\mathbf{a}}_\eta - \sum_{\eta'=1}^{4} \frac{\Delta\varphi_{\eta'}}{2\pi} \tilde{\mathbf{a}}_{\eta'}, \tag{7.64}$$

where η may take any of 1, 2, 3, and 4.

7.3.2 Screw 4*Q* State

In this subsection, we analyze the effect of phase shifts on the screw $4Q$ state composed of four proper screws given by

$$\mathbf{S}(\mathbf{r}) \propto \begin{pmatrix} \frac{1}{\sqrt{2}}\left((-c_1 + c_2 - c_3 + c_4) + \frac{1}{\sqrt{3}}(-s_1 + s_2 - s_3 + s_4)\right) \\ \frac{1}{\sqrt{2}}\left((c_1 - c_2 - c_3 + c_4) - \frac{1}{\sqrt{3}}(s_1 - s_2 - s_3 + s_4)\right) \\ \sqrt{\frac{2}{3}}(s_1 + s_2 + s_3 + s_4) + 2\,m \end{pmatrix}, \tag{7.65}$$

with $c_\eta = \cos\mathcal{Q}_\eta$ and $s_\eta = \sin\mathcal{Q}_\eta$, which is obtained from Eq. (1.11) by taking $N_Q = 4$ and

$$\psi_\eta^{\mathrm{c}} = \psi_\eta^{\mathrm{s}} = \frac{1}{2}, \tag{7.66}$$

$$\mathbf{e}_\eta^1 = \frac{\hat{\mathbf{z}} \times \mathbf{e}_\eta^0}{|\hat{\mathbf{z}} \times \mathbf{e}_\eta^0|}, \quad \mathbf{e}_\eta^2 = \mathbf{e}_\eta^0 \times \mathbf{e}_\eta^1, \quad \mathbf{e}_\eta^0 = \frac{\mathbf{q}_\eta}{q}, \tag{7.67}$$

$$\mathbf{m} = m\hat{\mathbf{z}}. \tag{7.68}$$

With the same manner to Tables 6.1–6.3, we summarize the symmetry operations and their decompositions for the screw $4Q$ state in Table 7.1. Similar to the previous arguments, from Table 7.1, we can obtain the symmetry operations which do not change the spin texture. In the 3D system, however, some of the symmetry operations are nonsymmorphic. For instance, the C_{4z} operation at $\tilde{\varphi} = \pi$ is reduced to the spatial translation by $\tilde{\mathbf{a}}_\eta$ as shown in Table 7.1, and hence, the operations of $\{C_{4z}| - \tilde{\mathbf{a}}_\eta\}$ does not change the spin texture, where $\{\mathcal{O}|\mathbf{t}\}$ denotes the translation by $\mathbf{t}$ after operating $\mathcal{O}$. This corresponds to a screw operation. Note that the situation is different from the

Table 7.1 Similar table to Tables 6.1–6.3 for the screw 4Q state in Eq. (7.65). The notations are common to those in Table 6.1. We take $\varphi_\eta = \frac{\tilde{\varphi}}{4}$ without loss of generality

Operation	Sum of phases	Translation	Magnetization
C_{4z}	$\tilde{\varphi} \to 2\pi - \tilde{\varphi}$	$\tilde{\mathbf{a}}_\eta$	$m \to m$
C_{2x}	$\tilde{\varphi} \to \tilde{\varphi}$	$\tilde{\mathbf{a}}_\eta + \tilde{\mathbf{a}}_{\eta'}$	$m \to -m$
$\mathcal{T}$	$\tilde{\varphi} \to \tilde{\varphi}$	$\tilde{\mathbf{a}}_\eta + \tilde{\mathbf{a}}_{\eta'}$	$m \to -m$

3Q cases, where a translation combined with rotation can be represented by a shift of the rotation axis. By considering such relations, we find that the screw 4Q state is symmetric under C_{4z}, $\mathcal{T}C_{2x}$, and their combinations for $\tilde{\varphi} = 0$, $\{C_{4z}| - \tilde{\mathbf{a}}_\eta\}$, $\mathcal{T}C_{2x}$, and their combinations for $\tilde{\varphi} = \pi$, otherwise C_{2z}, $\mathcal{T}C_{2x}$, and their combinations.

Hedgehogs in Hyperspace

To discuss the phase shift in the screw 4Q state in Eq. (7.65), we study the corresponding spin texture in the 4D hyperspace whose reciprocal space is spanned by the wave vectors $\mathbf{Q}_\eta$ in Eq. (7.58). As discussed in Sect. 7.3.1, a real-space spin configuration for a given phase summation $\tilde{\varphi}$ in the original 3D space corresponds to a hyperspace spin configuration on the hyperplane with $W = \frac{\tilde{\varphi}}{4}$. Following the procedure in the first subsubsection of Sect. 6.1.2, we first compute the positions of the topological defects in the 4D hyperspace. In the present case, however, the solutions for $\mathbf{S}(\mathbf{R}) = 0$ are given by lines rather than points in the 4D hyperspace. When $|m| < \frac{4}{\sqrt{6}} \sin W$, we obtain two solutions analytically:

$$(X^*, Y^*, Z^*) = L\left(0, 0, \frac{1}{2} \pm \frac{1}{2\pi} r_1^{\mathrm{scr}}(m, W)\right), \tag{7.69}$$

where

$$r_1^{\mathrm{scr}}(m, W) = \arccos\left(\frac{\sqrt{6}m}{4\sin W}\right). \tag{7.70}$$

Here, we show the solutions within $0 \le X, Y < \frac{L}{2}$ and $0 \le Z < L$, but the spatial translations of them with $\mathbf{a}_\eta$ in Eq. (7.57) also satisfy $\mathbf{S}(\mathbf{R}) = 0$. Meanwhile, when $|m| < \frac{4}{\sqrt{6}} \cos W$, we obtain other two solutions as

$$(X^*, Y^*, Z^*) = L\left(\frac{1}{4}, \frac{1}{4}, \frac{1}{2\pi} r_2^{\mathrm{scr}}(m, W)\right),\ L\left(\frac{1}{4}, \frac{1}{4}, \frac{1}{2} - \frac{1}{2\pi} r_2^{\mathrm{scr}}(m, W)\right), \tag{7.71}$$

where

$$r_2^{\mathrm{scr}}(m, W) = \arcsin\left(\frac{\sqrt{6}m}{4\cos W}\right). \tag{7.72}$$

The spatial translations also apply to this case. These solutions do not change their XY coordinates with m and $\tilde{\varphi}$. In addition to the above analytical solutions, we also find the other solutions by numerically solving the equation $\mathbf{S}(\mathbf{R}) = 0$. The two conditions $S_x(\mathbf{R}) = 0$ and $S_y(\mathbf{R}) = 0$ can be reduced to

$$\tan\left(\frac{q}{\sqrt{3}}Z^*\right) = \pm\sqrt{\frac{3\tan^2 W - 1}{3 - \tan^2 W}}, \tag{7.73}$$

and

$$\tan\left(\frac{q}{\sqrt{3}}Y^*\right) = \mp\sqrt{\frac{\left(\sqrt{3}\tan W + 1\right)\left(\sqrt{3} + \tan W\right)}{\left(\sqrt{3}\tan W - 1\right)\left(\sqrt{3} - \tan W\right)}}\tan\left(\frac{q}{\sqrt{3}}X^*\right). \tag{7.74}$$

Meanwhile, $S_z(\mathbf{R}) = 0$ is reduced to

$$\begin{aligned} &\cos\left(\frac{q}{\sqrt{3}}X^*\right)\cos\left(\frac{q}{\sqrt{3}}Y^*\right)\cos\left(\frac{q}{\sqrt{3}}Z^*\right)\sin W \\ &-\sin\left(\frac{q}{\sqrt{3}}X^*\right)\sin\left(\frac{q}{\sqrt{3}}Y^*\right)\sin\left(\frac{q}{\sqrt{3}}Z^*\right)\cos W + \frac{\sqrt{6}}{4}m = 0. \end{aligned} \tag{7.75}$$

We find that the numerical solutions exist when $\frac{1}{\sqrt{3}} < \tan W < \sqrt{3}$, namely $\frac{2\pi}{3} < \tilde{\varphi} < \frac{4\pi}{3}$. We note that the solutions in Eqs. (7.73)–(7.75) were not mentioned in the previous study for the screw $4Q$ state with $\tilde{\varphi} = \pi$ [16]. While the above solutions are points in the original 3D space, the line solutions are also obtained only for $m = 0$. When $W = 0$, $\frac{\pi}{6}$, and $\frac{\pi}{3}$, we find analytical solutions

$$(X^*, Y^*, Z^*) = (0, 0, *), \left(*, \frac{L}{4}, 0\right), \left(0, *, \frac{L}{4}\right), \tag{7.76}$$

where $*$ takes any value, respectively.

In the 4D hyperspace, the topological objects defined by the above solutions of $\mathbf{S}(\mathbf{R}) = 0$ except for Eq. (7.76) form closed loops, and the intersection of the loops by the hyperplane with $W = \frac{\tilde{\varphi}}{4}$ gives topological point defects in the original 3D space, which correspond to the hedgehogs and antihedgehogs discussed below. The hedgehog and antihedgehog always appear in pairs for each closed loop. The Dirac string connecting the hedgehog-antihedgehog pair in the 3D space is derived from the intersection of a 2D membrane in the 4D hyperspace whose edge and surface

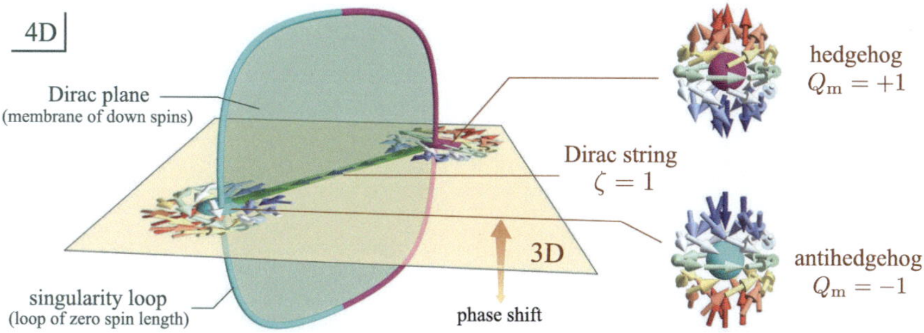

Fig. 7.17 Schematic figure for the correspondence between the original space and the hyperspace for the 3D $4Q$ case. The 3D $4Q$-HL is obtained as a 3D intersection (orange plane) of the 4D loop lattice in the 4D hyperspace. The cyan and magenta lines represent the singularity loop on which the spin length vanishes; the color distinguishes the source and sink of spins in the 3D intersection. The pale green plane represents the Dirac plane defined by downward spins. Other notations are common to those in Fig. 6.5. Reproduced from Ref. [19]. Copyright (2022) by the American Physical Society

are defined by $\mathbf{S}(\mathbf{R}) = 0$ and $\mathbf{S}(\mathbf{R}) = -\hat{\mathbf{z}}$, respectively. The 2D membrane can be regarded as an extension of the Dirac string in the higher dimension and hence, we may call it the Dirac plane. Thus, the hyperspace representation of the $4Q$ state is given by a 4D lattice composed of such closed loops, and the $4Q$ spin texture with the phase degree of freedom is defined as a 3D intersection of the 4D loop lattice. The correspondence of the topological properties between the original 3D space and the 4D hyperspace is schematically illustrated in Fig. 7.17.

Since it is difficult to visualize the 4D hyperspace, we present the topological objects in the original 3D space which are derived from the hyperspace representation above. Figure 7.18 shows the systematic change of the topological defects in the 3D space while changing m in Eq. (7.65) with $\tilde{\varphi} = \frac{\pi}{3}$ and π. Following the procedure in the second subsubsection of Sect. 6.1.2, we compute the monopole charge for the topological defects, Q_m in Eq. (1.7), and the vorticity of the Dirac strings which connect the hedgehogs and antihedgehogs, ζ in Eq. (1.4), in which the integral is taken along a closed path surrounding the Dirac string on the xy plane (see also the second subsubsection of Sect. 7.2.2). In order to distinguish the different topological phases, we also compute the total number of the topological defects, the hedgehogs and antihedgehogs, within the dashed cube shown in Fig. 7.16b, denoted by N_m; the total number per MUC is given by $\frac{N_m}{4}$.

First, we discuss the case of $\tilde{\varphi} = \frac{\pi}{3}$ shown in Fig. 7.18a, b, and c. For $m = 0$, there are 16 topological defects in total and half of them are hedgehogs with $Q_m = +1$ and the others are antihedgehogs with $Q_m = -1$, as shown in Fig. 7.18a. The hedgehogs and antihedgehogs derived from Eq. (7.69) are connected by the Dirac strings with $\zeta = -1$, while the other topological defects from Eq. (7.71) are connected by the Dirac strings with $\zeta = +1$. As discussed in Sect. 7.2.2, the length of the Dirac strings projected onto the z axis is directly related to the EMF. In this case, all the Dirac

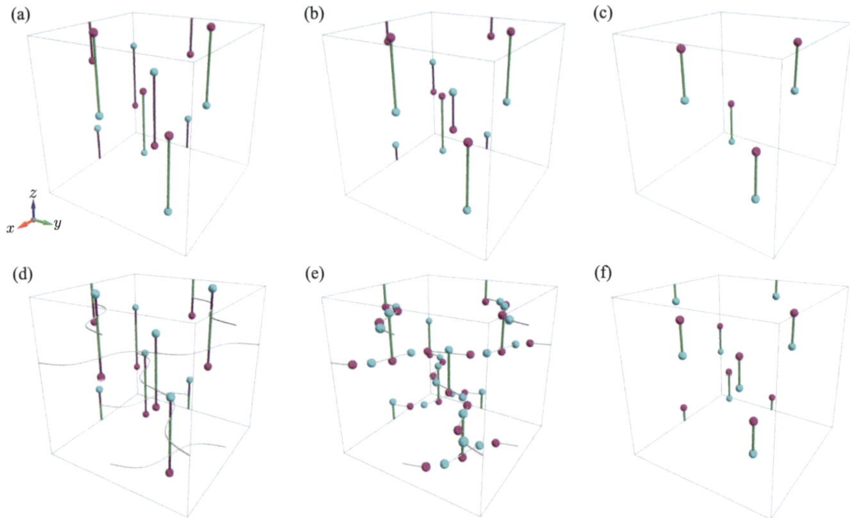

Fig. 7.18 Real-space distribution of the hedgehogs and antihedgehogs, and the Dirac strings within the L^3 cube [see Fig. 7.16b] for the screw $4Q$ state in Eq. (7.65) while changing m and $\tilde{\varphi}$: **a** $m = 0$, **b** $m = 0.3$, and **c** $m = 0.9$ at $\tilde{\varphi} = \pi/3$, and **d** $m = 0$, **e** $m = 0.7$, and **f** $m = 0.9$ at $\tilde{\varphi} = \pi$. The notations are common to those in Fig. 6.3, except for the Dirac strings running on the horizontal planes, whose vorticities are ill-defined, denoted by the white lines in (**d**) and (**e**). Reproduced from Ref. [19]. Copyright (2022) by the American Physical Society

strings run along the z axis and have the same length of $\frac{L}{2}$, resulting in the cancellation of the net EMF at $m = 0$ [see Eq. (7.78)]. When introducing m, the hedgehogs and antihedgehogs move toward their counterparts along the Dirac strings, as exemplified in Fig. 7.18b for $m = 0.3$. The Dirac strings with $\zeta = -1$ become shorter than those with $\zeta = +1$: The lengths change as $L\left(\frac{1}{2} - \frac{1}{\pi} r_2^{\mathrm{scr}}\left(m, \frac{\tilde{\varphi}}{4}\right)\right)$ and $\frac{L}{\pi} r_1^{\mathrm{scr}}\left(m, \frac{\tilde{\varphi}}{4}\right)$ for $\zeta = +1$ and -1, respectively. This differentiation gives rise to a net EMF, as will be discussed in the next subsubsection. By further increasing m, the hedgehogs and antihedgehogs connected by the Dirac strings with $\zeta = -1$ disappear with pair annihilation at $m = \frac{4}{\sqrt{6}} \sin \frac{\tilde{\varphi}}{4}$; namely N_{m} is reduced to 8, which defines a topological transition between the phases with different N_{m}. The remaining hedgehogs and antihedgehogs move toward each other while further increasing m, as shown in Fig. 7.18c. They also vanish with pair annihilation at $m = \frac{4}{\sqrt{6}} \cos \frac{\tilde{\varphi}}{4}$, which defines the other topological transition into a topologically trivial state with $N_{\mathrm{m}} = 0$.

Next, we discuss the case of $\tilde{\varphi} = \pi$ shown in Fig. 7.18d, e, and f. When $m = 0$, as shown in Fig. 7.18d, the system has eight pairs of the hedgehogs and antihedgehogs, similar to the case of $\tilde{\varphi} = \frac{\pi}{3}$. However, while the positions of the topological defects as well as their total number are same as those in Fig. 7.18a, half of the hedgehog and antihedgehog pairs are exchanged; all the Dirac strings have the hedgehogs at their lower edges in Fig. 7.18d, whereas only half of them do in Fig. 7.18a. Moreover, we find additional Dirac strings running on the planes perpendicular to the z axis, which are denoted by the white lines in the figure. Note that the vorticity is ill-defined for these horizontal Dirac strings since the integral is taken on a closed path on the xy plane. As shown in Fig. 7.18d, they intersect with the Dirac strings running along the z axis, whose vorticities ζ change their signs at the crossing points. While the vorticity ζ changes its sign along the Dirac strings with $\zeta = 1$ and -1 have the same length, leading to the cancellation of the net EMF, similar to the $\tilde{\varphi} = \frac{\pi}{3}$ case. By introducing m, the hedgehogs and antihedgehogs move toward each other along the vertical Dirac strings as in the case of $\tilde{\varphi} = \frac{\pi}{3}$, while the horizontal Dirac strings are intact. Through this motion, only the total length of the Dirac strings with $\zeta = -1$ is reduced [see Fig. 7.18d], leading to a large net EMF, as will be discussed in the next subsubsection. When m exceeds $\frac{2}{3}$, however, N_{m} increases from 16 to 48, as depicted in Fig. 7.18e. This is caused by a peculiar topological transition with the increase of N_{m} discussed in detail below. The additional defects are obtained from the numerical solutions with Eqs. (7.73), (7.74), and (7.75) for $m > \frac{2}{3}$, which appear in pairs on the horizontal Dirac strings; the horizontal Dirac strings are cut into pieces, both ends of which form hedgehogs or antihedgehogs. In other words, a cut results in pair creation of the hedgehog and antihedgehog. In this region, three hedgehogs and three antihedgehogs (a pair of hedgehog and antihedgehog with the vertical Dirac string, and hedgehog pair and antihedgehog pair with the horizontal Dirac strings) form a cluster like a twisted two-barred cross. While further increasing m, the hedgehogs and antihedgehogs move along the Dirac strings, and N_{m} decreases from 48 to 16 at $m = \sqrt{\frac{2}{3}}$. Here, two hedgehogs and one antihedgehog (or one hedgehog and two antihedgehogs) collide with each other at the same time in each cluster, leaving one (anti)hedgehog (see below). After this topological transition, all the Dirac strings have the hedgehogs at the upper edges, and hence, the vorticities are $\zeta = +1$, as shown in Fig. 7.18f. Finally, the remaining hedgehogs and antihedgehogs move toward their counterparts along the Dirac strings, and cause pair annihilation at $m = \frac{2}{\sqrt{3}}$, by which the system enters into a topologically trivial phase with $N_{\mathrm{m}} = 0$.

The evolution of the topological objects for the case of $\tilde{\varphi} = \pi$ includes, at least, two striking features. One is the increase of the total number of the hedgehogs and antihedgehogs while increasing m. This is highly nontrivial since usually the hedgehogs and antihedgehogs move toward each other along the Dirac string and cause pair annihilation, and hence, their number does not increase while increasing m [1, 9, 16, 18, 25]. It is worthy noting that the increase is caused by pair creation on the horizontal Dirac strings whose vorticities are ill-defined. The process is detailed in Fig. 7.19a and b. The other striking feature is the simultaneous fusion of three defects, as depicted in Fig. 7.19c and d. In this process, two hedgehogs (antihedgehogs) on the horizontal Dirac string and one antihedgehog (hedgehog) on the vertical Dirac string move toward the crossing point of the vertical and horizontal Dirac strings, and cause the fusion into a single hedgehog (antihedgehog). Note that the total monopole charge is conserved through the fusion. Hence, both striking features originate from the peculiar horizontal Dirac strings.

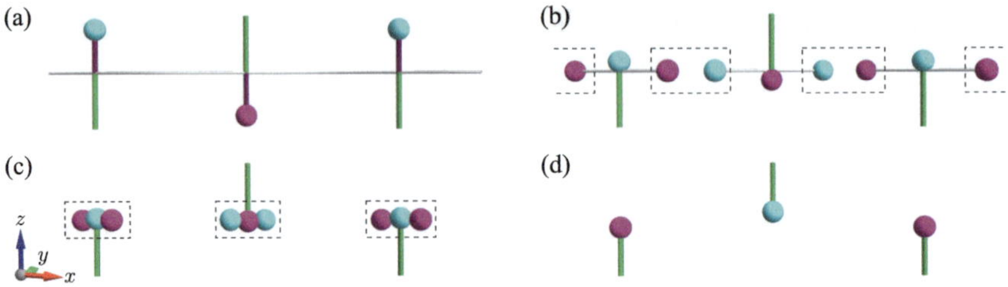

Fig. 7.19 Enlarged figures for the systematic evolution of the hedgehogs and antihedgehogs while changing m with $\tilde{\varphi} = \pi$: **a** $m = 0$, **b** $m = 0.7$, **c** $m = 0.8$, and **d** $m = 0.9$. The hedgehogs and antihedgehogs enclosed with the black dotted lines in (**b**) are generated by pair creation on the horizontal Dirac strings. Meanwhile, the three topological defects enclosed with the black dotted lines in (**c**) collide with each other simultaneously and cause a fusion into a hedgehog or an antihedgehog in (**d**). A part of the topological defects and the Dirac strings are depicted for better visibility. The notations are common to those in Fig. 7.18. Reproduced from Ref. [19]. Copyright (2022) by the American Physical Society

Thus far, we show the results only for $\tilde{\varphi} = \pi/3$ and π, but we note that the other cases fall into either behavior qualitatively. Specifically, the cases for $\frac{2\pi}{3} < \tilde{\varphi} < \frac{4\pi}{3}$ belong to the latter, while the others to the former. In the latter class, we find the additional numerical solutions as described above, which lead to the peculiar pair creation and fusion of the topological defects. We note that the fusion occurs simultaneously for all the created pairs when $\tilde{\varphi} = \pi$, while it takes place successively half by half for other values in $\frac{2\pi}{3} < \tilde{\varphi} < \frac{4\pi}{3}$, as discussed below.

Topological Phase Diagram

Figure 7.20 summarizes the topological phase diagram on the plane of m and $\tilde{\varphi}$ for the screw 4Q state in Eq. (7.65), determined by N_{m}. The result is periodic in the $\tilde{\varphi}$ direction with period of 2π and symmetric with respect to $\tilde{\varphi} = \pi$. We also plot the net EMF $\bar{b}_z$, which is defined as

$$\bar{b}_z = \frac{1}{4\pi L} \int d\mathcal{V} b_z(\mathbf{r}), \tag{7.77}$$

where the volume integration is taken within the L^3 cube in Fig. 7.16b. Following the arguments in Sect. 7.2.2 and Refs. [9, 16, 18, 25], $\bar{b}_z$ is rewritten into

$$\bar{b}_z = -\frac{1}{L} \sum_k \left(l_k^+ - l_k^- \right), \tag{7.78}$$

where l_k^+ and l_k^- denote the length of the kth Dirac strings with $\zeta = +1$ and $\zeta = -1$ projected onto the z axis, respectively, and the sum is taken for all the Dirac strings

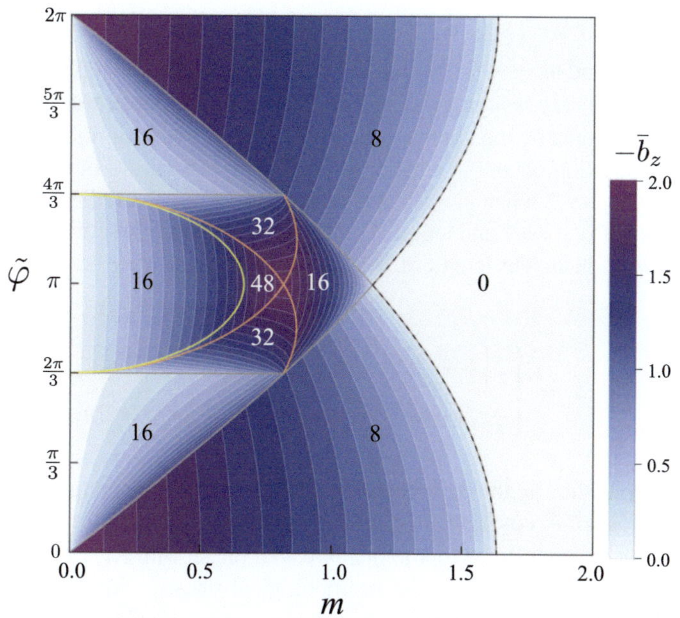

Fig. 7.20 Topological phase diagram for the screw 4Q state on the plane of m and $\tilde{\varphi}$, determined by the number of hedgehogs and antihedgehogs within the cube, N_m [see Fig. 7.16b]. The number displayed in each region represents N_m. The contour plot indicates the EMF $-\bar{b}_z$; the white lines denote the contours drawn every 0.1, and the black dashed lines denote the boundary to $\bar{b}_z = 0$. The gray solid lines denote pair annihilation of hedgehogs and antihedgehogs. The yellow line and the orange lines denote pair creation of hedgehogs and antihedgehogs and fusion of three topological defects, respectively, while increasing m. Reproduced from Ref. [19]. Copyright (2022) by the American Physical Society

involved in the cube. Note that the values of $-\bar{b}_z$ are plotted by contour in Fig. 7.20. The phase diagram for $m < 0$ is obtained in the same form with sign reversal of $-\bar{b}_z$ since the spin texture with (φ_η, m) is obtained by time-reversal operation on that with $(\varphi_\eta + \pi, -m)$.

In the phase diagram, we find the topological phases with $N_\mathrm{m} = 8, 16, 32$, and 48, in addition to the trivial phase with $N_\mathrm{m} = 0$ in the large m region. When $0 < \tilde{\varphi} < \frac{2\pi}{3}$ or $\frac{4\pi}{3} < \tilde{\varphi} < 2\pi$, the phases with $N_\mathrm{m} = 8$ and 16 appear. In the phase with $N_\mathrm{m} = 16$ for small m, 8 hedgehogs and 8 antihedgehogs are connected by the Dirac strings with $\zeta = +1$ and -1, and the Dirac strings with $\zeta = -1$ are shorter than those with $\zeta = +1$, as exemplified in Fig. 7.18b. While increasing m, the length difference between the long and short Dirac strings increases, leading to the increase of $-\bar{b}_z$. Specifically, the value of $-\bar{b}_z$ is given by

$$-\bar{b}_z = \pm 2\left[1 - \frac{2}{\pi}\left(r_1^{\mathrm{scr}}\left(m, \frac{\tilde{\varphi}}{4}\right) + r_2^{\mathrm{scr}}\left(m, \frac{\tilde{\varphi}}{4}\right)\right)\right], \tag{7.79}$$

where the upper (lower) sign is for $0 < \tilde{\varphi} < \frac{2\pi}{3}$ ($\frac{2\pi}{3} < \tilde{\varphi} < 2\pi$). At $m = \frac{4}{\sqrt{6}} \sin \frac{\tilde{\varphi}}{4}$ for $0 < \tilde{\varphi} < \frac{2\pi}{3}$ and $m = \frac{4}{\sqrt{6}} \cos \frac{\tilde{\varphi}}{4}$ for $\frac{4\pi}{3} < \tilde{\varphi} < 2\pi$, half of the topological defects connected by the Dirac strings with $\zeta = -1$ disappear with pair annihilation, leaving the defects connected by the Dirac stings with $\zeta = +1$, as exemplified in Fig. 7.18c. The pair annihilation occurs at smaller m when $\tilde{\varphi}$ approaches 0 or 2π, and $-\bar{b}_z$ is enhanced to $-\bar{b}_z \to 2$ when $m \to 0$ at $\tilde{\varphi} = 0$ or 2π (in the limit, there are only four Dirac strings with $\zeta = +1$ and length $\frac{L}{2}$). Meanwhile, in the phase with $N_{\mathrm{m}} = 8$, the increase of m reduces the length of the Dirac strings with $\zeta = +1$, leading to the decrease of $-\bar{b}_z$ as

$$-\bar{b}_z = \begin{cases} 2\left[1 - \frac{2}{\pi} r_2^{\mathrm{scr}}\left(m, \frac{\tilde{\varphi}}{4}\right)\right] & \text{for } 0 < \tilde{\varphi} < \frac{2\pi}{3} \\ \frac{4}{\pi} r_1^{\mathrm{scr}}\left(m, \frac{\tilde{\varphi}}{4}\right) & \text{for } \frac{4\pi}{3} < \tilde{\varphi} < 2\pi. \end{cases} \tag{7.80}$$

Finally, $-\bar{b}_z$ vanishes by the pair annihilation of the remaining hedgehogs and antihedgehogs at $m = \frac{4}{\sqrt{6}} \cos \frac{\tilde{\varphi}}{4}$ for $0 < \tilde{\varphi} < \frac{2\pi}{3}$ and $m = \frac{4}{\sqrt{6}} \sin \frac{\tilde{\varphi}}{4}$ for $\frac{4\pi}{3} < \tilde{\varphi} < 2\pi$, and the system enters into the topologically trivial phase with $N_{\mathrm{m}} = 0$ for larger m.

In contrast, when $\frac{2\pi}{3} < \tilde{\varphi} < \frac{4\pi}{3}$, the topological phases with $N_{\mathrm{m}} = 32$ and 48 appear additionally in the intermediate region of m, as shown in Fig. 7.20. In this range of $\tilde{\varphi}$, while increasing m from the phase with $N_{\mathrm{m}} = 16$, pair creation of the hedgehogs and antihedgehogs occurs on the phase boundary to the phase with $N_{\mathrm{m}} = 48$, which is denoted by the yellow line in the figure. No anomaly is found in $-\bar{b}_z$ at the topological transition since the pair-created topological objects move along the horizontal Dirac strings and this evolution does not contribute to $l_k^{\pm}$ in Eq. (7.78); see Fig. 7.18e. In these phases, however, the increase of m reduces the length of the Dirac strings with $\zeta = -1$, leading to the increase of $-\bar{b}_z$ as

$$-\bar{b}_z = 2\left[1 - \frac{2}{\pi}\left(r_1^{\mathrm{scr}}\left(m, \frac{\tilde{\varphi}}{4}\right) - r_2^{\mathrm{scr}}\left(m, \frac{\tilde{\varphi}}{4}\right)\right)\right]. \tag{7.81}$$

While further increasing m, half of the pair-created topological defects disappear through the fusion, which causes the topological transition from $N_{\mathrm{m}} = 48$ to 32 (orange lines in the figure). In the $N_{\mathrm{m}} = 32$ region, the value of $-\bar{b}_z$ is given by

$$-\bar{b}_z = 2\left[1 \mp \frac{2}{\pi}\left(r_1^{\mathrm{scr}}\left(m, \frac{\tilde{\varphi}}{4}\right) + r_2^{\mathrm{scr}}\left(m, \frac{\tilde{\varphi}}{4}\right)\right) \pm \frac{4}{\pi} \arccos\left(\sqrt{2\cos^2\frac{\tilde{\varphi}}{4} - \frac{1}{2}}\right)\right], \tag{7.82}$$

where the upper (lower) signs are for $\frac{2\pi}{3} < \tilde{\varphi} \leq \pi$ ($\pi \leq \tilde{\varphi} < \frac{4\pi}{3}$). With a further increase of m, the rest half of the pair-created topological defects cause the fusion and N_{m} is reduced from 32 to 16. In the $N_{\mathrm{m}} = 16$ phase, all of the topological defects are connected by the Dirac strings with $\zeta = +1$, as exemplified in Fig. 7.18f, where $-\bar{b}_z$ is given as

$$-\bar{b}_z = 2\left[1 + \frac{2}{\pi}\left(r_1^{\rm scr}\left(m, \frac{\tilde{\varphi}}{4}\right) - r_2^{\rm scr}\left(m, \frac{\tilde{\varphi}}{4}\right)\right)\right]. \tag{7.83}$$

Note that $-\bar{b}_z$ takes the maximum value of $-\bar{b}_z = 2$ at $(m, \tilde{\varphi}) = \left(\sqrt{\frac{2}{3}}, \pi\right)$, where all the pair-created defects cause the fusion simultaneously.

Figure 7.21 showcases typical spin configurations of the screw 4Q states for all the topological phases in Fig. 7.20, together with the hedgehogs and antihedgehogs, and

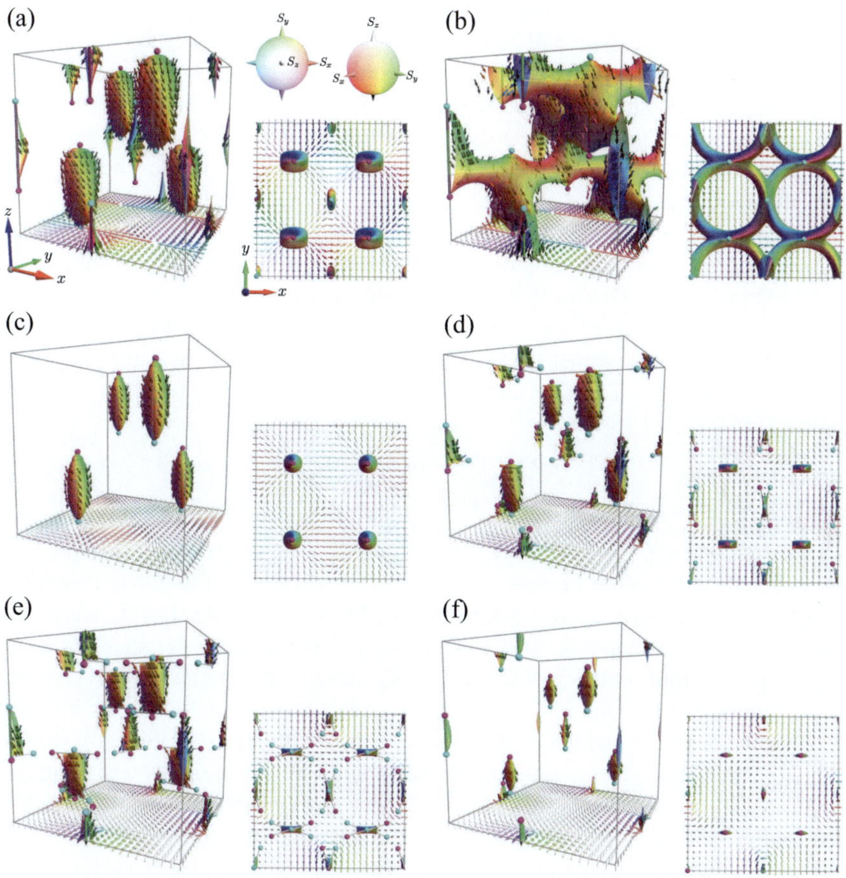

Fig. 7.21 Real-space spin configurations on the isosurfaces with $S_z(\mathbf{r}) = -0.9$ within the L^3 cube shown in Fig. 7.16b for the screw 4Q state in Eq. (7.65) for different topological phases in Fig. 7.20: **a** $m = 0$ and $\tilde{\varphi} = \frac{\pi}{3}$ ($N_{\rm m} = 16$), **b** $m = 0$ and $\tilde{\varphi} = \pi$ ($N_{\rm m} = 16$), **c** $m = 0.4$ and $\tilde{\varphi} = 0$ ($N_{\rm m} = 8$), **d** $m = 0.7$ and $\tilde{\varphi} = \frac{5\pi}{6}$ ($N_{\rm m} = 32$), **e** $m = 0.7$ and $\tilde{\varphi} = \pi$ ($N_{\rm m} = 48$), and **f** $m = 0.9$ and $\tilde{\varphi} = \pi$ ($N_{\rm m} = 16$). The spin configurations are also shown on the bottom planes of the cubes. The color of the arrows and the isosurfaces denote the xyz and xy components of $\mathbf{S}(\mathbf{r})$, respectively; see the inset in (**a**). In each figure, the right panel shows the top view. Reproduced from Ref. [19]. Copyright (2022) by the American Physical Society

the Dirac strings. The spin configurations are shown on the isosurfaces with $S_z(\mathbf{r}) = -0.9$ as well as the bottom plane of the L^3 cube. The isosurfaces, by definition, extend from the hedgehogs and antihedgehogs, and involve the Dirac strings inside. Figure 7.21a is for the $N_\mathrm{m} = 16$ state at $m = 0$ and $\tilde{\varphi} = \frac{\pi}{3}$. Figure 7.21b is for the different $N_\mathrm{m} = 16$ state at $m = 0$ and $\tilde{\varphi} = \pi$. The result demonstrates that the phase shift drastically changes the spin configurations as well as the real-space distributions of the topological objects, while N_m is the same as in Fig. 7.21a. In this state, the isosurfaces have complicated 3D networks due to the existence of the Dirac strings running on the horizontal planes. Figure 7.21c is for the $N_\mathrm{m} = 8$ phase at $m = 0.4$ and $\tilde{\varphi} = 0$. In this state, the isosurfaces become much simpler. Figure 7.21d and e are for the $N_\mathrm{m} = 32$ state at $m = 0.7$ and $\tilde{\varphi} = \frac{5\pi}{6}$ and $N_\mathrm{m} = 48$ states at $m = 0.7$ and $\tilde{\varphi} = \pi$, respectively. In both states, the spin configuration on a horizontal xy plane comprises a SkL with Bloch-type skyrmions, as exemplified on the bottom plane of the cube in the figure. We note that the skyrmions are deformed and elongated along the direction of the horizontal Dirac strings. Figure 7.21f is for the $N_\mathrm{m} = 16$ state at $m = 0.9$ and $\tilde{\varphi} = \pi$. In this state, while N_m takes the same value as that in Fig. 7.21a and b, the spin configuration is completely different from them. In all the cases, the spin configurations have twofold rotational symmetry about each vertical Dirac string. We note that the spin configurations with $\tilde{\varphi} = 0$ have fourfold rotational symmetry about each Dirac string, as exemplified in Fig. 7.21c, while those with $\tilde{\varphi} = \pi$ are symmetric for the screw operation $\{C_{4z} | -\tilde{\mathbf{a}}_\eta\}$, as exemplified in Fig. 7.21b, e, and f. These results are consistent with the symmetry arguments in Table 7.1.

Discussion

Let us briefly discuss the present results in comparison with the previous studies. A $4Q$ HL was experimentally discovered in $MnSi_{1-x}Ge_x$ [4]. In this study, the spin texture at zero magnetic field was interpreted as the screw $4Q$ state with four hedgehogs and four antihedgehogs ($N_\mathrm{m} = 8$), which corresponds to $\tilde{\varphi} = 0$ in our results, although the value of $\tilde{\varphi}$ was not examined experimentally. Meanwhile, the topological properties and the EMF in the screw $4Q$ state were theoretically studied at $\tilde{\varphi} = \pi$ while changing the external magnetic field [16], but the study was limited to the $N_\mathrm{m} = 16$ state as the solutions in Eqs. (7.73)–(7.75) were not included. Recently, some of the authors studied the evolution of the screw $4Q$ state on a 3D simple cubic lattice by variational calculations and simulated annealing, and found various types of topological transitions depending on the direction of the magnetic field [13]. We will discuss one of them, paying attention to the phase shift in Sect. 7.3.4.

7.3.3 Sinusoidal 4Q State

Next, we analyze the phase shift in the sinusoidal 4Q state given by

$$\mathbf{S}(\mathbf{r}) \propto \begin{pmatrix} \cos \mathcal{Q}_1 - \cos \mathcal{Q}_2 - \cos \mathcal{Q}_3 + \cos \mathcal{Q}_4 \\ \cos \mathcal{Q}_1 - \cos \mathcal{Q}_2 + \cos \mathcal{Q}_3 - \cos \mathcal{Q}_4 \\ \cos \mathcal{Q}_1 + \cos \mathcal{Q}_2 - \cos \mathcal{Q}_3 - \cos \mathcal{Q}_4 + 2\sqrt{3}m \end{pmatrix}, \tag{7.84}$$

which is obtained from Eq. (1.11) by taking $N_Q = 4$ and

$$\psi_\eta^{\mathrm{c}} = \frac{1}{2}, \quad \psi_\eta^{\mathrm{s}} = 0, \quad \mathbf{e}_\eta^1 = \frac{\mathbf{q}_\eta}{q}, \quad \mathbf{m} = m\hat{\mathbf{z}}. \tag{7.85}$$

Following the arguments in Sect. 7.3.2, we summarize the symmetry operations for the sinusoidal 4Q state in Table 7.2. We find that the spin texture is unchanged for $C_{4z}, \mathcal{I}, \mathcal{T}C_{2x}$, and their combinations for $\tilde{\varphi} = 0$, $\{C_{4z}| - \tilde{\mathbf{a}}_\eta\}, \mathcal{I}, \mathcal{T}C_{2x}$, and their combinations for $\tilde{\varphi} = \pi$, otherwise C_{2z}, S_{4z}, $\mathcal{T}C_{2x}$, and their combinations, where S_{4z} represents the fourfold improper rotation operation about the z axis.

Hedgehogs in Hyperspace

Following the same procedure in the first subsubsection of Sect. 7.3.2, we compute the positions of the topological defects in the 4D hyperspace for the spin texture corresponding to Eq. (7.84). Here, we show the solutions $\mathbf{S}(\mathbf{R}) = 0$ within $0 \leq X, Y < \frac{L}{2}$ and $0 \leq Z < L$, but the spatial translations of them with $\mathbf{a}_\eta$ in Eq. (7.57) also satisfy $\mathbf{S}(\mathbf{R}) = 0$. When $|m| < \frac{2}{\sqrt{3}} \sin W$, we obtain two solutions analytically:

$$(X^*, Y^*, Z^*) = L\left(0, 0, \frac{1}{4} \pm \frac{1}{2\pi} r_1^{\sin}(m, W)\right), \tag{7.86}$$

where

$$r_1^{\sin}(m, W) = \arccos\left(\frac{\sqrt{3}m}{2\sin W}\right). \tag{7.87}$$

Table 7.2 Similar table to Table 7.1 for the sinusoidal 4Q state in Eq. (7.84)

Operation	Sum of phases	Translation	Magnetization
C_{4z}	$\tilde{\varphi} \to 2\pi - \tilde{\varphi}$	$\tilde{\mathbf{a}}_\eta$	$m \to m$
C_{2x}	$\tilde{\varphi} \to \tilde{\varphi}$	0	$m \to -m$
$\mathcal{I}$	$\tilde{\varphi} \to 2\pi - \tilde{\varphi}$	$\tilde{\mathbf{a}}_\eta + \tilde{\mathbf{a}}_{\eta'}$	$m \to m$
$\mathcal{T}$	$\tilde{\varphi} \to \tilde{\varphi}$	$\tilde{\mathbf{a}}_\eta + \tilde{\mathbf{a}}_{\eta'}$	$m \to -m$

Meanwhile, when $|m| < \frac{2}{\sqrt{3}} \cos W$, we obtain other two analytical solutions as

$$(X^*, Y^*, Z^*) = L\left(\frac{1}{4}, \frac{1}{4}, \frac{3}{4} + \frac{1}{2\pi} r_2^{\sin}(m, W)\right), \quad L\left(\frac{1}{4}, \frac{1}{4}, \frac{1}{4} - \frac{1}{2\pi} r_2^{\sin}(m, W)\right), \tag{7.88}$$

where

$$r_2^{\sin}(m, W) = \arcsin\left(\frac{\sqrt{3}m}{2\cos W}\right). \tag{7.89}$$

The solutions in Eqs. (7.86) and (7.88) do not change their XY coordinates with m and $\tilde{\varphi}$, as Eqs. (7.69) and (7.71) for the screw $4Q$ case. Meanwhile, when $-\frac{2}{\sqrt{3}} \cos^2 W < m < \frac{2}{\sqrt{3}} \sin^2 W$, we obtain four solutions:

$$\begin{aligned}(X^*, Y^*, Z^*) =& L\left(\frac{1}{2\pi} r_{3+}^{\sin}(m, W), \frac{1}{2} - \frac{1}{2\pi} r_{3+}^{\sin}(m, W), \frac{1}{2} + \frac{W}{2\pi}\right), \\ & L\left(\frac{1}{2} - \frac{1}{2\pi} r_{3+}^{\sin}(m, W), \frac{1}{2\pi} r_{3+}^{\sin}(m, W), \frac{1}{2} + \frac{W}{2\pi}\right), \\ & L\left(\frac{1}{2\pi} r_{3+}^{\sin}(m, W), \frac{1}{2\pi} r_{3+}^{\sin}(m, W), \frac{1}{2} - \frac{W}{2\pi}\right), \\ & L\left(\frac{1}{2} - \frac{1}{2\pi} r_{3+}^{\sin}(m, W), \frac{1}{2} - \frac{1}{2\pi} r_{3+}^{\sin}(m, W), \frac{1}{2} - \frac{W}{2\pi}\right), \end{aligned} \tag{7.90}$$

where

$$r_{3+}^{\sin}(m, W) = \frac{1}{2} \arccos\left(\cos(2W) + \sqrt{3}m\right). \tag{7.91}$$

For $-\frac{2}{\sqrt{3}} \sin^2 W < m < \frac{2}{\sqrt{3}} \cos^2 W$, we obtain four solutions:

$$\begin{aligned}(X^*, Y^*, Z^*) =& L\left(\frac{1}{2\pi} r_{3-}^{\sin}(m, W), \frac{1}{2} - \frac{1}{2\pi} r_{3-}^{\sin}(m, W), \frac{W}{2\pi}\right), \\ & L\left(\frac{1}{2} - \frac{1}{2\pi} r_{3-}^{\sin}(m, W), \frac{1}{2\pi} r_{3-}^{\sin}(m, W), \frac{W}{2\pi}\right), \\ & L\left(\frac{1}{2\pi} r_{3-}^{\sin}(m, W), \frac{1}{2\pi} r_{3-}^{\sin}(m, W), 1 - \frac{W}{2\pi}\right), \\ & L\left(\frac{1}{2} - \frac{1}{2\pi} r_{3-}^{\sin}(m, W), \frac{1}{2} - \frac{1}{2\pi} r_{3-}^{\sin}(m, W), 1 - \frac{W}{2\pi}\right), \end{aligned} \tag{7.92}$$

where

$$r_{3-}^{\sin}(m, W) = \frac{1}{2} \arccos\left(\cos(2W) - \sqrt{3}m\right). \tag{7.93}$$

In contrast to Eqs. (7.86) and (7.88), the solutions in Eqs. (7.90) and (7.92) change their XY coordinates with m and $\tilde{\varphi}$, while the Z coordinates do not change. These solutions correspond to those obtained numerically in the screw 4Q case.

Figure 7.22 showcases the systematic change of the topological defects in the original 3D space while changing m in Eq. (7.84) with $\tilde{\varphi} = \frac{\pi}{3}$ and π. The monopole charge of the hedgehogs and antihedgehogs, $Q_{\rm m}$, and the vorticity of the Dirac strings, ζ, are calculated by the same procedure in the first subsubsection of Sect. 7.3.2.

First, we discuss the case of $\tilde{\varphi} = \frac{\pi}{3}$ shown in Fig. 7.22a, b, and c. When $m = 0$, there are 48 defects in total and half of them are hedgehogs with $Q_{\rm m} = +1$ and the others are antihedgehogs with $Q_{\rm m} = -1$, as shown in Fig. 7.22a. These topological defects are derived from Eqs. (7.86), (7.88), (7.90), and (7.92). There are two types of the Dirac strings connecting the hedgehogs and antihedgehogs: One is the vertical strings for the hedgehog-antihedgehog pairs, and the other is the horizontal ones for the hedgehog-hedgehog pairs or the antihedgehog-antihedgehog pairs. Similar to the screw 4Q case, the vorticity is ill-defined for the latter horizontal ones, and they intersect the former vertical ones whose vorticities change their signs at the crossing points; three hedgehogs and three antihedgehogs form a cluster like a

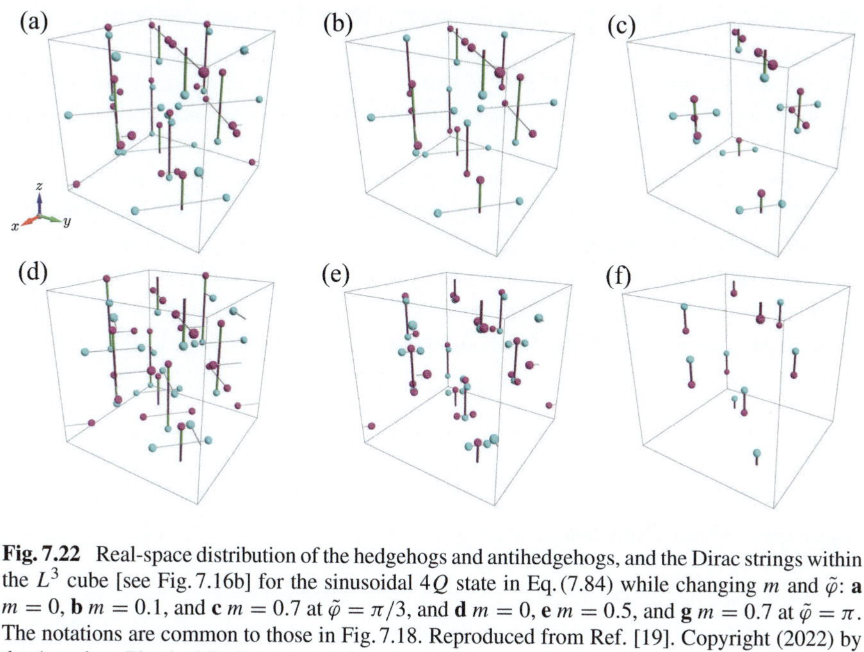

Fig. 7.22 Real-space distribution of the hedgehogs and antihedgehogs, and the Dirac strings within the L^3 cube [see Fig. 7.16b] for the sinusoidal 4Q state in Eq. (7.84) while changing m and $\tilde{\varphi}$: **a** $m = 0$, **b** $m = 0.1$, and **c** $m = 0.7$ at $\tilde{\varphi} = \pi/3$, and **d** $m = 0$, **e** $m = 0.5$, and **g** $m = 0.7$ at $\tilde{\varphi} = \pi$. The notations are common to those in Fig. 7.18. Reproduced from Ref. [19]. Copyright (2022) by the American Physical Society

twisted two-barred cross. Note that the horizontal ones are straight, in contrast to the curved ones in the screw $4Q$ case (see Fig. 7.18). Moreover, they are found for all $\tilde{\varphi}$, while they appear only for $\frac{2\pi}{3} < \tilde{\varphi} < \frac{4\pi}{3}$ in the screw $4Q$ case. All the vertical Dirac strings have the length of $\frac{L}{2}$ at $m = 0$, while the horizontal ones have a long or short length, resulting in the two types of clusters; the hedgehogs and antihedgehogs in the four clusters with longer horizontal strings are given by Eqs. (7.88) and (7.92), while those in the rest four with shorter ones are by Eqs. (7.86) and (7.90). Although the lengths of the $\zeta = 1$ and $\zeta = -1$ portions are different within each Dirac string, the total length for each vorticity is the same by considering all the Dirac strings, leading to a cancellation of the EMF at $m = 0$ (see the next subsubsection). By introducing m, the topological defects move along the Dirac strings, and two hedgehogs and one antihedgehog (or one hedgehog and two antihedgehogs) collide with each other, leaving one (anti)hedgehog, at $m = \frac{2}{\sqrt{3}} \sin^2 \frac{\tilde{\varphi}}{4}$ in the four clusters with shorter horizontal Dirac strings. The fusion process is similar to those in Fig. 7.19. At the topological transition by the fusion, N_{m} decreases from 48 to 32, and for larger m, the vertical Dirac strings left by the fusion, which have $\zeta = -1$, coexist with the clusters remaining, as shown in Fig. 7.22b. By further increasing m, the hedgehogs and antihedgehogs connected by the vertical Dirac strings cause the pair annihilation at $m = \frac{2}{\sqrt{3}} \sin \frac{\tilde{\varphi}}{4}$, leaving four clusters derived from Eqs. (7.88) and (7.92), as shown in Fig. 7.22c. At the topological transition by the pair annihilation, N_{m} is further reduced from 32 to 24. When $m = \frac{2}{\sqrt{3}} \cos^2 \frac{\tilde{\varphi}}{4}$, the fusion takes place in the remaining four clusters, which reduces N_{m} from 24 to 8, and leaves four vertical Dirac strings with $\zeta = -1$. Finally, the hedgehogs and antihedgehogs on the Dirac strings pair annihilate at $m = \frac{2}{\sqrt{3}} \cos \frac{\tilde{\varphi}}{4}$, where the system becomes topologically trivial with $N_{\mathrm{m}} = 0$.

Next, we discuss the case of $\tilde{\varphi} = \pi$ shown in Fig. 7.22d, e, and f. When $m = 0$, N_{m} is 48 as in the case of $\tilde{\varphi} = \frac{\pi}{3}$; the positions of the hedgehogs and antihedgehogs connected by the vertical Dirac strings are the same, but those connected by the horizontal Dirac strings are different, as shown in Fig. 7.22d. Similar to the $\tilde{\varphi} = \frac{\pi}{3}$ case, the total lengths of the Dirac strings with $\zeta = 1$ and -1 are the same. By introducing m, the topological defects move along the Dirac strings as shown in Fig. 7.22e, and the fusion occurs at $m = \frac{1}{\sqrt{3}}$ simultaneously in all the clusters. This leaves eight pairs of the hedgehogs and antihedgehogs connected by the vertical Dirac strings, as shown in Fig. 7.22f; N_{m} decreases from 48 to 16. In this state, all the Dirac strings have the length of $\frac{L}{\pi} r_1^{\mathrm{sin}} \left(m, \frac{\tilde{\varphi}}{4}\right) = L \left(\frac{1}{2} - \frac{1}{\pi} r_2^{\mathrm{sin}} \left(m, \frac{\tilde{\varphi}}{4}\right)\right)$ and the vorticity $\zeta = -1$. By further increasing m, the remaining pairs annihilate at $m = \sqrt{\frac{2}{3}}$.

Topological Phase Diagram

Performing similar calculations to those in the second subsubsection of Sect. 7.3.2 while changing $\tilde{\varphi}$ and m, we elaborate the phase diagram shown in Fig. 7.23. We also plot $-\bar{b}_z$ by the contour as before. Similar to Fig. 7.20, the result is again symmetric

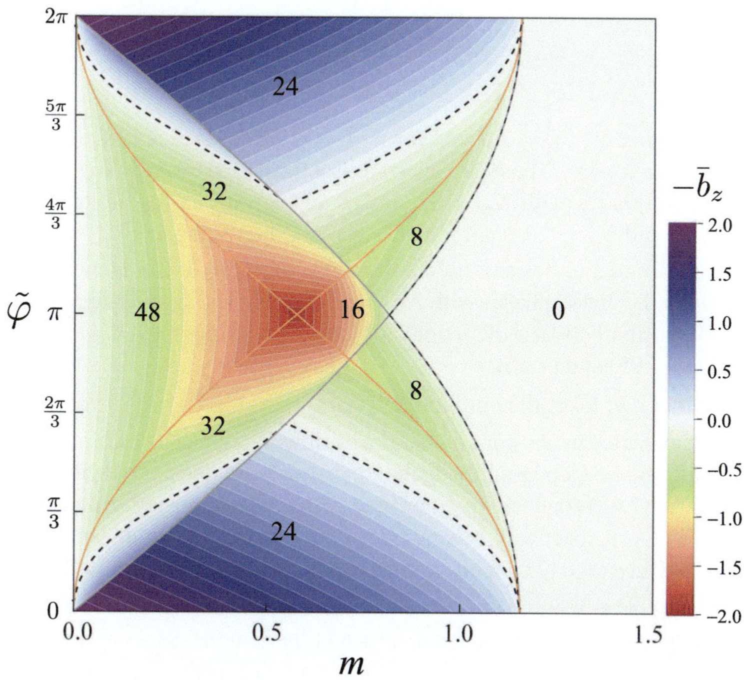

Fig. 7.23 Topological phase diagram for the sinusoidal 4Q state determined by N_{m}, with the contour plot of $-\bar{b}_z$, on the plane of m and $\tilde{\varphi}$. The notations are common to those in Fig. 7.20. Reproduced from Ref. [19]. Copyright (2022) by the American Physical Society

with respect to $\tilde{\varphi} = \pi$ and has 2π periodicity in $\tilde{\varphi}$, and the phase diagram for $m < 0$ is obtained in the same form with sign inversion of $-\bar{b}_z$. We find the topological phases with $N_{\mathrm{m}} = 8$, 16, 24, 32, and 48 in the phase diagram. There are two interesting features, in comparison with the result for the screw 4Q case in Fig. 7.20. One is that the topological phases with large values of N_{m}, such as $N_{\mathrm{m}} = 24$, 32, and 48, appear in the large portions of the phase diagram. In particular, the system has $N_{\mathrm{m}} = 48$ for all $\tilde{\varphi}$ in the small m limit, and turns into the $N_{\mathrm{m}} = 32$ state by the fusion on the orange lines in the phase diagram. Near $\tilde{\varphi} = 0$ or 2π, the $N_{\mathrm{m}} = 24$ regions extend widely in the intermediate m region, and the $N_{\mathrm{m}} = 32$ states appear in between. The value of $-\bar{b}_z$ in the phase with $N_{\mathrm{m}} = 48$ is given by

$$-\bar{b}_z = 2\left[1 + \frac{2}{\pi}\left(r_1^{\sin}\left(m, \frac{\tilde{\varphi}}{4}\right) - r_2^{\sin}\left(m, \frac{\tilde{\varphi}}{4}\right)\right)\right]. \tag{7.94}$$

Meanwhile, the values of $-\bar{b}_z$ in the $N_{\mathrm{m}} = 32$ and 24 phases are given as

$$-\bar{b}_z = 2\left[1 - \frac{2}{\pi}\left(r_1^{\sin}\left(m, \frac{\tilde{\varphi}}{4}\right) + r_2^{\sin}\left(m, \frac{\tilde{\varphi}}{4}\right)\right) - \frac{\tilde{\varphi}}{\pi}\right], \tag{7.95}$$

and

$$-\bar{b}_z = 2\left[1 - \frac{2}{\pi} r_2^{\sin}\left(m, \frac{\tilde{\varphi}}{4}\right) - \frac{\tilde{\varphi}}{\pi}\right], \tag{7.96}$$

respectively, when $0 \le \tilde{\varphi} \le \pi$; $-\bar{b}_z$ for $\pi \le \tilde{\varphi} \le 2\pi$ is obtained by the symmetry with respect to $\tilde{\varphi} = \pi$. These are in stark contrast to the screw $4Q$ case where the topological phases with $N_{\mathrm{m}} = 32$ and 48 appear only in the limited region for $\frac{2\pi}{3} < \tilde{\varphi} < \frac{4\pi}{3}$ and $N_{\mathrm{m}} = 24$ is not found. The other interesting feature is that the sign of $-\bar{b}_z$ is not limited to positive unlike the screw $4Q$ case; it can be negative while changing m and $\tilde{\varphi}$ in the phases with $N_{\mathrm{m}} = 32$ and 24. The sign changes are caused by the competition between the lengths of the Dirac strings with $\zeta = +1$ and -1; see also Eqs. (7.95) and (7.96).

We find that $-\bar{b}_z$ takes the minimum value of -2 at $(m, \tilde{\varphi}) = \left(\frac{1}{\sqrt{3}}, \pi\right)$, where all the defects connected by the horizontal Dirac strings cause the fusion simultaneously. This state has the hedgehog-antihedgehog pairs given by the analytical solutions in Eqs. (7.86) and (7.88) connected by the Dirac strings whose vorticities and lengths are all $\zeta = -1$ and $\frac{L}{4}$, respectively. Meanwhile, $-\bar{b}_z$ takes the maximum value of 2 in the limit of $m \to 0$ at $\tilde{\varphi} = 0\,(2\pi)$. This is understood as follows. When $\tilde{\varphi} = 0$, the system has the hedgehogs and antihedgehogs given by Eqs. (7.88) and (7.92) connected by the Dirac strings with the vorticities $\zeta = 1$ and the lengths $L\left(\frac{1}{2} - \frac{1}{\pi} r_2^{\sin}\left(m, \frac{\tilde{\varphi}}{4}\right)\right)$. This leads to

$$-\bar{b}_z = 2\left[1 - \frac{2}{\pi} r_2^{\sin}\left(m, \frac{\tilde{\varphi}}{4}\right)\right]. \tag{7.97}$$

Similarly, when $\tilde{\varphi} = 2\pi$, the hedgehogs and antihedgehogs are given by Eqs. (7.86) and (7.90), and $-\bar{b}_z$ is given by

$$-\bar{b}_z = \frac{4}{\pi} r_1^{\sin}\left(m, \frac{\tilde{\varphi}}{4}\right). \tag{7.98}$$

Note that Eqs. (7.97) and (7.98) are related with each other by the symmetry with respect to $\tilde{\varphi} = \pi$. Hence, $-\bar{b}_z$ goes to 2 in the limit of $m \to 0$ for both $\tilde{\varphi} = 0$ and 2π.

Figure 7.24 showcases typical spin configurations of the sinusoidal $4Q$ states for all the topological phases in Fig. 7.23, in a similar manner to Fig. 7.21. Figure 7.24a is for the $N_{\mathrm{m}} = 48$ state at $m = 0$ and $\tilde{\varphi} = \pi$. In this state, all the isosurfaces have the same shape and volume. Figure 7.24b is for the $N_{\mathrm{m}} = 32$ state at $m = 0.2$ and $\tilde{\varphi} = \frac{\pi}{3}$. The isosurfaces including the pairs of the hedgehog and antihedgehog connected by the Dirac strings with $\zeta = -1$ are small (hardly seen in the figure) compared to those including the twisted two-barred crosses, and they disappear by pair annihilation while increasing m, as shown in Fig. 7.24c for the $N_{\mathrm{m}} = 24$ state at $m = 0.6$ and $\tilde{\varphi} = \frac{\pi}{3}$. In these states, any horizontal xy plane intersecting the Dirac strings with $\zeta = -1$ gives a SkL with antiskyrmions, as exemplified on the bottom planes of the

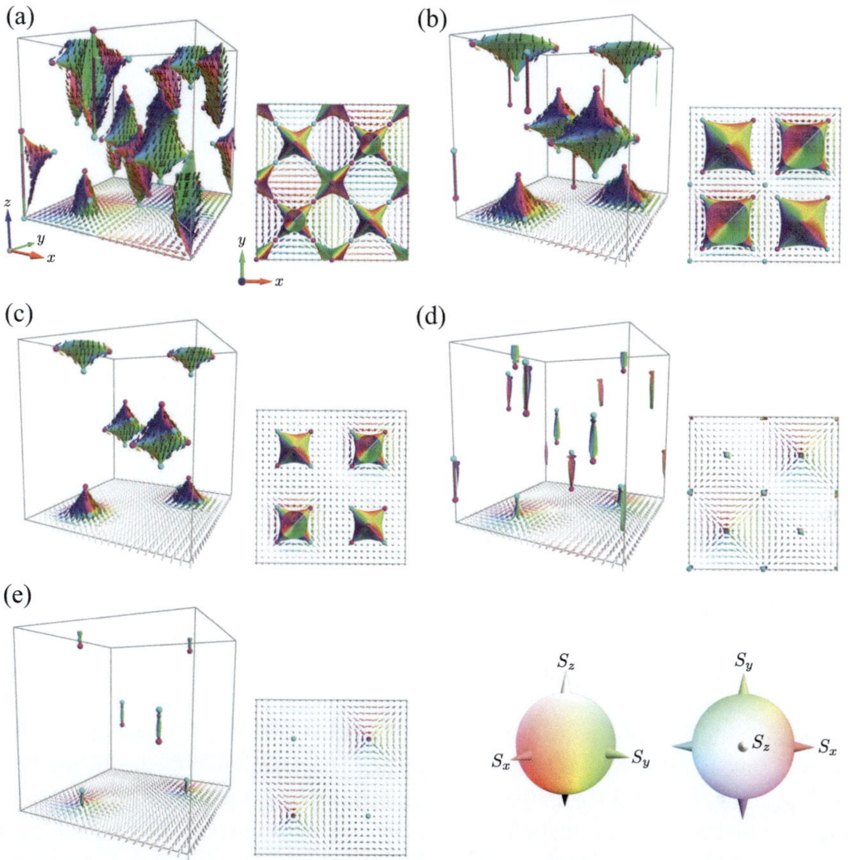

Fig. 7.24 Real-space spin configurations on the isosurfaces with $S_z(\mathbf{r}) = -0.9$ for the sinusoidal $4Q$ state in Eq. (7.84) for different topological phases in Fig. 7.23: **a** $m = 0$ and $\tilde{\varphi} = \pi$ ($N_\mathrm{m} = 48$), **b** $m = 0.2$ and $\tilde{\varphi} = \frac{\pi}{3}$ ($N_\mathrm{m} = 32$), **c** $m = 0.6$ and $\tilde{\varphi} = \frac{\pi}{3}$ ($N_\mathrm{m} = 24$), **d** $m = 0.6$ and $\tilde{\varphi} = \pi$ ($N_\mathrm{m} = 16$), and **e** $m = 0.8$ and $\tilde{\varphi} = \frac{5\pi}{6}$ ($N_\mathrm{m} = 8$). The notations are common to those in Fig. 7.21. Reproduced from Ref. [19]. Copyright (2022) by the American Physical Society

cubes. Figure 7.24d is for the $N_\mathrm{m} = 16$ state at $m = 0.6$ and $\tilde{\varphi} = \pi$. In this state, the horizontal Dirac strings disappear, and all the isosurfaces become small, while they retain knoblike features as the remnant of the horizontal Dirac strings. In this case also, the horizontal xy plane is a SkL with antiskyrmions, and the antiskyrmions appear on the xy planes with almost all z coordinates. Figure 7.24e is for the $N_\mathrm{m} = 8$ state at $m = 0.8$ and $\tilde{\varphi} = \frac{5\pi}{6}$. In this state, the number of the Dirac strings is halved and both Dirac strings and surrounding isosurfaces are shrunk. Hence, the antiskyrmions are found on the horizontal planes with limited z coordinates, in contrast to the case in Fig. 7.24d. In all the above cases, the spin configurations have fourfold improper rotational symmetry about each vertical Dirac string, as shown in Fig. 7.24. Moreover, the spin configurations with $\tilde{\varphi} = \pi$ are symmetric for the screw operation $\{C_{4z}| - \tilde{\mathbf{a}}_\eta\}$ and the spatial-inversion operation, as exemplified in Fig. 7.24a and d. These results are consistent with the symmetry arguments in Table 7.2.

7.3.4 Numerical Analysis of Phase Shifts in the Screw 4Q State

Thus far, we elucidated the topological properties of 3D 4Q-HLs by systematically changing the phases of their constituent waves as well as the magnetization. Here, to study how the actual phases evolve, we perform similar analysis to that in Sect. 6.1.4 for the 3D 4Q states on a 3D simple cubic lattice obtained in the previous study [13] for the effective spin model in Eq. (3.26) with $J = 1$, $K = 0.6$, $D = 0.3$, and $N = 16^3$. A set of the wave vectors denoted by $\mathbf{Q}_\eta$ in Eq. (3.26) is set as $\mathbf{q}_\eta$ shown in Fig. 7.16a and $q = |\mathbf{q}_\eta| = \frac{\sqrt{3}\pi}{4}$, i.e., $L = 8$. In this case, it was shown that the model stabilizes the 4Q states for $0 \leq h \lesssim 1.395$, by using the variational calculation for $h = 0$ and the simulated annealing for $h \geq 0$ [13] (see Sect. 3.4.2). When $h = 0$, the ground state is given by a superposition of four proper screws, similar to the one discussed in Sect. 7.3.2. When $0 \leq h \lesssim 0.575$, the ground state obtained by the simulated annealing is a 4Q-HL with $N_\mathrm{m} = 16$, as shown in Fig. 3.4c. By increasing h, the hedgehogs and antihedgehogs move toward each other and half of the hedgehog-antihedgehog pairs cause pair annihilation at $h \simeq 0.575$, which changes the spin state to a different 4Q-HL with $N_\mathrm{m} = 8$. The positions of the hedgehogs and antihedgehogs before and after the topological phase transition are shown in Fig. 3.4d and e, respectively. The remaining hedgehogs and antihedgehogs pair annihilate at $h \simeq 1.395$ and the system turns into a topologically trivial 4Q state with $N_\mathrm{m} = 0$, and finally, into the forced ferromagnetic state for $h \gtrsim 1.395$. In the following, we focus on the region for $0 \leq h \leq 1.2$ where the spin configurations can be well described by Eq. (7.65); the spin configurations for $h \gtrsim 1.2$ are largely modulated around the hedgehogs and antihedgehogs, and the topological properties are not well captured by Eq. (7.65).

Following the procedure in Sect. 4.3, we extract the phases of the constituent waves in the two 4Q-HLs by assuming that the ground state is well approximated by a superposition of proper screws like in Eq. (7.65). In this case, we estimate the cost function defined by

$$U(\varphi_1, \varphi_2, \varphi_3, \varphi_4, m) = \frac{1}{N} \sum_i \left(1 - \mathbf{S}_i^\mathrm{SA} \cdot \mathbf{S}_i^{\mathrm{scr}4Q}(\varphi_1, \varphi_2, \varphi_3, \varphi_4, m)\right), \tag{7.99}$$

where $\mathbf{S}_i^\mathrm{SA}$ is the spin configuration obtained by the simulated annealing and $\mathbf{S}_i^{\mathrm{scr}4Q}$ is that generated from Eq. (7.65) with $\mathbf{e}_\eta^2 \to -\mathbf{e}_\eta^2$ [in the previous study [13], D was taken to be positive in Eq. (3.26), which prefers left-handed screws]. We define the sum of φ_η^* in the same form as in Eq. (6.36):

$$\tilde{\varphi}^* = \pi - \left| \pi - \mathrm{Mod}\left[\sum_{\eta=1}^{4} \varphi_\eta^*, 2\pi \right] \right|. \tag{7.100}$$

The upper panel of Fig. 7.25 shows the h dependence of the optimal values of U. In the whole range of h, the optimal U is less than 0.021, indicating that the optimal $\{\mathbf{S}_i^{\mathrm{scr}4Q}\}^*$ reproduces well the numerically obtained ground state $\{\mathbf{S}_i^{\mathrm{SA}}\}$. The lower panel of Fig. 7.25 shows the h dependences of the energy per site calculated from $\{\mathbf{S}_i^{\mathrm{scr}4Q}\}^*$ and $\{\mathbf{S}_i^{\mathrm{SA}}\}$, $E^{\mathrm{scr}4Q}$ and E^{SA}, respectively, and their difference, $\Delta E = E^{\mathrm{scr}4Q} - E^{\mathrm{SA}}$. We find that the energy is well reproduced for all h (the relative error is less than 1.9 %.) Both U and ΔE show rapid changes when approaching $h \sim 0.6$, which is related to the topological transition discussed below. We note that U as well as ΔE shows a hump at $h \simeq 0.58$; we return to this point in the end of this subsection.

Figure 7.26a shows the h dependences of $\tilde{\varphi}^*$ and N_m calculated from the optimal $\{\mathbf{S}_i^{\mathrm{scr}4Q}\}^*$. The results of N_m well reproduce those for $\{\mathbf{S}_i^{\mathrm{SA}}\}$ plotted by the orange points in the figure. For the $N_\mathrm{m} = 16$ state at $h = 0$, we obtain $\tilde{\varphi}^* \simeq \frac{\pi}{3}$, while it is equivalent to $\tilde{\varphi}^* \simeq \pi$ and $\frac{4\pi}{3}$ because of the threefold rotational symmetry about the

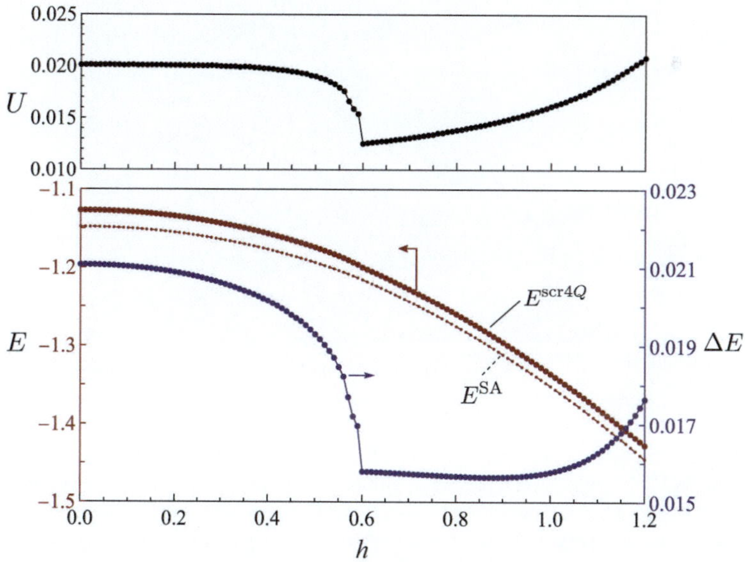

Fig. 7.25 Magnetic field dependences of (top) the cost function U in Eq. (7.99) and (bottom) the energy per site calculated from $\{\mathbf{S}_i^{\mathrm{SA}}\}$ and $\{\mathbf{S}_i^{\mathrm{scr}4Q}\}^*$, and the difference of the energy, ΔE. Reproduced from Ref. [19]. Copyright (2022) by the American Physical Society

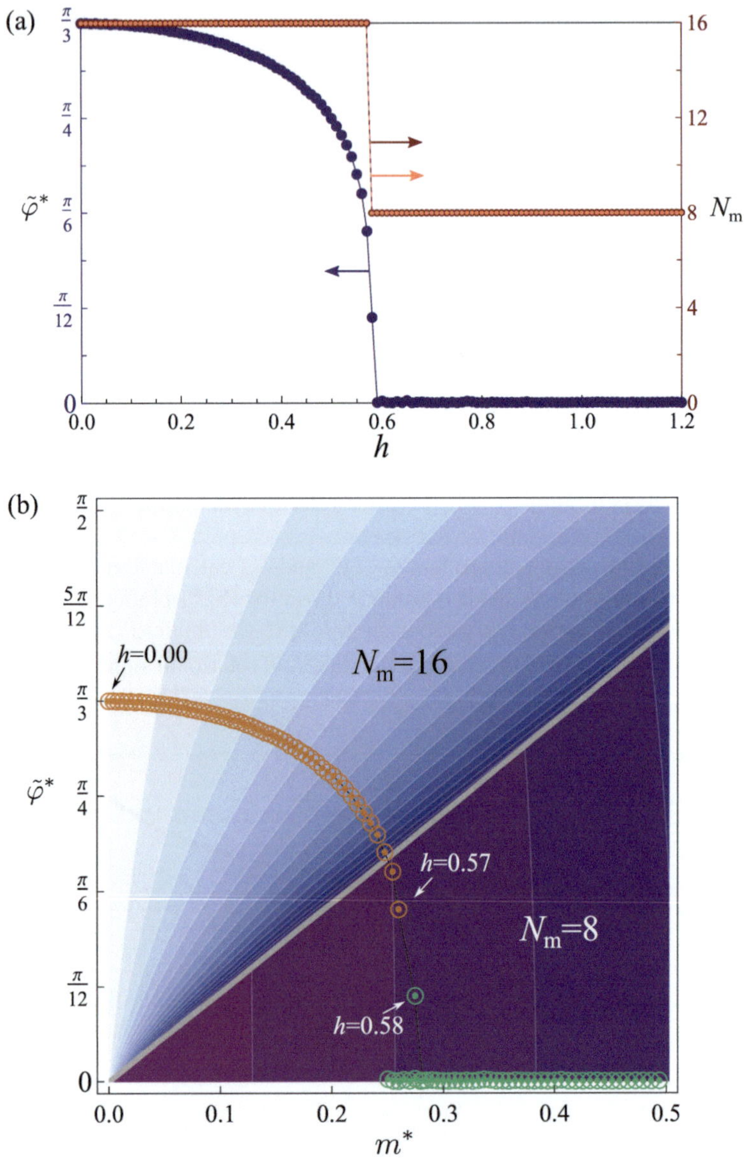

Fig. 7.26 **a** Magnetic field dependences of $\tilde{\varphi}^*$ [Eq. (7.100)] and N_{m} calculated from the optimal spin textures $\{\mathbf{S}_i^{\mathrm{scr}4Q}\}^*$, which are represented by blue and red points, respectively. For comparison, N_{m} obtained by the simulated annealing is plotted by the orange points. **b** Evolution of m^* and $\tilde{\varphi}^*$ on the phase diagram in Fig. 7.20. The background is an enlarged figure of the phase diagram in Fig. 7.20. The orange and green circles for each data represent $N_{\mathrm{m}} = 16$ and $N_{\mathrm{m}} = 8$, respectively. Reproduced from Ref. [19]. Copyright (2022) by the American Physical Society

[111] axis. By increasing h, $\tilde{\varphi}^*$ gradually decreases from $\simeq \frac{\pi}{3}$, but rapidly reduces to $\simeq 0$ when approaching $h \simeq 0.585$; $\tilde{\varphi}^* \simeq 0$ for $h \gtrsim 0.585$. The rapid change of $\tilde{\varphi}^*$ appears to occur as a precursor of the topological change from $N_{\mathrm{m}} = 16$ to 8.

Figure 7.26b shows the evolution of m^* and $\tilde{\varphi}^*$ obtained from the optimal $\{\mathbf{S}_i^{\mathrm{scr}4Q}\}^*$, plotted on the phase diagram in Fig. 7.20. The orange and green circles represent the phases with $N_{\mathrm{m}} = 16$ and 8 in Fig. 7.26a, respectively. The result indicates that N_{m} calculated from $\{\mathbf{S}_i^{\mathrm{scr}4Q}\}^*$ is almost consistent with the phase diagram calculated in Sect. 7.3.2 and the topological transition from $N_{\mathrm{m}} = 16$ to 8 is correlated with the rapid phase shift to $\simeq 0$. We note that two orange points with $N_{\mathrm{m}} = 16$ are obtained in the $N_{\mathrm{m}} = 8$ region of the phase diagram. As in the $3Q$ case in Sect. 6.1.4, we speculate that the discrepancy is presumably due to the insufficiency of Eq. (7.65). Another possible reason is that the phase diagram in Fig. 7.20 is calculated in continuous space, while the analysis in this section is done for the discrete lattice system; the phase boundary between $N_{\mathrm{m}} = 16$ and 8 may be shifted by the discretization.

Near the topological transition, we note that m^* increases monotonically up to $h = 0.58$, but slightly decreases at $h \simeq 0.59$ after $\tilde{\varphi}^*$ reduces to $\simeq 0$, and increases again for larger h, as shown in Fig. 7.26b. In the previous study, it was pointed out that the system shows the first-order phase transition at $h \simeq 0.595$ accompanied by a jump of the net magnetization [13]. Hence, the nonmonotonic behavior of m^* implies that $\{\mathbf{S}_i^{\mathrm{scr}4Q}\}^*$ is insufficient to approximate the sudden change of the spin texture through the first-order transition. We speculate that the hump in U in the top panel of Fig. 7.25 appears to be related with this issue.

7.4 Current-Induced Dynamics of $3Q$-HL

In this section, we investigate the effect of an electric current on 3D topological spin crystals. The topological defects introduced in Sect. 1.1 exhibit stable particle-like nature due to their topological protection. Utilizing such spin structures as nano- to micro-scale information carriers would lead to the miniaturization and increased capacity of magnetic memory devices. In such applications, it is essentially important to establish an efficient way to transmit the topological defects as carriers of information. One of the fundamental methods for achieving this is the application of an electric current. Thus far, previous studies for such a current-induced drift motion have primarily focused on skyrmions, with little attention paid to other topological defects such as hedgehogs. Unlike skyrmions, hedgehogs have singular points in the 3D spin structures (see Sect. 1.1.3), and hence they are expected to bring about qualitatively different current-induced dynamics on a discrete lattice. Furthermore, as discussed in Sects. 7.2 and 7.3, the HLs exhibit a variety of periodic arrangements of the hedgehogs and antihedgehogs, which can be controlled through the magnetic field. This, in turn, offers control of the HLs by an electric current under the magnetic field, making it a crucial issue in the exploration of the spin moiré engineering.

Here, we study the drift motions of the $3Q$-HLs composed of three waves under an electric current. Through the numerical simulations based on the LLG equation, we investigate the velocity of the drift motion while changing the amplitude and direction of the applied current. In Sect. 7.4.1, we present the ground-state phase diagram under the magnetic field for the model in Eq. (3.21). We also show the real-space distribution of the topological defects for the two types of the $3Q$-HLs with and without the DM interaction. In Sect. 7.4.2, we study the current-velocity relation for these HLs under an electric current by numerically solving the LLG equation in Eq. (4.9). In Sect. 7.4.3, we discuss the origin of the threshold current to cause the drift motion of the hedgehogs by evaluating the potential barrier from lattice discretization.

7.4.1 Phase Diagram

Before going into the current-induced motion of the hedgehogs, we study the ground state of the Hamiltonian in Eq. (3.21) assuming the three wave vectors $\mathbf{Q}_1 = (Q, 0, 0)$, $\mathbf{Q}_2 = (0, Q, 0)$, and $\mathbf{Q}_3 = (0, 0, Q)$, by numerically solving the LLG equation in Eq. (4.6) with $\alpha = 0.04$ and $\Delta t = 0.2$. Hereafter, we set $Q = \pi/6$, $\Delta = 0.3$, $D = 0$ or 0.3, $\gamma = 0.6Q$, and $r_c = 16$ in Eq. (3.21) under the PBC and apply the magnetic field in the z direction as $\mathbf{H} = (0, 0, H)$. We perform the following calculations for the $N = 12^3$-site system, which corresponds to the single MUC. We confirmed that the system size effect is small by comparing the results with those for the 60^3-site system. To determine the magnetic phases from the numerically optimized spin configurations, we compute the spin structure factor $\mathcal{S}_{\mathbf{q}}$ defined by

$$\mathcal{S}_{\mathbf{q}} = \frac{1}{N}\sum_{i,j} \mathbf{S}_i \cdot \mathbf{S}_j e^{i\mathbf{q}\cdot(\mathbf{r}_i - \mathbf{r}_j)}. \tag{7.101}$$

In addition, to characterize their topological property, we compute real-space positions of hedgehogs and antihedgehogs marked by nonzero monopole charge Q_{m} defined by Eq. (7.6). For the present model, Q_{m} takes 0, −1, or 1. We further compute the total number of hedgehogs and antihedgehogs within the MUC N_{m} by Eq. (7.8). In addition, we also calculate the average EMF $\bar{\mathbf{b}}$ by taking a spatial average of the EMF at site i, $\mathbf{b}_i = (b_{i,x}, b_{i,y}, b_{i,z})$, defined as Eq. (7.9). In addition, we also identify the skyrmion strings connecting the hedgehog and antihedgehog, by evaluating the vorticity ζ for six faces composing each unit cube. The vorticity for the face of the i-th unit cube on the $\mu\nu$ plane is given by

$$\zeta = \frac{1}{2\pi}\left(\arg\left[\frac{S^+(\mathbf{r}_i + \hat{\boldsymbol{\mu}})}{S^+(\mathbf{r}_i)}\right] + \arg\left[\frac{S^+(\mathbf{r}_i + \hat{\boldsymbol{\mu}} + \hat{\boldsymbol{\nu}})}{S^+(\mathbf{r}_i + \hat{\boldsymbol{\mu}})}\right]\right.$$
$$\left. + \arg\left[\frac{S^+(\mathbf{r}_i + \hat{\boldsymbol{\nu}})}{S^+(\mathbf{r}_i + \hat{\boldsymbol{\mu}} + \hat{\boldsymbol{\nu}})}\right] + \arg\left[\frac{S^+(\mathbf{r}_i)}{S^+(\mathbf{r}_i + \hat{\boldsymbol{\nu}})}\right]\right), \tag{7.102}$$

with $S^{+}(\mathbf{r}_i) = S_x(\mathbf{r}_i) + \mathrm{i}S_y(\mathbf{r}_i)$. Here, the Dirac string is supposed to run perpendicular to a plaquette when ζ is nonzero and the average of $S_{i,z}$ for the four spins constituting the plaquette is negative. For each Dirac string connecting the hedgehog and antihedgehog, we compute the length projected onto the y axis, $d_{\mathrm{m},y}$ [see Fig. 7.27b and c].

$D = 0$ Case

First, let us discuss the results for $D = 0$. Figure 7.27a shows the field dependences of $\mathcal{S}_{\mathbf{q}}$, $\bar{b}_z$, N_{m}, and $d_{\mathrm{m},y}$. $\mathcal{S}_{\mathbf{q}}$ displayed in the top panel shows that the system is in a 3Q state for $0 \leq H \lesssim 0.745$, and turns into a 2Q state for $H \gtrsim 0.745$. In the 3Q state, three $\mathcal{S}_{\mathbf{Q}_\eta}$ are identical at zero field and $0 < \mathcal{S}_{\mathbf{Q}_2} \leq \mathcal{S}_{\mathbf{Q}_1} = \mathcal{S}_{\mathbf{Q}_3}$ for nonzero field. At the phase transition from the 3Q state to the 2Q state, $\mathcal{S}_0$, corresponding to the uniform magnetization square, and $\mathcal{S}_{\mathbf{Q}_\eta}$ show discontinuous changes, implying the first-order phase transition. In the 2Q state, $\mathcal{S}_{\mathbf{Q}_2} = 0$ and $0 < \mathcal{S}_{\mathbf{Q}_1} = \mathcal{S}_{\mathbf{Q}_3}$. As shown in the middle panel of Fig. 7.27a, N_{m} is 8 in the 3Q state, and hence this state is the 3Q-HL with four pairs of hedgehogs and antihedgehogs in the MUC. Meanwhile,

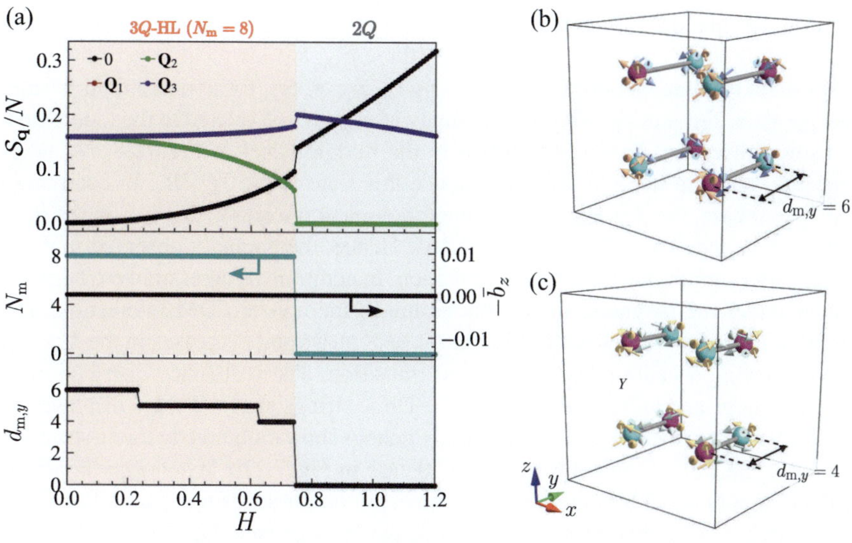

Fig. 7.27 **a** H dependences of the spin structure factor $\mathcal{S}_{\mathbf{q}}$ (top), the total emergent magnetic field $\bar{b}_z$, and the number of hedgehogs and antihedgehogs within the MUC, N_{m} (middle), and the length of the Dirac string projected onto the y axis, $d_{\mathrm{m},y}$ (bottom) for the model in Eq. (3.21) with $D = 0$. **b** and **c** Real-space distribution of the hedgehogs and antihedgehogs, the Dirac strings, and spins around the hedgehogs and antihedgehogs within the MUC for the ground state at **b** $H = 0$ and **c** $H = 0.7$. The magenta and cyan spheres represent the hedgehog and antihedgehog with $Q_{\mathrm{m}} = 1$ and -1, respectively. The Dirac strings running in the y direction are represented by gray lines. The arrows and their colors represent the direction and S_z component of spins, respectively

$\bar{b}_z$ is zero in the entire field region, which can be understood from the real-space distribution of the Dirac string discussed below.

The real-space distribution of the hedgehogs and antihedgehogs, and the Dirac strings at $H = 0$ is shown in Fig. 7.27b. The hedgehogs and antihedgehogs form the NaCl structure and they are connected by the Dirac strings running in the y direction with the length of $d_{\mathrm{m},y} = 6$. By increasing H, the hedgehogs and antihedgehogs move toward each other by shortening the Dirac strings, as shown in Fig. 7.27c at $H = 0.7$. H dependence of $d_{\mathrm{m},y}$ is plotted in the bottom panel of Fig. 7.27a. $d_{\mathrm{m},y}$ decreases from 6 to 4 successively, but suddenly becomes zero for the $2Q$ state by increasing H. In continuous space, $\bar{b}_z$ corresponds to the net EMF in Eq. (7.77), which can be calculated from the difference in the projected length along the z axis of the Dirac strings with $\zeta = 1$ and -1. Meanwhile, in the present case, all the Dirac strings run along the y axis, and accordingly, ζ in Eq. (1.4) defined on the closed loop on the xy plane is ill-defined, leading to $\bar{b}_z = 0$.

$D = 0.3$ Case

Next, we discuss the results for $D = 0.3$. The field dependences of the physical quantities are presented in Fig. 7.28 in the same manner as Fig. 7.27a. $\mathcal{S}_{\mathbf{q}}$ in the top panel shows that the system is in the $3Q$ state for $0 \leq H \lesssim 0.935$, and turns into the $2Q$ state for $H \gtrsim 0.935$. In the $3Q$ state, three $\mathcal{S}_{\mathbf{Q}_\eta}$ are identical at zero field, but all of them become inequivalent as $0 < \mathcal{S}_{\mathbf{Q}_2} \leq \mathcal{S}_{\mathbf{Q}_3} \leq \mathcal{S}_{\mathbf{Q}_1}$ for nonzero field. Within the $3Q$ state, $\mathcal{S}_{\mathbf{Q}_\eta}$ and $\mathcal{S}_0$ exhibit an anomaly at $H \simeq 0.465$ related to the topological transition discussed below. As plotted in the middle panel of Fig. 7.28, N_{m} takes nonzero values in the $3Q$ state, and hence this state is the $3Q$-HL. In contrast to the $D = 0$ case, the $3Q$-HL exhibits the topological transition; N_{m} is 8 in the low-field region and turns into 4 at $H \simeq 0.465$. Hence, the anomaly observed in $\mathcal{S}_{\mathbf{q}}$ at $H \simeq 0.465$ signals this topological transition. In addition, $\bar{b}_z$ takes nonzero negative values under the magnetic field. This is due to the nonzero DM interaction. The amplitude of $\bar{b}_z$ increases with H in the $N_{\mathrm{m}} = 8$ phase and decreases in the $N_{\mathrm{m}} = 4$ phase, leaving a peak at the topological transition. The real-space distribution of the hedgehogs and antihedgehogs, and the Dirac strings at zero field is displayed in Fig. 7.29a, where the hedgehogs and antihedgehogs show a slightly distorted structure from the NaCl arrangement for the $D = 0$ case in Fig. 7.27b. When $D = 0$, all the hedgehog-antihedgehog pairs are connected by the Dirac strings parallel to the y axis with $d_{\mathrm{m},y} = 6$. For $D = 0.3$, however, two of four pairs have $d_{\mathrm{m},y} = 5$, while the remaining two pairs show $d_{\mathrm{m},y} = 7$. Furthermore, the z coordinates of the paired hedgehogs and antihedgehogs are different by one, leading to the Dirac strings with kinks, as shown in Fig. 7.29a.

Similar to the $D = 0$ case, $d_{\mathrm{m},y}$ decreases by increasing H, as shown in the bottom panel of Fig. 7.28. $d_{\mathrm{m},y}$ starting from 5 becomes zero by the topological transition at $H \simeq 0.465$ and then, that starting from 7 becomes zero by the phase transition to the $2Q$ state at $H \simeq 0.935$. Figure 7.29b displays the real-space distribution of the topological objects at $H = 0.45$, just before the topological transition. By increasing

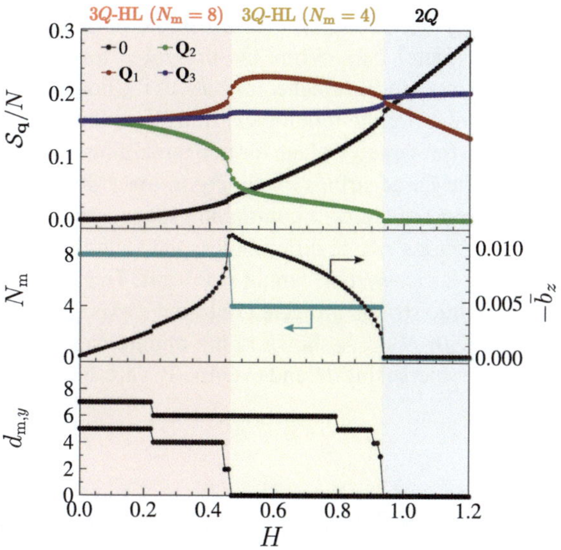

Fig. 7.28 Similar figure to Fig. 7.27a for the model in Eq. (3.21) with $D = 0.3$. The other notations are common to those in Fig. 7.27a

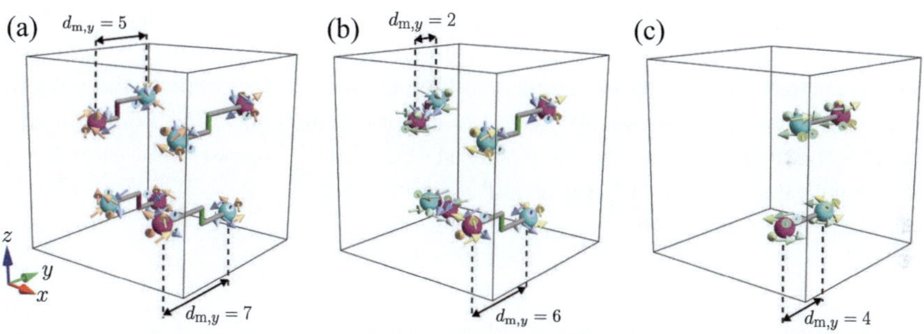

Fig. 7.29 Real-space distribution of the hedgehogs and antihedgehogs, the Dirac strings, and spins around the hedgehogs and antihedgehogs within the MUC for the ground state of the model in Eq. (3.21) with $D = 0.3$: **a** $H = 0$, **b** $H = 0.45$, and **c** $H = 0.92$. The Dirac strings running along the z axis with $\zeta = 1$ and -1 are represented by the green and purple lines, respectively. The other Dirac strings are represented by gray lines. The other notations are common to those in Fig. 7.27b and c

H, the hedgehogs and antihedgehogs move toward each other along the Dirac string by reducing $d_{\mathrm{m},y}$ while keeping the kink structure of the Dirac string. The pair with $d_{\mathrm{m},y} = 2$ in Fig. 7.29b disappears at $H \simeq 0.465$, which leads to the topological transition. The displacement of the hedgehogs and antihedgehogs in the z direction remains under the magnetic field and it brings about the length of the projected Dirac string along the z axis to have a nonzero value. Considering the corresponding spin texture in continuous space, the Dirac strings change their length in the field direction continuously by increasing H, leading to the net EMF depending on H through Eq. (7.78) (see Sect. 7.2.2). While the presented results are for a discrete lattice system, the remnants of this effect contribute to the nonzero $\bar{b}_z$. As illustrated in Fig. 7.29c at $H = 0.92$, after the topological transition, the paired hedgehog and antihedgehog move onto the same xy plane for two remaining pairs, and accordingly, they are connected by the Dirac strings extending in the y direction without kinks. In the corresponding spin structure in continuous space, the z coordinates of the hedgehogs and antihedgehogs can be different less than a lattice constant, while the simulation for a discrete lattice system cannot resolve it. This difference leads to the nonzero length of the Dirac strings projected onto the z axis, giving rise to nonzero $\bar{b}_z$ in the 3Q-HL state with $N_{\mathrm{m}} = 4$ shown in the middle panel of Fig. 7.28. These pairs also reduce $d_{\mathrm{m},y}$ by increasing H and eventually vanish at the transition to the 2Q state at $H \simeq 0.935$.

7.4.2 Longitudinal and Transverse Motions of Hedgehog

Now, we investigate the drift motion of hedgehogs driven by an electric current. We numerically solve the LLG equation in Eq. (4.9) with $\alpha = 0.04$, $\beta = 0.02$, and $\Delta t = 0.2$ for $\hat{\mathbf{j}}_e \parallel \hat{\mathbf{x}}, \hat{\mathbf{y}}$, or $\hat{\mathbf{z}}$ while changing the normalized current density $\tilde{j}_e$. Starting from the ground-state spin configuration obtained in Sect. 7.4.1, we switch on an electric current at $t = 0$ and examine the time evolution. To calculate the velocity of the drift motion in the steady state, we adopt the method in the Fourier space. Specifically, by using the real-time evolution of the Fourier transform of spins $\mathbf{S}_{\mathbf{q}}(t)$, we compute the velocity in the x, y, and z directions as

$$v_x = -\frac{1}{Q(t_f - t_i)} \int_{t_i}^{t_f} \frac{d}{dt} \arg \left(S_{\mathbf{Q}_1,y}(t) \right) dt, \tag{7.103}$$

$$v_y = -\frac{1}{Q(t_f - t_i)} \int_{t_i}^{t_f} \frac{d}{dt} \arg \left(S_{\mathbf{Q}_2,z}(t) \right) dt, \tag{7.104}$$

$$v_z = -\frac{1}{Q(t_f - t_i)} \int_{t_i}^{t_f} \frac{d}{dt} \arg \left(S_{\mathbf{Q}_3,x}(t) \right) dt, \tag{7.105}$$

respectively. Here, we use the spin component of $\mathbf{S}_{\mathbf{Q}_\eta}$ whose amplitude is the largest among the three spin components for each $\mathbf{Q}_\eta$.[7] Note that the velocities calculated with the other spin components essentially give the same results. The integrals in Eqs. (7.103)–(7.105) are taken from $t_i = 500$ to $t_f = 1000$.

---

[7] The largest spin component is determined by the exchange anisotropy in the model in Eq. (3.21); see Fig. 3.3b. For instance, at $\mathbf{q} = \mathbf{Q}_1$, the exchange anisotropy is strong in the S_y direction, as represented by the long axis of the blue ellipsoid, leading to the largest amplitude in $S_{\mathbf{Q}_1,y}$.

$D = 0$ Case: Longitudinal Drift Motion of Hedgehogs

First, we discuss the results for the model in Eq. (3.21) with $D = 0$. In this case, the drift motion of the spin texture occurs only along the current direction; the drift motion to the transverse direction does not occur for any $\hat{\mathbf{j}}_e$ since $\bar{b}_z = 0$. In Fig. 7.30a, we show the velocity of the longitudinal drift motion for $\hat{\mathbf{j}}_e \parallel \hat{\mathbf{x}}$, v_x, on the plane of the magnetic field H and the normalized current density $\tilde{j}_e$. In this case, the electric current is applied in the direction perpendicular to the Dirac strings, as depicted in the inset (see also Fig. 7.27); essentially the same results are obtained for $\hat{\mathbf{j}}_e \parallel \hat{\mathbf{z}}$. We find that for all H, the longitudinal drift motion occurs above a certain threshold value $\tilde{j}_{e,\mathrm{c}}$ represented by the white line. Above the threshold, v_x increases monotonically with $\tilde{j}_e$. The threshold current depends on H, reflecting the change of real-space configurations of the hedgehogs and antihedgehogs. By increasing H from zero, $\tilde{j}_{e,\mathrm{c}}$ monotonically increases and shows a cusp at $H \simeq 0.235$ denoted by the gray dashed line, which corresponds to the discontinuous change of $d_{\mathrm{m},y}$ from 6 to 5 [see the bottom panel of Fig. 7.27a]. By further increasing H, $\tilde{j}_{e,\mathrm{c}}$ slightly decreases first, and then, again starts to increase toward the other cusp at $H \simeq 0.625$ with the Dirac string shortening from $d_{\mathrm{m},y} = 5$ to 4. $\tilde{j}_{e,\mathrm{c}}$ keeps increasing toward the phase transition to the 2Q state accompanied by the change of N_{m} to zero at $H \simeq 0.745$ denoted by the solid gray line. A nonzero $\tilde{j}_{e,\mathrm{c}}$ is also observed in the 2Q state for $H \gtrsim 0.745$, while it decreases by increasing H. We will discuss the reason for the H dependences of the threshold current in the 3Q-HL and 2Q states in Sect. 7.4.3.

In Fig. 7.30b, we show v_y for $\hat{\mathbf{j}}_e \parallel \hat{\mathbf{y}}$, where the current direction is parallel to the Dirac strings. Similar to the results for $\hat{\mathbf{j}}_e \parallel \hat{\mathbf{x}}$, a nonzero threshold current is required to cause the longitudinal drift motion of the spin texture. However, $\tilde{j}_{e,\mathrm{c}}$ exhibits a different trend in the H dependence. We find that $\tilde{j}_{e,\mathrm{c}}$ decreases by increasing H, but

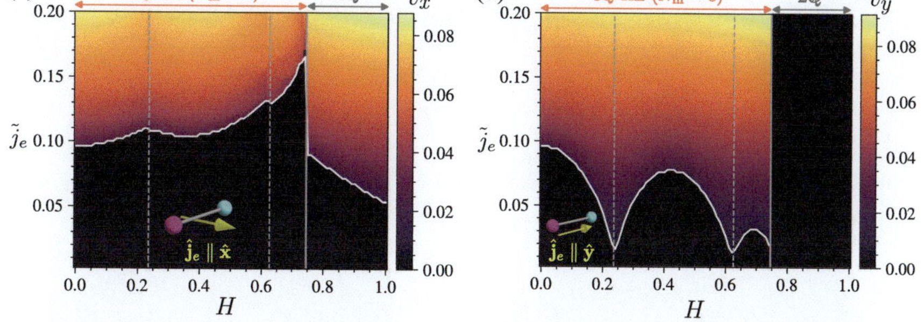

Fig. 7.30 The velocity of the longitudinal motion of the spin texture, v_μ, under the electric current for the model in Eq. (3.21) with $D = 0$ on the plane of the magnetic field H and the current density $\tilde{j}_e$: **a** $\hat{\mathbf{j}}_e \parallel \hat{\mathbf{x}}$ and **b** $\hat{\mathbf{j}}_e \parallel \hat{\mathbf{y}}$. The white lines represent the threshold current for the drift motion, $\tilde{j}_{e,\mathrm{c}}$. The dashed and solid gray lines denote the discontinuous changes in $d_{\mathrm{m},y}$ and N_{m}, respectively. The relationship between the current direction and the Dirac string connecting the hedgehog and antihedgehog is schematically shown in the inset. See also Fig. 7.27

suddenly turns to increase with the discontinuous change of $d_{\mathrm{m},y}$ from 6 to 5, leaving a sharp dip with the minimum of $\tilde{j}_e \simeq 0.016$ at $H \simeq 0.235$. By further increasing H, $\tilde{j}_{e,\mathrm{c}}$ again decreases and shows another dip at $H \simeq 0.625$ corresponding to the further change of $d_{\mathrm{m},y}$ from 5 to 4. Finally, $\tilde{j}_{e,\mathrm{c}}$ shows a drop toward the phase transition to the $2Q$ state at $H \simeq 0.745$. In contrast to the case with $\hat{\mathbf{j}}_e \parallel \hat{\mathbf{x}}$, the drift motion does not occur in the $2Q$ state since this state exhibits the spin modulations only in the x and z directions, as represented by $\mathcal{S}_{\mathbf{Q}_2} = 0$ in Fig. 7.27a. We note that, in addition to the dip structures with each reduction in $d_{\mathrm{m},y}$, the maximum value of $\tilde{j}_{e,\mathrm{c}}$ in each dome decreases with the increase of H.

Let us summarize the qualitative behavior of the threshold current density $\tilde{j}_{e,\mathrm{c}}$ and its relation to the real-space arrangement of hedgehogs and antihedgehogs, and the Dirac strings. When the current is perpendicular to the Dirac string, i.e., $\hat{\mathbf{j}}_e \parallel \hat{\mathbf{x}}$ or $\hat{\mathbf{z}}$, $\tilde{j}_{e,\mathrm{c}}$ exhibits a dominant trend to increase along with the reduction of $d_{\mathrm{m},y}$ in the $3Q$-HL phase. In addition, $\tilde{j}_{e,\mathrm{c}}$ exhibits cusp structures corresponding to the discrete changes in $d_{\mathrm{m},y}$. In the case of the current parallel to the Dirac string, i.e., $\hat{\mathbf{j}}_e \parallel \hat{\mathbf{y}}$, $\tilde{j}_{e,\mathrm{c}}$ decreases sharply with the discrete changes in $d_{\mathrm{m},y}$ alongside a decreasing trend with the reduction of $d_{\mathrm{m},y}$ toward the phase transition, leading to the multiple dip and dome structures. Thus, $\tilde{j}_{e,\mathrm{c}}$ not only serves as a probe for the change of the topological properties but also signals the changes in the real-space configuration of the hedgehogs and antihedgehogs.

$D = 0.3$ Case: Hedgehog Hall Effect

Next, we discuss the results for $D = 0.3$. In this case, in contrast to the case with $D = 0$, not only the longitudinal drift motion but also the transverse one can occur since $\bar{b}_z \neq 0$ under the magnetic field, as shown in the middle panel of Fig. 7.28. Let us begin with the longitudinal one. In Fig. 7.31a, we show the longitudinal velocity v_x for $\hat{\mathbf{j}}_e \parallel \hat{\mathbf{x}}$, where an electric current is perpendicular to the Dirac string. Similar

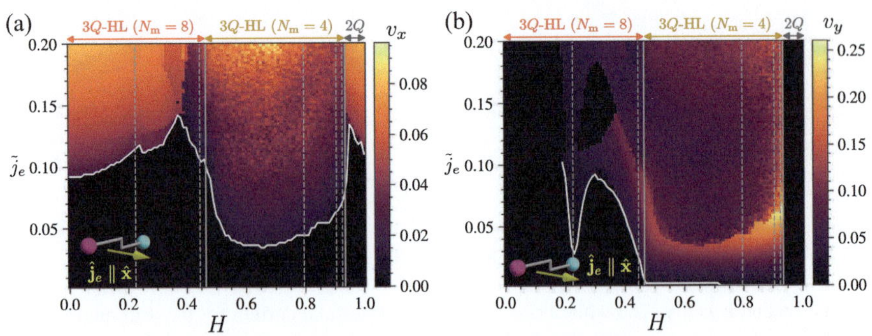

Fig. 7.31 The velocities for the (**a**) longitudinal and (**b**) transverse motions for $D = 0.3$ with $\hat{\mathbf{j}}_e \parallel \hat{\mathbf{x}}$. The notations are common to those in Fig. 7.30

to the $D = 0$ case, a nonzero threshold current density $\tilde{j}_{e,c}$ is required to cause the longitudinal drift motion, and v_x increases with $\tilde{j}_e$ above the threshold value, as shown in Fig. 7.31a. In addition, $\tilde{j}_{e,c}$ increases by increasing H and shows a cusp around the discontinuous change of $d_{m,y}$ from 7 and 5 to 6 and 4, respectively, at $H \simeq 0.225$, denoted by the gray dashed line (see the bottom panel of Fig. 7.28). These behaviors are similar to the $D = 0$ case in Fig. 7.30a. By further increasing H, $\tilde{j}_{e,c}$ increases, but starts to decrease at $H \simeq 0.375$ before reaching the topological transition at $H \simeq 0.465$ denoted by the gray solid line. Notably, in the 3Q-HL state with $N_m = 4$ after the topological transition, the threshold value is reduced and roughly half of the typical value in the $N_m = 8$ phase. For each H, v_x seemingly increases with $\tilde{j}_e$ above the threshold, but shows nonmonotonic fine oscillations represented by the mosaic patterns in the color map of Fig. 7.31a. In this $N_m = 4$ phase, by increasing H from 0.465, $\tilde{j}_{e,c}$ decreases, shows a minimum at $H \simeq 0.655$, and starts to increase toward the phase transition to the 2Q state accompanied by the change of N_m to zero at $H \simeq 0.935$, denoted by the other gray solid line. In the 2Q state, v_x takes a nonzero threshold value, which decreases by increasing H, similar to the $D = 0$ case.

We then discuss the transverse drift motion. Such a transverse motion is known for the skyrmions and is called the skyrmion Hall effect [7, 24] [see Fig. 1.11a]. In the following, we refer to the transverse drift motion of the hedgehogs as the hedgehog Hall effect. In Fig. 7.31b, we present the result for the transverse velocity v_y for $\hat{\mathbf{j}}_e \parallel \hat{\mathbf{x}}$. At zero field, v_y is zero for all $\tilde{j}_e$, namely the transverse motion is not observed since $\bar{b}_z = 0$. By introducing H, $\bar{b}_z$ becomes nonzero, and accordingly, we find that the transverse drift motion appears and becomes visible around the first discontinuous change of $d_{m,y}$ at $H \simeq 0.225$. The threshold current density $\tilde{j}_{e,c}$ shows a dip at $H \simeq 0.225$ and decreases to zero toward the topological transition at $H \simeq 0.465$, as represented by the white line in Fig. 7.31b. Intriguingly, $\tilde{j}_{e,c}$ for v_y is smaller than that for v_x in Fig. 7.31a. This means that in the range of $\tilde{j}_e$ above the threshold for v_y and below that for v_x, the transverse motion occurs without the longitudinal one; the hedgehog Hall effect maximizes the Hall angle [6]

$$\theta^{\text{Hall}} = \arctan\left(\frac{v_y}{v_x}\right) = \frac{\pi}{2}, \tag{7.106}$$

which we call the perfect hedgehog Hall effect. In this region, by increasing $\tilde{j}_e$ above the threshold, v_y first increases, and then becomes zero, as represented by the black island for $0.2 \lesssim H \lesssim 0.4$ in Fig. 7.31b. The value of $\tilde{j}_e$ at which v_y drops to zero corresponds to the threshold current density for v_x. By further increasing $\tilde{j}_e$, v_y starts to increase again above the black island. After the topological transition for $H \gtrsim 0.465$, the hedgehog Hall effect is more clearly observed in the 3Q-HL phase with $N_m = 4$. Notably, the threshold value $\tilde{j}_{e,c}$ is vanishingly small, and accordingly, the perfect hedgehog Hall effect can be universally driven by very small field in this region. Moreover, the typical transverse velocity is 0.1–0.2, which is about four times larger than that for the longitudinal one. We will discuss the origin of the perfect hedgehog Hall effect in Sect. 7.4.3. In this phase, v_y monotonically increases with $\tilde{j}_e$

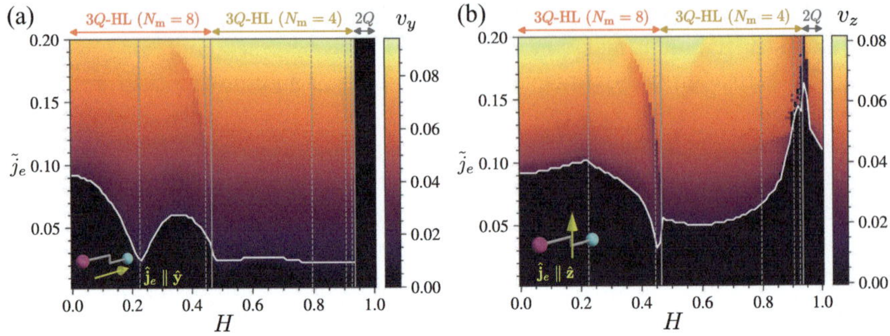

Fig. 7.32 The velocities for the longitudinal motions for $D = 0.3$ with **a** $\hat{\mathbf{j}}_e \parallel \hat{\mathbf{y}}$ and **b** $\hat{\mathbf{j}}_e \parallel \hat{\mathbf{z}}$. The notations are common to those in Fig. 7.30

below $\tilde{j}_{e,c}$ of v_x, above which v_y is reduced first and then, slowly increases with $\tilde{j}_e$ accomapnied by the nonmonotonic fine oscillations, similar to v_x in Fig. 7.31a.

In Fig. 7.32a, we present the results of v_y for $\hat{\mathbf{j}}_e \parallel \hat{\mathbf{y}}$, where the electric current is applied in the direction of the major portions of the Dirac string as shown in the inset (see also Fig. 7.29). Intriguingly, the hedgehog Hall effect does not appear and only the longitudinal motion occurs above the threshold value in this parameter regime. Although the perfect hedgehog Hall effect as well as the hedgehog Hall effect widely appears in the wide region of the phase diagram in the $\hat{\mathbf{j}}_e \parallel \hat{\mathbf{x}}$ case, the transverse motion of the hedgehogs is largely suppressed by changing the direction of the electric current. This is due to the anisotropic potential barrier discussed in Sect. 7.4.3. In this case, similar to the $D = 0$ case, $\tilde{j}_{e,c}$ shows dips at the discontinuous changes of $d_{m,y}$ in the $N_m = 8$ phase at $H \simeq 0.225$ and $H \simeq 0.465$. In the $N_m = 4$ phase, $\tilde{j}_{e,c}$ exhibits a weak H dependence and the effect from the Dirac string shortening is small. We note that the drift motion does not occur in the $2Q$ state, since $\mathcal{S}_{\mathbf{Q}_2} = 0$ as shown in Fig. 7.28. In Fig. 7.32b, we show the results of v_z for $\hat{\mathbf{j}}_e \parallel \hat{\mathbf{z}}$, where the electric current is perpendicular to the major portions of the Dirac strings. In this case, the hedgehog Hall effect does not occur since $\bar{b}_x$ and $\bar{b}_y$ are zero, and thus only the longitudinal motion appears. The threshold value $\tilde{j}_{e,c}$ increases with H and shows a cusp at the discontinuous change of $d_{m,y}$ at $H \simeq 0.225$. In contrast to the $\hat{\mathbf{j}}_e \parallel \hat{\mathbf{x}}$ case in Fig. 7.31a, $\tilde{j}_{e,c}$ decreases by increasing H after the cusp and shows a sharp dip around the topological transition at $H \simeq 0.465$. The decrease of $\tilde{j}_{e,c}$ is presumably due to the kink structures of the Dirac strings shown in Fig. 7.29. Around the topological transition, the Dirac strings with $d_{m,y} = 2$ connect the hedgehog and antihedgehog separated by 2 in the y direction and 1 in the z direction. The Dirac strings extending in the z direction might dominantly contribute to $\tilde{j}_{e,c}$ for $\hat{\mathbf{j}}_e \parallel \hat{\mathbf{z}}$ and play a similar role to the Dirac string parallel to the y direction for the $\hat{\mathbf{j}}_e \parallel \hat{\mathbf{y}}$ case, leading to the decrease of $\tilde{j}_{e,c}$. After the topological transition, N_m turns into 4 and the kink structures in the Dirac strings disappear. In this phase, $\tilde{j}_{e,c}$ exhibits a rapid increase with H toward the phase transition to the $2Q$ state at $H \simeq 0.935$, similar to the $\hat{\mathbf{j}}_e \parallel \hat{\mathbf{x}}$ case in Fig. 7.31a.

Our results show that the hedgehog Hall effect is much more flexible than the skyrmion Hall effect. It can be controlled by the amplitudes and directions of the magnetic field and the electric current. Most strikingly, hedgehogs can exhibit two extreme motions: purely longitudinal motion without any transverse one and purely transverse motion without any longitudinal one. By choosing appropriate parameters, one can even achieve switching between these two extremes. For instance, at $H = 0.4$, one can achieve the hedgehog Hall effect by applying the current in the x direction with $\tilde{j}_e = 0.2$ (Supplementary Video 1). On the one hand, by reducing the current density, one can switch it to the perfect hedgehog Hall effect at $\tilde{j}_e = 0.2$ (Supplementary Video 2). On the other hand, by changing the current direction to the y direction at $\tilde{j}_e = 0.2$, one can suppress the Hall effect, leading to purely longitudinal motion (Supplementary Video 3).

7.4.3 *Potential Barrier from Lattice Discretization*

In Sect. 7.4.2, we find that the drift motions of the hedgehogs and antihedgehogs are caused by an electric current. The threshold value of the current density $\tilde{j}_{e,c}$ depends on H, reflecting the real-space configurations of the hedgehogs and antihedgehogs, especially the discontinuous changes of the lengths of the Dirac strings. Furthermore, we find the perfect hedgehog Hall effect, which is a transverse motion without a longitudinal motion, in the presence of the DM interaction. These results imply that the lattice discretization brings about a potential barrier affecting the drift motions of hedgehogs and this barrier undergoes significant modulation as the hedgehogs and antihedgehogs move along the Dirac strings. Furthermore, it is expected that the potential barrier depends on the drift direction under the magnetic field.

Here, we compute the potential barrier that the hedgehogs and antihedgehogs experience during the drift motion due to the spatial changes in their positions. When the superposition in Eq. (1.11) is given, a spatial translation by $\Delta\mathbf{r}$ can be achieved through a phase shift from φ_η from $\varphi_\eta + \Delta\varphi_\eta$ with Eq. (4.16) and $\mathbf{a}_\eta = \frac{2\pi}{Q}\hat{\mathbf{Q}}_\eta$. Specifically, for the present 3Q-HL case, a spatial translation by $\Delta\mathbf{r} = (\Delta x, \Delta y, \Delta z)$ is represented by a phase shift as

$$\Delta\mathbf{r} = \frac{1}{Q}\left(\Delta\varphi_1, \Delta\varphi_2, \Delta\varphi_3\right). \tag{7.107}$$

Thus, the potential barrier can be computed by calculating the energy change while varying the phase φ_η. The 3Q-HL is represented by the superpositions of three elliptical helices, and hence we assume

$$\mathbf{S}_i^{\text{ellip}3Q} \propto \begin{pmatrix} \psi_2\epsilon_2 \sin \mathcal{Q}_2 + \psi_3 \cos \mathcal{Q}_3 \\ \psi_1 \cos \mathcal{Q}_1 + \psi_3\epsilon_3 \sin \mathcal{Q}_3 \\ \psi_1\epsilon_1 \sin \mathcal{Q}_1 + \psi_2 \cos \mathcal{Q}_2 + m \end{pmatrix}, \tag{7.108}$$

with $\mathcal{Q}_\eta = \mathbf{Q}_\eta \cdot \mathbf{r}_i + \varphi_\eta$. Equation (7.108) is obtained from Eq. (1.11) by taking

$$\psi_\eta^{\text{c}} = \psi_\eta, \quad \frac{\psi_\eta^{\text{s}}}{\psi_\eta^{\text{c}}} = \epsilon_\eta, \tag{7.109}$$

$$\mathbf{e}_1^1 = \hat{\mathbf{y}}, \ \mathbf{e}_2^1 = \hat{\mathbf{z}}, \ \mathbf{e}_3^1 = \hat{\mathbf{x}}, \tag{7.110}$$

$$\mathbf{e}_1^2 = \hat{\mathbf{z}}, \ \mathbf{e}_2^2 = \hat{\mathbf{x}}, \ \mathbf{e}_3^2 = \hat{\mathbf{y}}. \tag{7.111}$$

$$\mathbf{m} - m\hat{\mathbf{z}}. \tag{7.112}$$

We note that Eq. (7.108) can represent the $2Q$ state in the high-field region by taking $\psi_2 = 0$. Following the procedure in Sect. 4.3, we approximate the spin structures obtained by the LLG simulation by Eq. (7.108), and estimate the variational parameters by minimizing the cost function

$$U(\{\psi_\eta, \epsilon_\eta, \varphi_\eta\}, m) = \frac{1}{N}\sum_i \left(1 - \mathbf{S}_i^{\text{LLG}} \cdot \mathbf{S}_i^{\text{ellip}3Q}(\{\psi_\eta, \epsilon_\eta, \varphi_\eta\}, m)\right), \tag{7.113}$$

where $\mathbf{S}_i^{\text{LLG}}$ is the ground-state spin configuration obtained by numerically solving the LLG equation and $\mathbf{S}_i^{\text{ellip}3Q}$ is that generated from Eq. (7.108). The optimal state $\{\mathbf{S}_i^{\text{ellip}3Q}\}^*$ shows the cost function less than 6×10^{-4} for all H, and hence $\{\mathbf{S}_i^{\text{LLG}}\}$ is well reproduced by Eq. (7.108).

By using the optimal parameters obtained by the above procedure and varying only the phase φ_η to represent a spatial translation by $\Delta\mathbf{r}$ given by Eq. (7.107), we compute the energy change per site for the model in Eq. (3.21) measured from that for the optimal spin configuration $\{\mathbf{S}_i^{\text{ellip}3Q}\}^*$, denoted by ΔE. In Fig. 7.33a, we show ΔE for $D = 0$ and $H = 0$ on the plane of Δx and Δy with $\Delta z = 0$. ΔE is zero at $(\Delta x, \Delta y) = (n_1, n_2)$ (n_1 and n_2 are integers), and it takes the maximum value at $(\Delta x, \Delta y) = (0.5 + n_1, 0.5 + n_2)$. This indicates that in the lowest-energy state, hedgehogs and antihedgehogs are located at the interstitial positions of the lattice sites, and the energy increases when they approach the faces of the unit cubes, whose vertices are the lattice sites. When $(\Delta x, \Delta y, \Delta z) = (0.5, 0.5, 0)$, the hedgehogs and antihedgehogs are positioned on the sides of the unit cubes, leading to the largest ΔE. By further increasing Δz to 0.5, the hedgehogs and antihedgehogs come to the lattice sites, leading to the maximum of ΔE. These results imply that the $3Q$-HL at zero field undergoes a potential barrier approximately given by

$$\Delta E(\Delta\mathbf{r}) = \frac{V_x}{2}(1 - \cos(2\pi\Delta x)) + \frac{V_y}{2}(1 - \cos(2\pi\Delta y)) + \frac{V_z}{2}(1 - \cos(2\pi\Delta z)). \tag{7.114}$$

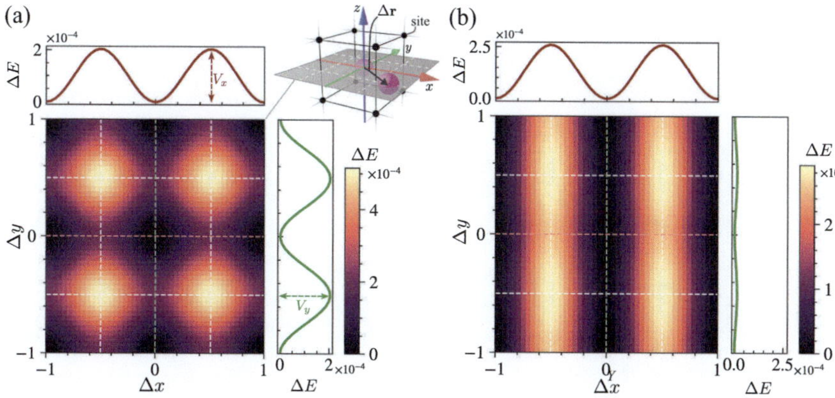

Fig. 7.33 Energy change per site, ΔE, for a spatial translation of the 3Q-HL by $\Delta\mathbf{r} = (\Delta x, \Delta y, 0)$ measured from that for the optimal spin configuration $\{\mathbf{S}_i^{\mathrm{ellip}3Q}\}^*$ for the model in Eq. (3.21) with $D = 0$ at **a** $H = 0$ and **b** $H = 0.25$. The white dashed lines denote the intersection of the xy plane and the unit cubes with the corners at the lattice sites. See also the schematic illustration for the translation of the hedgehog shown in the inset of (**a**). ΔE along the red and green dashed lines are presented in the top and right panels, respectively

In fact, ΔE along the x and y axes shown in the top and right panels of Fig. 7.33a behave like cosine curves. We note that $V_x = V_y = V_z$ by symmetry at zero field.

In Fig. 7.33b, we show ΔE for $D = 0$ at $H = 0.25$. Interestingly, the potential barrier along the y axis, V_y, is largely reduced and ΔE is nearly flat in the y direction. The decrease of the barrier along the y direction is attributed to the changes in the positions of hedgehogs and antihedgehogs associated with the Dirac string shortening from 6 to 5 at $H \simeq 0.235$. At this change, the hedgehogs and antihedgehogs are positioned on the faces of the unit cubes on the xz plane, reducing the energy barrier in the y direction compared to when they are in the interstitial positions of the lattice sites.

To investigate the H dependences of the potential barrier, we compute its amplitude in the direction of μ ($\mu = x, y, z$), V_μ, in Eq. (7.114) as

$$V_\mu = \Delta E(\Delta\mathbf{r} = \hat{\boldsymbol{\mu}}/2) - \Delta E(\Delta\mathbf{r} = 0), \tag{7.115}$$

for each H. In Fig. 7.34a, we show V_μ for $D = 0$ while varying H. At zero field, $V_x = V_y = V_z$ due to symmetry. By increasing H, V_x and V_z increase keeping the relation $V_x = V_z$, and show the cusps at the discontinuous changes of $d_{\mathrm{m},y}$ at $H \simeq 0.235$ and $H \simeq 0.625$. They increase further toward the phase transition to the 2Q state at $H \simeq 0.745$, but decrease in the 2Q state after the sharp drop at the phase transition. In stark contrast, V_y decreases with increasing H and shows the dips at the discontinuous changes of $d_{\mathrm{m},y}$ at $H \simeq 0.235$ and $H \simeq 0.625$, and the peak value of V_y in each dome decreases toward the phase transition. Eventually, V_y becomes zero in the 2Q state since the spin texture shows the modulations only in the x and z

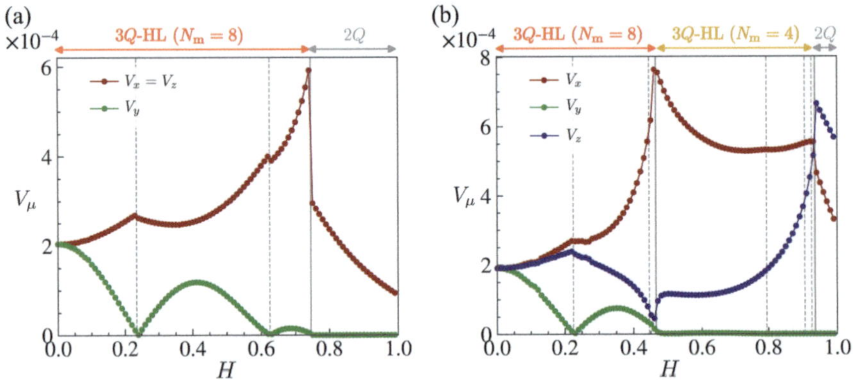

Fig. 7.34 H dependences of V_x, V_y, and V_z denoted by the red, green, blue dots, respectively, for **a** $D = 0$ and **b** $D = 0.3$. The dashed and solid gray lines denote the discontinuous changes in $d_{\mathrm{m},y}$ and N_{m}, respectively

directions. These H dependences of V_μ are similar to those of the threshold current density for the longitudinal drift, namely $\tilde{j}_{e,\mathrm{c}}$ of v_μ for $\hat{\mathbf{j}}_e \parallel \hat{\boldsymbol{\mu}}$ shown in Fig. 7.30, indicating the correlation between V_μ and $\tilde{j}_{e,\mathrm{c}}$; the drift motion is caused when the electric current becomes large enough to overcome the potential barrier V_μ. We note that V_y is vanishingly small at $H \simeq 0.235$ and $H \simeq 0.625$, while $\tilde{j}_{e,\mathrm{c}}$ shown in Fig. 7.30b takes nonzero values. This is presumably due to the insufficient simulation time; $\tilde{j}_{e,\mathrm{c}}$ might be overestimated since conducting longer simulations is required to resolve small velocities more accurately.

In Fig. 7.34b, we show V_μ for $D = 0.3$. Similar to the $D = 0$ case, H dependences of V_μ explain $\tilde{j}_{e,\mathrm{c}}$ for the longitudinal drift motion. First, let us discuss V_x. In the 3Q-HL with $N_{\mathrm{m}} = 8$ for $0 \leq H \lesssim 0.465$, V_x increases toward the topological transition at $H \simeq 0.465$ by increasing H, and in the 3Q-HL with $N_{\mathrm{m}} = 4$ for $0.465 \lesssim H \lesssim 0.935$, it first decreases and then gradually increases by increasing H. These trends resemble those of $\tilde{j}_{e,\mathrm{c}}$ shown in Fig. 7.31a, while the magnitude relationship of V_x in the $N_{\mathrm{m}} = 8$ and $N_{\mathrm{m}} = 4$ phases is opposite to that of $\tilde{j}_{e,\mathrm{c}}$. The reason for the smaller $\tilde{j}_{e,\mathrm{c}}$ despite the larger V_x in the $N_{\mathrm{m}} = 4$ phase is attributed to the hedgehog Hall effect observed in this phase; in addition to the driving force directly arising from the current, the spin structure experiences a Magnus force in the x direction due to the transverse motion in the y direction [7, 24]. Thus, the longitudinal drift motion in the x direction can be caused by a smaller current density, despite encountering a larger potential barrier. We note that the decrease of V_x with increasing H in the 2Q state is also compatible with the decrease of $\tilde{j}_{e,\mathrm{c}}$. Next, we discuss the result of V_z. The H dependence of V_z well resembles that of $\tilde{j}_{e,\mathrm{c}}$ in Fig. 7.32b, showing a cusp at $H \simeq 0.225$ and a dip at $H \simeq 0.465$, and then a rapid increase toward the phase transition to the 2Q state. In this case, the Magnus force does not affect the drift motion of the hedgehogs and antihedgehogs since the force in the z direction is zero. We note that V_z decreases by increasing H in the 2Q state, which is also compatible with $\tilde{j}_{e,c}$. Lastly, we discuss V_y. Similar to $\tilde{j}_{e,c}$ of v_y for $\hat{\mathbf{j}}_e \parallel \hat{\mathbf{y}}$ shown in Fig. 7.32a, $\tilde{j}_{e,c}$ shows the dip at $H \simeq 0.225$ in the 3Q-HL with $N_m = 8$. Intriguingly, V_y becomes vanishingly small in the 3Q-HL with $N_m = 4$ for $H \gtrsim 0.465$. This agrees well with the small $\tilde{j}_{e,c}$ of v_y for $\hat{\mathbf{j}}_e \parallel \hat{\mathbf{x}}$ and $\hat{\mathbf{j}}_e \parallel \hat{\mathbf{y}}$ shown in Figs. 7.31b and 7.32a, respectively. Thus, the large difference of V_x and V_y as well as nonzero $\bar{b}_z$ is important to cause the perfect hedgehog Hall effect.

We discuss the origin of the vanishingly small V_y in the 3Q-HL phase with $N_m = 4$ for $0.465 \lesssim H \lesssim 0.935$ in Fig. 7.34b. This appears to be related to the small higher harmonics of the spin structure factor as follows. Figure 7.35a shows H dependences of $\mathcal{S}_{n\mathbf{Q}_\eta}$ ($n = 1, 2, 3$) for $D = 0.3$ calculated from $\{\mathbf{S}_i^{\mathrm{LLG}}\}$ and $\{\mathbf{S}_i^{\mathrm{ellip}3Q}\}^*$, both of which show good agreement. Notably, $\mathcal{S}_{2\mathbf{Q}_2}$ and $\mathcal{S}_{3\mathbf{Q}_2}$ are vanishingly small compared to the others in the 3Q-HL with $N_m = 4$. The higher harmonics represent rapidly oscillating components in real space, and hence they can be largely affected by the lattice discretization. This is schematically shown in Fig. 7.35b. When the wave number is smaller than $6Q = \pi$, the higher harmonics restored from the values on the lattice sites before and after the translation are identical. However, the original wave cannot be restored for the case with $6Q$; the inverse Fourier transform of the gray dots taking zero cannot reproduce the gray lines, while the red dots can do. This corresponds to the Nyquist wave number. The presence of such wave components leads to a change in $\mathbf{S}_\mathbf{q}$ by the translation, resulting in a potential barrier through Eq. (3.14). Thus, smaller amplitudes of the higher harmonics above the Nyquist wave number suppress the effects from lattice discretization. Due to the norm condition

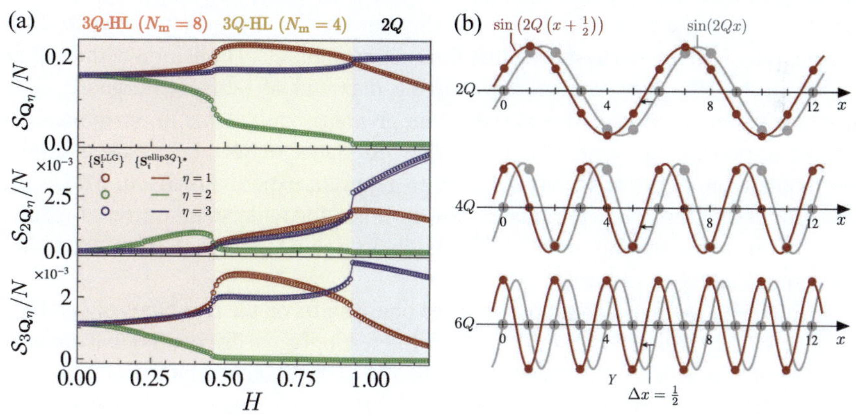

Fig. 7.35 **a** H dependences of $\mathcal{S}_{\mathbf{Q}_\eta}$ (top), $\mathcal{S}_{2\mathbf{Q}_\eta}$ (middle), and $\mathcal{S}_{3\mathbf{Q}_\eta}$ (bottom) for the model in Eq. (3.21) with $D = 0.3$. The results calculated from $\{\mathbf{S}_i^{\mathrm{LLG}}\}$ and $\{\mathbf{S}_i^{\mathrm{ellip}3Q}\}^*$ are represented by the circles and solid lines, respectively. **b** Schematic figures for the effect from the spatial translation by a half lattice constant $\Delta x = \frac{1}{2}$ for the higher harmonics. The lines and dots represent the higher harmonics in continuous space and their values on the lattice sites, respectively. The original higher harmonics and those after the translation are denoted by the gray and red colors, respectively. Presented here are the higher harmonics with the wave numbers of $2Q$ (top), $4Q$ (middle), and $6Q$ (bottom) with $Q = \frac{\pi}{6}$

$|\mathbf{S}(\mathbf{r})|^2 = 1$, $\mathcal{S}_{\mathbf{q}}$ and $\mathcal{S}_{2\mathbf{q}}$ are tightly related to each other. In the present case, $\mathcal{S}_{3\mathbf{Q}\eta}$ is vanishingly small only for $\eta = 2$, and accordingly, the modulation with the Nyquist wave vector $6\mathbf{Q}_\eta$ is small only in the y direction, leading to vanishingly small V_y.

7.5 Concluding Remarks of this Chapter

To conclude this chapter, we discuss the implications of our results. In Sect. 7.1, we investigated the control of the spin textures and their topological properties for the $3Q$- and $4Q$-HLs by changing the amplitude of spin moirés through the spatial anisotropy. Experimentally, two types of the HLs as well as the $3Q$-SkL are realized in $MnSi_{1-x}Ge_x$ [3, 22], as introduced in Sect. 1.4.3. Although the microscopic origin is under debate, a scenario was proposed based on the itinerant nature of electrons included in the present model in Eq. (3.27) with isotropic interactions [13]. In this model, the anisotropy in the effective long-range interactions would be introduced through the change of the electronic state of itinerant electrons, and hence, our results suggest that the interesting transitions found here might be caused by modifying the electronic band structure. Such band modification can be achieved, e.g., by external pressure and chemical substitution, and the systematic experimental studies to confirm our scenario are desirable.

In Sect. 7.2, we clarified the control of the topological properties and the EEMF in the $3Q$-HL by changing the twist angle. This would shed light on novel dynamical electromagnetic responses, which may be useful to detect the topological properties. For instance, in the case denoted by the blue lines in Fig. 7.14, the topological Hall conductivity is predicted to show a maximal at the topological transition with $N_{\text{pair}} = 4 \to 1$ in the HL phase accompanied by the maximal of $-\bar{b}_z$. Furthermore, $\bar{e}_z$ is also maximized at the topological transition, giving rise to the maximum dynamical electric responses, e.g., emergent inductance introduced in Sect. 1.3.3. Interestingly, the modulation of the twist angle has been reported experimentally in MnGe [8, 10]. Our results not only provide a starting point for understanding these results but also serve as a guiding map to explore the intriguing emergent electromagnetic phenomena and their control.

In Sect. 7.3, we investigated the effect of phase shifts on the two types of $4Q$-HLs composed of helices and sinusoidal waves. Interestingly, we discovered that unusual Dirac strings appear on the planes perpendicular to the magnetization direction, and their evolution leads to unconventional topological phenomena: pair creation and fusion. These topological phenomena caused by the horizontal Dirac strings have not been found in the $3Q$ HLs [9, 25]. Our finding indicates the importance of the phase degree of freedom for such unexplored topological objects. Furthermore, $\bar{b}_z$ is always negative in the superposition of helices, while it can be both negative and positive depending on the phase and the magnetization in the superposition of sinusoidal waves. This result may be related to the sign change of the topological Hall conductivity with the increase of the magnetic field reported for the $4Q$-HL in $MnSi_{1-x}Ge_x$ [5] (see Fig. 1.15).

In Sect. 7.4, we revealed that the threshold current density for the current-induced drift motion of the $3Q$-HL depends on the magnetic field, reflecting the real-space configurations of the hedgehogs and antihedgehogs. Our finding implies that the threshold can be used as a probe for the topological transitions. The real-space spin structure of the $3Q$-HL can be experimentally studied by the LTEM [20]. While detailed information about the exact spatial arrangement may be hard to resolve, the relative positions of the hedgehogs and antihedgehogs can be inferred through the relatively straightforward method of electrically measuring the threshold current. Furthermore, we revealed the anisotropic potential barrier brings about the perfect hedgehog Hall effect exhibiting the Hall motion with a maximally large Hall angle. Such an anisotropic barrier can also arise in 2D spin textures. It is intriguing to extend the discussion to the SkL and find a similar perfect Hall effect.

References

1. Binz, B., & Vishwanath, A. (2006). Theory of helical spin crystals: Phases, textures, and properties. *Physical Review B, 74*, Article 214408.
2. Binz, B., Vishwanath, A., & Aji, V. (2006). Theory of the helical spin crystal: a candidate for the partially ordered state of MnSi. *Physical Review Letters, 96*, Article 207202.
3. Fujishiro, Y., Kanazawa, N., Nakajima, T., Yu, X. Z., Ohishi, K., Kawamura, Y., Kakurai, K., Arima, T., Mitamura, H., Miyake, A., Akiba, K., Tokunaga, M., Matsuo, A., Kindo, K., Koretsune, T., Arita, R., & Tokura, Y. (2019). Topological transitions among skyrmion- and hedgehog-lattice states in cubic chiral magnets. *Nature Communications, 10*(1), 1059.
4. Fujishiro, Y., Kanazawa, N., Shimojima, T., Nakamura, A., Ishizaka, K., Koretsune, T., Arita, R., Miyake, A., Mitamura, H., Akiba, K., Tokunaga, M., Shiogai, J., Kimura, S., Awaji, S., Tsukazaki, A., Kikkawa, A., Taguchi, Y., & Tokura, Y. (2018). Large magneto-thermopower in MnGe with topological spin texture. *Nature Communications, 9*(1), 408.
5. Fujishiro, Y., Kanazawa, N., & Tokura, Y. (2020). Engineering skyrmions and emergent monopoles in topological spin crystals. *Applied Physics Letters, 116*(9), Article 090501.
6. Jiang, W., Zhang, X., Yu, G., Zhang, W., Wang, X., Benjamin Jungfleisch, M., Pearson, J. E., Cheng, X., Heinonen, O., & Wang, K. L., et al. (2017). Direct observation of the skyrmion Hall effect. *Nature Physics, 13*(2), 162–169.
7. Jonietz, F., Mühlbauer, S., Pfleiderer, C., Neubauer, A., Münzer, W., Bauer, A., Adams, T., Georgii, R., Böni, P., Duine, R. A., Everschor, K., Garst, M., & Rosch, A. (2010). Spin transfer torques in MnSi at ultralow current densities. *Science, 330*(6011), 1648–1651.
8. Kanazawa, N., Kitaori, A., White, J. S., Ukleev, V., Rønnow, H. M., Tsukazaki, A., Ichikawa, M., Kawasaki, M., & Tokura, Y. (2020). Direct observation of the statics and dynamics of emergent magnetic monopoles in a chiral magnet. *Physical Review Letters, 125*, Article 137202.
9. Kanazawa, N., Nii, Y., Zhang, X.-X., Mishchenko, A. S., De Filippis, G., Kagawa, F., Iwasa, Y., Nagaosa, N., & Tokura, Y. (2016). Critical phenomena of emergent magnetic monopoles in a chiral magnet. *Nature Communications, 7*(1), 11622.
10. Kanazawa, N., White, J. S., Rønnow, H. M., Dewhurst, C. D., Morikawa, D., Shibata, K., Arima, T., Kagawa, F., Tsukazaki, A., Kozuka, Y., Ichikawa, M., Kawasaki, M., & Tokura, Y. (2017). Topological spin-hedgehog crystals of a chiral magnet as engineered with magnetic anisotropy. *Physical Review B, 96*, 220414(R).
11. Kitaori, A., Kanazawa, N., Ishizuka, H., Yokouchi, T., Nagaosa, N., & Tokura, Y. (2021). Enhanced electrical magnetochiral effect by spin-hedgehog lattice structural transition. *Physical Review B, 103*, L220410.

12. Nagaosa, N., & Tokura, Y. (2013). Topological properties and dynamics of magnetic skyrmions. *Nature Nanotechnology, 8*(12), 899–911.
13. Okumura, S., Hayami, S., Kato, Y., & Motome, Y. (2020). Magnetic hedgehog lattices in noncentrosymmetric metals. *Physical Review B, 101*, Article 144416.
14. Okumura, S., Hayami, S., Kato, Y., & Motome, Y. (2020). Tracing monopoles and anti-monopoles in a magnetic hedgehog lattice. *JPS Conference Proceedings, 30*, Article 011010.
15. Ozawa, R., Hayami, S., Barros, K., Chern, G. W., Motome, Y., Batista, C. D. (2016). Vortex crystals with chiral stripes in itinerant magnets. *Journal of the Physical Society of Japan, 85*(10), 103703.
16. Jin-Hong Park and Jung Hoon Han. (2011). Zero-temperature phases for chiral magnets in three dimensions. *Physical Review B, 83*, Article 184406.
17. Shimizu, K., Okumura, S., Kato, Y., & Motome, Y. (2021). Phase transitions between helices, vortices, and hedgehogs driven by spatial anisotropy in chiral magnets. *Physical Review B, 103*, Article 054427.
18. Shimizu, K., Okumura, S., Kato, Y., & Motome, Y. (2021). Spin moiré engineering of topological magnetism and emergent electromagnetic fields. *Physical Review B, 103*, Article 184421.
19. Shimizu, K., Okumura, S., Kato, Y., & Motome, Y. (2022). Phase degree of freedom and topology in multiple-q spin textures. *Physical Review B, 105*, Article 224405.
20. Tanigaki, T., Shibata, K., Kanazawa, N., Yu, X., Onose, Y., Park, H. S., Shindo, D., & Tokura, Y. (2015). Real-space observation of short-period cubic lattice of skyrmions in MnGe. *Nano Letters, 15*(8), 5438–5442.
21. Tatara, G. (2019). Effective gauge field theory of spintronics. *Physica E, 106*, 208–238.
22. Tokunaga, Y., Yu, X. Z., White, J. S., Rønnow, H. M., Morikawa, D., Taguchi, Y., & Tokura, Y. (2015). A new class of chiral materials hosting magnetic skyrmions beyond room temperature. *Nature Communications, 6*(1), 7638.
23. Volovik, G. E. (1987). Linear momentum in ferromagnets. *Journal of Physics C: Solid State Physics, 20*(7), L83–L87.
24. Zang, J., Mostovoy, M., Han, J. H., & Nagaosa, N. (2011). Dynamics of skyrmion crystals in metallic thin films. *Physical Review Letters, 107*, 136804.
25. Zhang, X.-X., Mishchenko, A. S., De Filippis, G., & Nagaosa, N. (2016). Electric transport in three-dimensional skyrmion/monopole crystal. *Physical Review B, 94*, Article 174428.

Chapter 8
Summary

In this thesis, we have theoretically studied the generation, annihilation, and control of the periodic arrangements of topological spin textures, called topological spin crystals, in magnets. To facilitate the systematic discussion on the control of the topological spin crystals, we proposed the "spin moiré" picture by taking particular note of the observation that the topological spin crystals are often represented as superpositions of multiple spin density waves. By exploiting the lessons for controlling the superstructures in conventional moirés, we have investigated various ways of controlling the spin textures and resultant magnetic and electronic phenomena—spin moiré engineering. Specifically, we have studied the spin moiré engineering for several different parameters in moiré patterns, such as the number, twist angles, and relative phases of the superposed waves, for several topological spin crystals: chiral soliton lattices (CSLs) in one dimension, skyrmion lattices (SkLs) in two dimensions, and hedgehog lattices (HLs) in three dimensions. Through the control of these parameters, we have elucidated the modulations and evolutions of the spin textures, topological properties, and associated emergent electromagnetic fields (EEMFs).

In Chap. 1, we introduced the fundamentals of topological defects in magnetism, topological spin crystals, the EEMFs arising from the Berry phase, and the resultant transport and optical phenomena. We also briefly reviewed the experimental studies on the topological spin crystals, focusing on the CSLs, SkLs, and HLs. In addition, we discussed three stabilization mechanisms for such topological spin crystals. In the end of this chapter, we presented the motivation of our study.

In Chap. 2, we introduced the main concept of this thesis, spin moiré and its engineering. After reviewing flexible control of the moiré patterns widely known and used in optics and atomic layered materials, we showed that such controls can be straightforwardly extended to the topological spin crystals composed of multiple spin helices. We also discussed several advantages of the spin moiré to the structural moiré in the layered materials.

K. Shimizu, *Spin Moiré Engineering and Emergent Electromagnetism in Topological Spin Crystals*, Springer Theses, https://doi.org/10.1007/978-981-96-5992-0_8

In Chap. 3, we introduced the theoretical models used in this thesis. First, we introduced a phenomenological model based on the Ginzburg-Landau theory for describing the thermodynamic properties of chiral magnets. Next, we introduced a minimal spin model, which describes the fundamental magnetic properties of chiral magnets both in and out of equilibrium. Then, we presented a spin-charge coupled model for a metallic magnet in which the effects from itinerant electrons are explicitly taken into account. Finally, we introduced the effective spin model derived by the perturbation expansion for the spin-charge coupled model.

In Chap. 4, we presented the analytical and numerical methods that we use in this thesis. First, we introduced the variational method to obtain the lowest-energy state for a given classical-spin Hamiltonian. We also introduced a more elaborate way to optimize the spin configuration based on a gradient based optimization method. Next, we presented an approach to extracting parameters in the constituent waves in spin moirés from the numerically obtained real-space spin configuration. We also introduced the Landau-Lifshitz-Gilbert (LLG) equation describing the time evolution by the classical-spin Hamiltonian including time-dependent external fields. Then, we presented a numerical method based on the linear spin-wave theory to compute the magnon spectrum. Finally, we introduced a theoretical framework to systematically analyze the effect of phase shifts in spin moirés by developing the hyperspace representation.

In Chap. 5, we presented the results of the spin moiré engineering in one dimension. Specifically, we focus on a one-dimensional chiral magnet, where a helical state at zero field acquires the solitonic feature to form the CSL by the magnetic field perpendicular to the chiral axis. The CSL is a one-dimensional spin moiré represented by a superposition with higher harmonic waves. We studied the flexible control of the topological properties and the emergent electric field (EEF) through not only static external magnetic fields but also dynamically oscillating fields. First, we studied the fundamental excitation properties by calculating the magnon spectra that exhibit multiple gaps opened by the solitonic feature of the CSL. In the system with edges, additional excitation modes appear within the magnon band gaps. By numerical simulations based on the LLG equation, we showed that intriguing soliton penetration dynamics is caused by a circularly-polarized AC magnetic field with the frequency of the edge mode. We found that the total number of infiltrated solitons starts to incrementally increase in a sequential manner by increasing the amplitude of the rotating magnetic field above the threshold. We also revealed that the threshold field depends on the solitonic feature of the spin texture. Then, we investigated the control of the EEF while changing the static and linearly-polarized AC magnetic fields by using the LLG simulation. For the bulk response calculated under the periodic boundary condition, we clarified that the net EEF becomes prominent at the magnetic resonance frequencies. Notably, we found that the EEF is enlarged along with the solitonic feature of the CSL. In addition, we found that the higher frequency resonance modes are more clearly visible in the frequency profile of the EEF response than the magnetic one. We also found that the static magnetic field parallel to the chiral axis drives the EEF in the field direction accompanied by the spin-wave propagation. Moreover, for the EEF response in the presence of the edges calculated under the open boundary condition, we find that the contributions from the additional edge modes become comparable to or larger than the bulk responses in realistic system sizes.

In Chap. 6, we showed the results of the spin moiré engineering in two dimensions. Among many parameters in spin moirés, we focused on the phase degree of freedom in the SkLs. Specifically, we investigated the control of the spin textures through phase shifts. We studied the effect of phase shifts on the SkLs composed of three waves. The phase degree of freedom plays a prominent role when the number of superposed waves is larger than the spatial dimension; this is the case for the $3Q$-SkLs in two dimensions. Using the hyperspace representation developed in Chap. 4, we systematically investigate the effect of phase shifts on two types of SkLs composed of three helices and three sinusoidal waves. We clarified the topological phase diagram while changing the phases and the uniform magnetization, distinguished by the skyrmion number N_{sk} ranging from -2 to 2. In the superposition of three helices, we found that the SkLs with $N_{\mathrm{sk}} = \pm 1$ are dominant, and the SkLs with higher topological numbers $N_{\mathrm{sk}} = \pm 2$ appear with nonzero magnetization. In contrast, in the superposition of three sinusoidal waves, the SkLs with $N_{\mathrm{sk}} = \pm 2$ are dominant and $N_{\mathrm{sk}} = \pm 1$ phases appear with nonzero magnetization. In addition, by analyzing the numerical data obtained in the previous study, we showed that the phase shift is induced by the magnetic field, resulting in a topological transition with a change in N_{sk} from 2 to 1.

In Chap. 7, we discussed the spin moiré engineering in three dimensions. First, we presented the results of controlling spin moirés with the amplitudes of superposed waves through spatial anisotropy. We performed a variational calculation for the effective spin model for noncentrosymmetric itinerant magnets by changing the anisotropy in the interactions. We showed that the two types of 3D HLs stabilized in the isotropic cases are gradually modulated by the spatial anisotropy with changing amplitudes of the superposed helices. The $3Q$-HL turns into the 2D spin structures by increasing the hard-axis anisotropy along one of the helical directions, while it turns into another type of 2D one and finally a 1D helical state for the easy-axis anisotropy. Meanwhile, the $4Q$-HL changes into a $3Q$-HL and a 1D helical state by increasing the hard- and easy-axis anisotropy, respectively. In addition to these magnetic phase transitions, we found topological transitions caused by pair annihilation of hedgehogs and antihedgehogs, by calculating the real-space positions of the topological defects and the flows of the emergent magnetic field. Next, we presented the results of controlling spin moirés by the twist angles between the helical directions. For the 3D spin moiré composed of three helices in continuous space, we systematically clarified the evolution of the topological objects and the EEMFs while changing the angle spanning the three helical directions as well as the uniform magnetization. We found two topological transitions by pair annihilation of the hedgehogs and antihedgehogs, where the number of the hedgehog-antihedgehog pairs changes from 4 to 1 and to 0. We showed that the net EEMFs are maximized at the former topological transition with the vanishment of three pairs. Furthermore, by variational calculations based on the Ginzburg-Landau free energy, we clarified how the helical angle changes when the magnetization is increased by an external magnetic field. We showed two representative cases starting from the $3Q$-HL: In one case, the helical angle increases, and the HL turns into a 2D SkL, and in the other case, the helical angle decreases and the HL turns into a 1D conical state. We found that the EEMFs exhibit characteristic changes through these transitions. Next, we presented the results of the effects of phase shifts in the $4Q$-HLs. Similar to the $3Q$-SkL cases in Chap. 6, using the hyperspace representation, we systematically analyzed the topological phase diagram for the two types of HLs composed of four helices and sinusoidal waves. In the case of the helices, we found various topological phases classified by the number of hedgehogs and antihedgehogs per unit cube, N_{m}: $N_{\mathrm{m}} = 8, 16, 32,$ and 48 while changing the phase and the uniform magnetization. Meanwhile, for the case of the sinusoidal waves, we obtained the topological phases with $N_{\mathrm{m}} - 8, 16, 24, 32,$ and 48. In thc formcr, thc largc portions of thc phase diagram are occupied by the phases with $N_{\mathrm{m}} = 8$ and 16, while for the latter, the major portions are occupied by the phases with larger N_{m}. The emergent magnetic field (EMF) arising from the noncoplanar spin configurations is always negative for the former, but it changes the sign depending on the phase and magnetization for the latter. Interestingly, we discovered unconventional evolution of topological objects: pair creation of hedgehogs and antihedgehogs and fusion of three hedgehogs and antihedgehogs, the latter of which maximizes the amplitude of the EMF. Furthermore, we studied phase shifts in the $4Q$-HL caused by the magnetic field by analyzing the numerical data obtained in the previous study. We clarified that by increasing the magnetic field for the $4Q$-HL, the phase shift occurs, causing a topological transition with a reduction of N_{m}. Finally, we investigated the control of the drift motions of the $3Q$-HL caused by an electric current with the LLG simulation. When the Dzyaloshinskii-Moriya interaction is absent, we clarified that an electric current causes the drift motion of the $3Q$-HL longitudinal to the current direction above a certain threshold $\tilde{j}_{e,\mathrm{c}}$. We found that the value of $\tilde{j}_{e,\mathrm{c}}$ depends on the magnetic field and the direction of the electric current relative to the Dirac strings. In the presence of the Dzyaloshinskii-Moriya interaction, we found that the $3Q$-HL shows the drift motion transverse to the current direction that we call the hedgehog Hall effect. Specifically, we revealed that the threshold current of the transverse motion is smaller than that of the longitudinal motion, resulting in the perfect hedgehog Hall effect with the maximally large Hall angle of $\frac{\pi}{2}$. Furthermore, we clarified that these behaviors can be explained by the potential barrier experienced by the hedgehogs and antihedgehogs on the discrete lattice.

To conclude, we have proposed the spin moiré picture to systematically discuss the control of the spin textures, their topological properties, and the EEMFs, and demonstrated these controls in a variety of topological spin textures. A moiré is a general phenomenon in physics that brings about not only unexpected superstructures but also intriguing physical properties, as illustrated in the diverse progress in atomic layered materials such as the twisted bilayer graphene. As we have demonstrated in this thesis, the spin moiré provides yet another novel platform for intriguing physics, with several advantages to the structural moiré in the layered materials: (1) While the structural moiré is mostly limited to two dimensions, the spin moiré can be extended from one to three dimensions, (2) since the constituent waves are vector (spinor) fields, the spin moiré exhibits a richer superstructure compared to the scalar potential field in the structural moiré, (3) the spin moiré allows flexible control through external fields such as magnetic fields, electric currents, and temperature, although it is usually difficult to modulate the structural moiré after the fabrication, and (4) the spin moiré exhibits topologically nontrivial spin textures, giving rise to novel emergent electromagnetic phenomena. Our results based on the spin moiré picture not only establish the versatile tool to study the control of topological magnetism but also pave the way for further exploration of topological spin textures and the resultant emergent electromagnetic phenomena. Our present studies have been mostly limited to the analyses of magnetic and electric properties based on the EEMFs. Meanwhile, especially in the spin moiré with a small period, the electric response arising from the band topology in momentum space becomes crucial, in addition to the real-space topological properties. Examining the electronic states by extending the discussion to explicitly incorporate itinerant electrons are essential for establishing a more solid foundation for spin moiré engineering, particularly in understanding the electric and magnetic properties, and quantum transport phenomena.

Appendix A
Soliton Penetration from Chirality Domain Boundary

In this Appendix, we demonstrate that soliton penetration dynamics similar to that discussed in Sect. 5.1.3 takes place even in the bulk system exhibiting the different chirality domains. Here, we consider the system where crystal structures with distinct chiralities form domains. Such systems have indeed been observed in experiments, wherein micro-sized samples exhibit the spin texture coexisting with right-handed and left-handed CSLs due to the opposite chirality, by using LTEM [1]. To mimic such situations, we consider a 1000-site system under the PBC, where the sign of the DM interaction is reversed in the right half of the system such that the left half exhibits a right-handed CSL, while the right half displays a left-handed CSL, as schematically depicted in Fig. A.1a. The Hamiltonian reads

$$\mathcal{H}(t) = \sum_l \left[-J\mathbf{S}_l(t) \cdot \mathbf{S}_{l+1}(t) - \mathbf{D}_l \cdot (\mathbf{S}_l(t) \times \mathbf{S}_{l+1}(t)) + \mathbf{h}(t) \cdot \mathbf{S}_l(t) \right], \quad \text{(A.1)}$$

where $D = |\mathbf{D}_l| = \tan(\pi/10)$ and

$$\mathbf{D}_l = \begin{cases} D\hat{\mathbf{x}} & (0 \le l < 500), \\ -D\hat{\mathbf{x}} & (500 \le l < 1000). \end{cases} \quad \text{(A.2)}$$

We calculate the spin dynamics by numerically solving the LLG equation in Eq. (4.6) with the same parameters used in Sect. 5.1.3.

Although the PBC is used here, the system exhibits the mid-gap modes with the same frequency as the edge mode in Sect. 5.1.2. Furthermore, these modes are exponentially localized at the boundaries of the different chirality, i.e., $l \simeq 0$ and $l \simeq 500$. In Fig. A.1b, we show the real-space and real-time distribution of $S_{l,z}(t)$ with $(h_x^{\mathrm{stat}}, h_z^{\mathrm{stat}}) = (0, 0.05)$ for the counterclockwise rotating magnetic field with $\omega = 0.076$ and $h^{\mathrm{rot}} = 0.013$. After applying the rotating magnetic field, the system exhibits the precursor motion wherein the bulk solitons in the left (right) half are compressed toward $l = 0$ $(l = L - 1)$, and accordingly, the sparse-soliton region is realized around $l = 500$. Subsequently, additional solitons penetrate into the system from the boundary of the different chirality at $t \simeq 4000$, as indicated by the green

K. Shimizu, *Spin Moiré Engineering and Emergent Electromagnetism in Topological Spin Crystals*, Springer Theses, https://doi.org/10.1007/978-981-96-5992-0

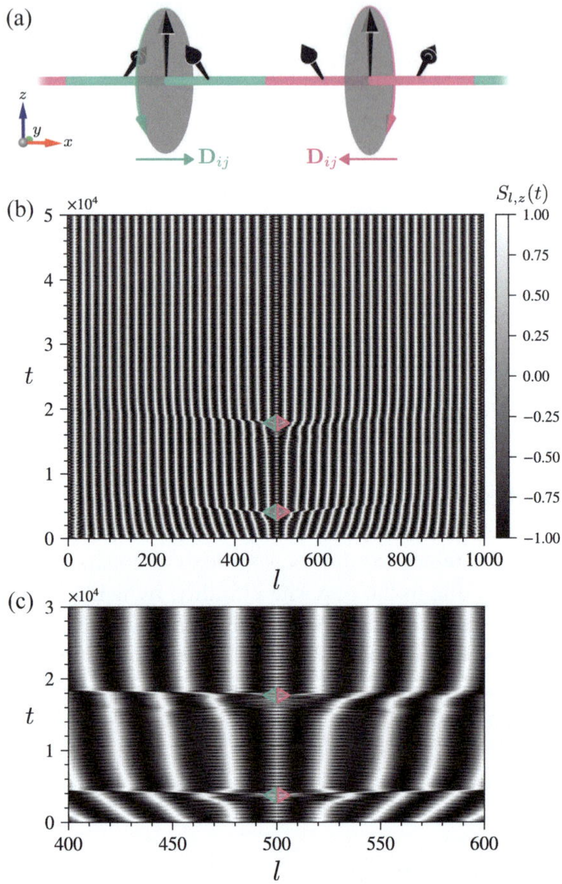

Fig. A.1 **a** Schematic picture of the setup for the system with the domain of the opposite chirality. For the 1000-site system under the PBC, the DM vector is given by $\mathbf{D}_l = D\hat{\mathbf{x}}$ $(-D\hat{\mathbf{x}})$ in the left (right) 500 bonds of the system. The black arrows represent the spins with the opposite swirling directions between the green and pink regions. **b** Real-space and real-time distribution of the z component of spins $S_{l,z}(t)$ for the 1000-site system with the counterclockwise rotating magnetic field for $(h_x^{\mathrm{stat}}, h_z^{\mathrm{stat}}) = (0, 0.05)$, $\omega = 0.076$, and $h^{\mathrm{rot}} = 0.013$. **c** The same results shown for 200 sites at the center of the system with a shorter time window. The green and pink triangles represent the time when solitons penetrate simultaneously into the left and right domains of the system, respectively

and pink triangles. Notably, solitons do not penetrate into just one chirality domain but both domains, as seen in the enlarged figure in Fig. A.1c. This is linked to spatial inversion symmetry of the system at the boundary. We also find another soliton penetration at $t \simeq 1.8 \times 10^4$ in a similar manner.

Reference

[1] Togawa, Y., Koyama, T., Nishimori, Y., Matsumoto, Y., McVitie, S., McGrouther, D., Stamps, R. L., Kousaka, Y., Akimitsu, J., Nishihara, S., Inoue, K., Bostrem, I. G., Sinitsyn, VI. E., Ovchinnikov, A. S., & Kishine, J. (2015). *Physical Review B, 92*, 220412(R).

Appendix B
Surface Barrier

In this Appendix, we investigate a potential barrier encountered when introducing a single chiral soliton from an edge of the system into the bulk of the FFM state. A similar situation was examined in a continuum model [1] and the discretized model [2, 3]. We first prepare a single chiral soliton capped by polarized spins on both sides. Specifically, we prepare the system of size $L + 1$ with $\mathbf{S}_0 = \mathbf{S}_L = (0, 0, -1)^\mathsf{T}$ and optimize the spin texture based on the Hamiltonian in Eq. (3.4). The optimization is conducted while varying L from 20 to 40 one by one, and we identify the lowest-energy single chiral soliton state. Subsequently, we embed this single chiral soliton state to be centered at $l = l_\mathrm{c}$ into the fully-polarized FFM state with $\mathbf{S}_l = (0, 0, -1)^\mathsf{T}$ for the $L = 1000$-site system ($l = 0, 1, \ldots, L - 1$) and calculate the total energy E_{l_c} while varying l_c from -20 to 20. If a part of the soliton extends beyond the edge of the system when l_c approaches zero or is negative, we compute the energy by disregarding the protruding portion. These analyses facilitate calculating the potential barrier encountered for the single soliton to penetrate from the edge into the FFM system.

Figure B.1 shows the l_c dependence of the total energy measured from the fully-polarized FFM state, namely, $\Delta E = E_{l_\mathrm{c}} - E_{l_\mathrm{c}=-20}$, for several h_z^stat with $h_x^\mathrm{stat} = 0$.

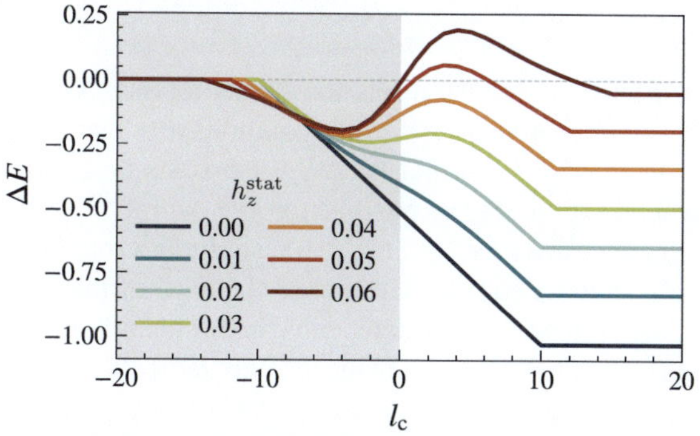

Fig. B.1 Energy for a single chiral soliton measured from the fully-polarized FFM state, ΔE, while changing the center position of the soliton l_c with $h_x^\mathrm{stat} = 0$ for several values of h_z^stat

K. Shimizu, *Spin Moiré Engineering and Emergent Electromagnetism in Topological Spin Crystals*, Springer Theses, https://doi.org/10.1007/978-981-96-5992-0

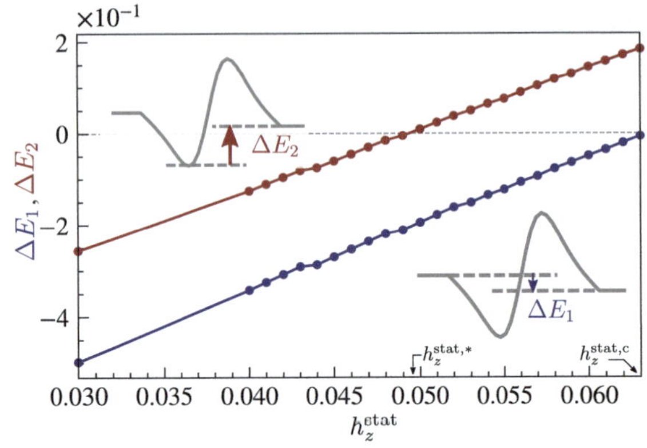

Fig. B.2 The energy gain ΔE_1 achieved by introducing a single soliton into the FFM state, and the optimal energy difference ΔE_2 when placing the soliton either outside or inside the FFM state, while varying $h_z^{\rm stat}$, represented by the blue and red lines, respectively. See Eqs. (B.1) and (B.2) for the definitions of ΔE_1 and ΔE_2, respectively

For $h_z^{\rm stat} = 0$, ΔE decreases by increasing $l_{\rm c}$, meaning that the chiral soliton can smoothly penetrate into the system without any potential barrier, and finally, the system gains the energy of $\Delta E \simeq -1.03$ for sufficiently large $l_{\rm c}$. By increasing $h_z^{\rm stat}$, the amount of this energy gain decreases, and the slope of ΔE at $l_{\rm c} = 0$ increases. The slope becomes zero at $h_z^{\rm stat} \simeq 0.025$ and then turns to be positive by further increasing $h_z^{\rm stat}$, resulting in a dip and a hump for $l_{\rm c} < 0$ and $l_{\rm c} > 0$, respectively. The hump corresponds to the potential barrier experienced by the single chiral soliton. Although the system still gains the energy by introducing the chiral soliton as $\Delta E_{l_{\rm c}=20}$ is negative, the soliton is required to overcome the potential barrier around the edge. The energy gain achieved by the soliton penetration turns to be positive for $h_z^{\rm stat} \gtrsim h_z^{\rm stat,c}$ as the FFM state is the ground state. Our results for the discretized system are qualitatively consistent with the previous study in continuum space [1].

In Fig. B.2, we show the $h_z^{\rm stat}$ dependence of the energy gain accompanied by introducing a soliton into the fully-polarized FFM state, ΔE_1, and the optimal energy difference between the cases where the soliton is placed outside ($l_{\rm c} \leq 0$) and inside ($l_{\rm c} \geq 0$) the system, ΔE_2, defined by

$$\Delta E_1 = E_{l_{\rm c}=20} - E_{l_{\rm c}=-20}, \tag{B.1}$$

$$\Delta E_2 = \min_{l_{\rm c}\geq 0} \Delta E_{l_{\rm c}} - \min_{l_{\rm c}\leq 0} \Delta E_{l_{\rm c}}, \tag{B.2}$$

respectively. See the insets for the definitions. As discussed above, the energy gain ΔE_1 is negative at $h_z^{\rm stat} = 0$, monotonically increases with $h_z^{\rm stat}$, and becomes zero at $h_z^{\rm stat} = h_z^{\rm stat,c}$. Similarly, ΔE_2 also starts from negative and increases with $h_z^{\rm stat}$, whereas it turns to be positive for $h_z^{\rm stat} > h_z^{\rm stat,*} \simeq 0.0495$. In the soliton penetration dynamics discussed in Sect. 5.1.3, applying a rotating magnetic field brings about the bulk solitons compressed toward the edge of the system, leaving the sparse region at the opposite edge where spins predominantly point downward. For $h_z^{\rm stat} < h_z^{\rm stat,*}$, placing a chiral soliton inside the bulk rather than outside decreases the energy, leading to the penetration, as discussed in Sect. 5.1.3. In contrast, for $h_z^{\rm stat} > h_z^{\rm stat,*}$, it is more favorable to place a chiral soliton outside the system ($l_{\rm c} \simeq -3$), hence preventing the systematic soliton penetration. In this region, the penetration of solitons becomes more difficult, while the application of the rotating magnetic field with a larger amplitude may cause a soliton penetration due to the large disturbances in the nonequilibrium state.

The above discussions have been limited to the situation where placing a single chiral soliton in the fully-polarized FFM state. Nevertheless, considering the problem with multiple solitons, placing them inside the system still yields a larger energy gain than placing a single soliton outside. By introducing n solitons deeply inside the system, the system gains the energy of $n\Delta E_1$. For instance, at $h_z^{\rm stat} = 0.06$, $n\Delta E_1 \leq \Delta E_2$ is realized for $n \geq 3$. Hence, placing multiple chiral solitons inside the bulk rather than placing a single chiral soliton outside decreases the energy for $h_z^{\rm stat} > h_z^{\rm stat,*}$. To cause such multiple penetrations, a substantial rotating magnetic field is required, leading to the larger threshold field of $h^{\rm rot}$, as shown in Fig. 5.6a.

References

[1] Shinozaki, M., Masaki, Y., Aoki, R., Togawa, Y., & Kato, Y. (2018). *Physical Review B, 97*, 214413.
[2] Masaki, Y., Aoki, R., Togawa, Y., & Kato, Y. (2018). *Physical Review B, 98*, 100402(R).
[3] Masaki, Y. (2020). *Physical Review B, 101*, 214424.

Appendix C
Some Analytical Formulae for $\chi_{E,\mu}(\omega)$

In this Appendix, we discuss the analytical formula of the complex admittance of the EEF $\chi_{E,\mu}(\omega)$ calculated in Sect. 5.2. In Sect. C.1, we discuss $\chi_{E,\mu}(\omega)$ on resonance and obtain a simple formula to understand the behavior of the resonance peak. In Sect. C.2, we discuss the behavior of $\chi_{E,\mu}(\omega)$ in the large-ω limit.

C.1 $\chi_{E,\mu}(\omega)$ on Resonance

In this section we discuss $\chi_{E,\mu}(\omega)$ at resonance frequencies analytically in the continuum limit. In continuous space, the EEF at position x and time t in Eq. (5.9) is written as

$$E^{\text{em}}(x,t) = \frac{\partial \mathbf{S}(x,t)}{\partial t} \cdot \mathbf{C}^{\text{vc}}(x,t), \tag{C.1}$$

where $\mathbf{C}^{\text{vc}}(x,t)$ is the vector spin chirality given by

$$\mathbf{C}^{\text{vc}}(x,t) = \mathbf{S}(x,t) \times \frac{\partial \mathbf{S}(x,t)}{\partial x}. \tag{C.2}$$

The Fourier component of the EEF is obtained as

$$E^{\text{em}}(q,\omega) = -i \int dq' \int d\omega'\, \omega' \mathbf{S}(q',\omega') \cdot \mathbf{C}^{\text{vc}}(q-q',\omega-\omega'). \tag{C.3}$$

Then, at a resonance frequency, the total EEF, $\bar{E}^{\text{em}}(\omega) = E^{\text{em}}(q=0,\omega)$, is well approximated by

$$\bar{E}^{\text{em}}(\omega) \simeq -i\omega \int dq\, \mathbf{S}(q,\omega) \cdot \mathbf{C}^{\text{vc}}(-q,\omega=0). \tag{C.4}$$

K. Shimizu, *Spin Moiré Engineering and Emergent Electromagnetism in Topological Spin Crystals*, Springer Theses, https://doi.org/10.1007/978-981-96-5992-0

For the noncollinear spin textures discussed in the main text, $\mathbf{C}^{\text{vc}}(q, 0)$ is dominated by the $q = mQ$ components, where m is an integer and Q is the ordering wave number. Thus, $\chi_{E,\mu}(\omega)$ for the resonance dynamics is approximately given by

$$\chi_{E,\mu}(\omega) \simeq -i\omega \sum_m \frac{\mathbf{S}(mQ, \omega) \cdot \mathbf{C}^{\text{vc}}(-mQ, \omega = 0)}{h_\mu^{\text{AC}}(\omega)}. \tag{C.5}$$

In the helical state with $h_x^{\text{stat}} = h_z^{\text{stat}} = 0$, $\mathbf{C}^{\text{vc}}(q, \omega = 0)$ consists of only the $q = 0$ component ($m = 0$). Thus, when h_x^{stat} and h_z^{stat} are small and the modulation from the helical spin texture is weak, Eq. (C.5) is further approximated by

$$\chi_{E,\mu}(\omega) \simeq -i\omega\chi_{S,\nu\mu}(\omega)\bar{C}_\nu^{\text{vc}}, \tag{C.6}$$

where $\chi_{S,\nu\mu}(\omega)$ is the dynamical spin susceptibility in Eq. (5.12) and $\bar{C}_\nu^{\text{vc}}$ is the ν component of the vector spin chirality for the ground state given by

$$\bar{\mathbf{C}}^{\text{vc}} = \frac{1}{L} \int dx\, \mathbf{C}^{\text{vc}}(x, t = 0). \tag{C.7}$$

In Eq. (C.6), $\chi_{E,\mu}(\omega)$ is linear in ω. This explains why the resonance peaks with high frequencies appear more clearly in $\chi_{E,\mu}(\omega)$ than in $\chi_{S,\mu\nu}(\omega)$, as found in the second subsubsection of Sect. 5.2.2.

C.2 $\chi_{E,\mu}(\omega)$ in Large-ω Region

In this section, we show that $\chi_{E,\mu}(\omega)$ for large ω can be understood from the vector spin chirality, as in Eq. (5.17). For simplicity, we consider the continuum limit again. By plugging Eq. (4.6) into Eq. (5.9), we obtain

$$\begin{aligned} E^{\text{em}}(x, t) &= \frac{1}{1+\alpha^2} \Big[-\mathbf{S}(x, t) \times \mathbf{h}^{\text{eff}}(x, t) \\ &\quad + \alpha \mathbf{S}(x, t) \times \big(\mathbf{S}(x, t) \times \mathbf{h}^{\text{eff}}(x, t)\big) \Big] \cdot \left(\mathbf{S}(x, t) \times \frac{\partial \mathbf{S}(x, t)}{\partial x} \right) \\ &= -\frac{1}{1+\alpha^2} \mathbf{h}^{\text{eff}}(x, t) \cdot \left(\frac{\partial \mathbf{S}(x, t)}{\partial x} + \alpha \mathbf{S}(x, t) \times \frac{\partial \mathbf{S}(x, t)}{\partial x} \right). \end{aligned} \tag{C.8}$$

Since $\mathbf{h}^{\text{eff}}(x, t) - \mathbf{h}(t)$ is always perpendicular to $\frac{\partial \mathbf{S}(x,t)}{\partial x}$, the total EEF is calculated as

$$\begin{aligned}\bar{E}^{\mathrm{em}}(t) &= \frac{1}{L}\int_0^L dx\ E^{\mathrm{em}}(x,t) \\ &= -\frac{1}{L(1+\alpha^2)}\left[\mathbf{h}(t)\cdot\int_0^L dx\ \frac{\partial \mathbf{S}(x,t)}{\partial x}\right. \\ &\quad \left. +\alpha\int_0^L dx\ \mathbf{h}^{\mathrm{eff}}(x,t)\cdot\left(\mathbf{S}(x,t)\times\frac{\partial \mathbf{S}(x,t)}{\partial x}\right)\right].\end{aligned} \tag{C.9}$$

The first term in the square bracket of Eq. (C.9) vanishes in the PBC case, while it is nonzero in the order of $1/L$ in the OBC case. Hence, in the following, we focus on the second term in Eq. (C.9). Note that it vanishes when $\alpha = 0$, indicating that the dissipative dynamics is essential for generating the bulk EEF [1].

For sufficiently large ω, the spins hardly follow the AC magnetic field, and hence, it is a good approximation to replace $\mathbf{S}(x,t)$ with the ground-state spin configuration at $t = 0$, $\mathbf{S}(x,0)$. In this approximation, $\mathbf{h}^{\mathrm{eff}}(x,t)$ can be decomposed as

$$\mathbf{h}^{\mathrm{eff}}(x,t) = \mathbf{h}^{\mathrm{AC}}(t) + \mathbf{h}^{\mathrm{eff}}(x,0). \tag{C.10}$$

By substituting Eq. (C.10) into Eq. (C.9), the EEF is given by

$$\begin{aligned}\bar{E}^{\mathrm{em}}(t) &\simeq -\frac{\alpha}{L(1+\alpha^2)}\mathbf{h}^{\mathrm{AC}}(t)\cdot\int_0^L dx\ \mathbf{S}(x,0)\times\frac{\partial \mathbf{S}(x,0)}{\partial x} \\ &= -\frac{\alpha}{1+\alpha^2}\mathbf{h}^{\mathrm{AC}}(t)\cdot\bar{\mathbf{C}}^{\mathrm{vc}}.\end{aligned} \tag{C.11}$$

Note that the contribution from $\mathbf{h}^{\mathrm{eff}}(x,0)$ vanishes as $\bar{E}^{\mathrm{em}}(0) = 0$ and $\mathbf{h}^{\mathrm{AC}}(0) = 0$. Equation (C.11) leads to Eq. (5.17).

Reference

[1] Kishine, J. -I., Bostrem, I. G., Ovchinnikov, A. S., & Sinitsyn, Vl. E. (2012). *Physical Review B, 86*, 214426.